Nitrogen Capture

Anthony S. Travis

Nitrogen Capture

The Growth of an International Industry (1900–1940)

Springer

Anthony S. Travis
Sidney M. Edelstein Center for the History
and Philosophy of Science,
Technology and Medicine
The Hebrew University of Jerusalem
Jerusalem, Israel

ISBN 978-3-030-09859-9 ISBN 978-3-319-68963-0 (eBook)
https://doi.org/10.1007/978-3-319-68963-0

Printed on acid-free paper

This Springer imprint is published by the registered company Springer International Publishing AG part of Springer Nature.
The registered company address is: Gewerbestrasse 11, 6330 Cham, Switzerland

Preface

The supply of food and its scientific relation to agriculture has a history going back to the 1840s, with the growth of studies into plant nutrition and soil fertility based on products containing the three major nutrients: nitrogen, phosphorus and potassium. Unlike phosphorus and potassium, the availability of nitrogen products involved many challenges, geographical, political, economic and technological. This study is devoted mainly to the story of the industrial capture of atmospheric nitrogen and its transformation into stable compounds for use as fertilizers. The endeavour led to the most far-reaching development in industrial chemistry during the early twentieth century, the Haber–Bosch process, inaugurated in 1913 by BASF, a German firm endowed with superior technical capabilities. The process, drawing on a method devised by the academic chemist Fritz Haber in 1909 and developed into a manufacturing process by Carl Bosch and his team at BASF, used unprecedented brute force conditions—a combination of high pressure, a novel catalyst, and elevated temperature—to bring about the combination, or fixation, of the unreactive gas nitrogen with hydrogen, as ammonia.

The Haber–Bosch process, at the peak of pre-World War I chemical technology, arose from a vast industrial research effort based on emerging technologies. The first processes, developed by 1910, drew on hydro-electric power to capture nitrogen by using electric arcs, as an oxide, and by another electrothermal reaction as calcium cyanamide. The driving force in all cases was the need for self-sufficiency in nitrogen products, in particular to reduce the overwhelming reliance in Europe and the United States on Chile saltpetre (sodium nitrate, or "nitrates") imported from South America. There was also the supposition that the mineral nitrate would be exhausted before the mid-twentieth century.

Haber–Bosch synthetic ammonia was converted into the fertilizer ammonium sulphate. In marketing terms, success rested in large part on the fact that ammonium sulphate—from gas works, coke ovens and, more recently, Mond producer gas plants, and calcium cyanamide—was already familiar to farmers. Along with the nitrates, ammonium sulphate contributed towards ensuring food security for expanding populations.

While compounds of nitrogen are essential as fertilizers, they are also required in the manufacture of explosives. This had tremendous implications after the outbreak of World War I in August 1914. Previously, the European nations that would become the Allies, or Entente Powers, and the opposing Central Powers were dependent on Chile saltpetre for both agriculture and production of the nitric acid required to produce explosive nitro compounds, such as TNT. For the Central Powers, the availability of the vital nitrate ceased following the Battle of the Falkland Islands in December 1914, when the British Royal Navy sent the Kaiser's Far Eastern Squadron to the bottom of the Atlantic. The resulting nitrate shortage in Germany and the long stalemate on the Western Front from early 1915 stimulated technical improvement and massive expansion—with government support—of synthetic nitrogen processes for calcium cyanamide and ammonia, as well as major developments in the production from ammonia of nitric acid by catalytic oxidation. During the war, ammonium nitrate, for explosives, took precedence over ammonium sulphate. Nitrogen for explosives and fertilizers became a key sector in the emerging academic–industrial–military complexes that came into existence from late 1914, first in Germany and then in Britain; similar arrangements arose in France, Italy, and Japan and also from June 1916 in the United States. The Haber–Bosch high-pressure ammonia process came to the fore from mid-1916, mainly as a result of the Hindenburg programme of state-led industrial expansion.

Following the Armistice in November 1918, the BASF ammonia process inaugurated a new era in industrial chemistry, with a global reach. This happened after BASF, in the belief that its process could not be imitated, refused to license the novel technology, except on terms that it dictated, none of which were acceptable. Research was encouraged outside Germany, based on wartime investigations, into rival high-pressure synthetic ammonia processes. By the early 1920s, the huge technical challenges were overcome in Italy and France, while soon after British investigators successfully imitated the German process. The new high-pressure ammonia technologies were of immense significance for the growth of chemical industries in nations that had previously been left behind in the establishment of science-based industry based on coal. Hydro-electricity, for example, became important in the production by electrolysis of pure hydrogen, as well as a source of power. This, and coke-based processes for generating hydrogen, had significant implications following the reshaping of Europe in the aftermath of the war. In the geopolitical arena, synthetic ammonia was high among large-scale strategic self-sufficiency and state-sponsored programmes in Italy, Russia, and Japan—at the very time that the new high-pressure processes became widely available. As a result, the chemical industries of these nations, under the influences of fascism, communism and colonial modernization projects, began moving into the top ranks. Notwithstanding the widespread availability of high-pressure processes, extensive war-built calcium cyanamide factories were also available for peacetime fertilizer production. Calcium cyanamide remained significant, particularly in Italy and Japan, until the late 1930s. By then, there were new and urgent needs for nitrogen products.

Here, I attempt to illustrate the impacts of various discoveries and inventions related to the supply of stable nitrogen products that contribute to the world food supply and the way in which the new processes fostered further applications. The story almost neatly falls into place during three distinct periods of modern history: (1) pre-World War I, from around 1890, with the growth of the by-product coke industry, a source of ammonia, the development of electrochemical and electrothermal processes, and studies at the frontier of scientific knowledge into gas reactions carried out at high pressures in the presence of catalysts, culminating with the introduction of the astonishing Haber–Bosch process of BASF; (2) World War I, 1914–1918, during which time military and civilian needs created the unprecedented growth of factories for the capture of atmospheric nitrogen; and (3) the inter-war years, two decades divided by the Wall Street crash of 1929, the first of which saw the emergence and global spread of rival high-pressure ammonia technologies, invariably to satisfy technological enthusiasm, if not momentum, as well as strategic needs, and the second of which witnessed the growing ambitions, with significant dependence on self-sufficiency in nitrogen requirements, of authoritarian states, notably the Soviet Union and Imperial Japan. From the perspective of industrial chemistry, the 1930s also featured the widespread introduction of techniques arising out of adaptations of the high-pressure ammonia process, including coal-to-oil conversion by hydrogenation, and research that led to the discovery of polythene and other novel materials when organic compounds were subjected to high pressures.

The aim is thus to present a broad overview of the nitrogen industry, and where it led, until the late 1930s, by which time technical developments in ammonia synthesis had almost come to a standstill. In doing so, this study looks way beyond the Haber–Bosch process to survey in some detail the impact of rival processes, none more so than that of Luigi Casale, which enabled early and rapid international expansion. Of particular relevance, however, is the infringement on the monopoly of BASF by British investigators, described here through recourse to rarely used archival sources.

My own interest in nitrogen capture began in the early 1980s, when working with an offshoot of the British Schools Council Industry Project, the Brent Schools and Industry Project, a programme of the London Borough of Brent aimed at introducing the application of science into the classroom. At that time, the few studies of the history of the ammonia process and of related early twentieth-century nitrogen fixation processes were written for specialist audiences. The outcome of my research was the publication in 1984 of *The High Pressure Chemists*. Usefully, the project was aided by the presence in Brent, at North Wembley, of the research laboratories of the (British) General Electric Company (GEC), with which Fritz Haber's skilful co-inventor Robert Le Rossignol was associated. It was through former GEC head chemist Ralph C. Chirnside, a close friend of Le Rossignol, that I was to learn about some of the associations with Haber. No less important was the manufacturing and research facility of Johnson Matthey Metals Ltd, located on the site of the former Wembley British Empire Exhibition grounds. There I was able to view the weaving of platinum wire gauzes required in the oxidation of ammonia to nitric acid, a vital step in munitions manufacture that was brought to near perfection under conditions of war after 1914. Johnson Matthey also operated a nitric acid pilot plant at the Wembley facility.

Since the early 1990s, several published biographical accounts have appeared of the inventor of the high-pressure ammonia synthesis method, the Nobel laureate Fritz Haber. Few of these, however, deal in a balanced way with the technical story of the process with which Haber and Carl Bosch were so intimately associated and the rival nitrogen fixation processes: the electric arc and Frank-Caro (calcium cyanamide) processes, as well as the high-pressure processes of Casale, Claude, Fauser, Uhde-Mont Cenis, General Chemical Corporation and Nitrogen Engineering Corporation. It is in order to make up for this lacuna that I here present the result of an extensive reworking of my earlier research, reflecting recent scholarship and incorporating the studies of several colleagues, including participants in the European Science Foundation's Evolution of Chemistry in Europe 1789–1939 programme. I would especially like to acknowledge the Historical Group of the Royal Society of Chemistry for inviting me to give the 2014 Wheeler Award Lecture on the topics dealt with in the first part of this study and the Society for the History of Alchemy and Chemistry for the privilege of receiving its 2015 Morris Award. It was the enthusiastic response from the audience to the lecture accompanying the Morris Award event at the London Science Museum's new research centre that stimulated a study into the story of nitrogen fixation after 1920. This has enabled me to greatly expand on my Springer Brief, *The Synthetic Nitrogen Industry in World War I: Its Emergence and Expansion*, published in 2015, and to produce the present volume; it draws on a range of archival sources and includes revisions and amendments. There is, moreover, a special emphasis on global developments leading up to and following the end of hostilities in November 1918. This provides an opportunity to introduce a more integrated account of certain people, processes and places that have been little studied and that are blanks even to many historians.

I wish to thank Luca Bianchi of Casale SA, who kindly answered a number of questions and provided much useful background information. Thanks are also due to David Aubin, Marcello Benegiamo, William H. Brock, Robert Bud, Zehava Cohen, Joseph Gal, Igal Galili, Karl Hall, Andrea Hohmeyer, Ernst Homburg, Michael Jewess, Jeffrey A. Johnson, Frank Mecklenburg, Nick Lera, Chaya Meier Herr, Kenneth P. Magee, Hideko Tamaru Oyama, S. Ted Oyama, Peter Reed, Robin C. Travis, Bob Weintraub and Hubert Weitensfelder. A special thanks to Peter J. T. Morris, formerly of the London Science Museum, with whom I have shared an interest in the history of chemical technology for well over two decades, and at whose suggestion I undertook the writing of the Springer Brief. I am grateful to Dr Morris for an extensive review of an earlier version of the manuscript and for kindly providing information based on his own research. I would also like to thank Sofia Costa, associate editor for chemistry at Springer, for her ongoing encouragement and several suggestions for improvement of the manuscript.

The Chemical Heritage Foundation, Philadelphia, provided a travel grant that enabled me to draw upon the rich archival and library resources at its Othmer Library. These supplement those to be found at the Sidney M. Edelstein Library for the History and Philosophy of Science, Technology and Medicine at the Israel National Library, Jerusalem, and the Imperial College Archives and Corporate Records Unit, London. The staff of these organizations as well as of the Leo Baeck Institute, Center for Jewish History, New York; the Harman Science Library, The Hebrew University of Jerusalem; the Wellcome Collection, London; and the Bundesarchiv Berlin are thanked for great assistance. Finally thanks to colleagues at the Sidney M. Edelstein Center for the History and Philosophy of Science, Technology and Medicine, The Hebrew University of Jerusalem.

In order to aid reader familiarity with the main thrust of this volume, the first four chapters are presented as portraits, all broadly thematic, that, by allowing some repetition within different contexts, enable an appreciation of important events in the development of the first modern fertilizers required in food production. This includes discussion of the roles of superphosphates (the source of phosphorus) and Stassfurt potash (the source of potassium), which, through their introductions into agriculture well before the South American nitrates, emphasized the emerging scientific endeavour into increasing crop yields. Phosphorus and potassium played an important role in the nitrogen story when from the late 1920s they were made available mixed with nitrogen products for use as compound fertilizers. Also included are the early laboratory experiments with atmospheric nitrogen which, though they could not be put to practical use, certainly inspired late nineteenth-century investigators. Particularly important is the role of ammonium sulphate, first obtained from coal gas and coke works and then from the synthetic nitrogen processes. From around 1920 it was the main nitrogen fertilizer until the late 1930s.

What follows is an account of events that were as much a prelude to World War I as were the stories of the buildup of fleets of battleships among the "Great Powers" in the decade or so prior to 1914 (Chaps. 5 and 6). This is later reflected, at the climax of hostilities in November 1918, through the vast nitrogen factories undergoing expansion, under construction or planned in Germany, Britain, France and the United States (Chaps. 7 and 8). In the aftermath came the struggles until the mid-1920s to imitate or develop rivals to the Haber–Bosch process (Chaps. 9 and 10). Then there is a focus on developments across broad geographical areas encompassing the United States and Europe (Chaps. 11 and 12). This is the period when there emerged—often through involvement in high-pressure nitrogen chemistry—giant industrial chemical corporations through mergers (Allied Chemical, 1921; IG Farben, 1925; and ICI, 1926) and acquisitions (Montecatini), and national research laboratories. Then came the problem of overproduction and the severe economic crisis at the end of the 1920s (Chap. 13). The global context is completed with the particular situations in the Soviet Union (Chap. 14) and Japan (Chap. 15). The wider impact of synthetic ammonia technology is emphasized with a summary of inventions resulting from the success of high-pressure processes (Chap. 16). We then return to the main inventor, Fritz Haber, and the post-war situation in Germany (Chap. 17). This is followed with a homage to many of those involved in industrial nitrogen capture and the role of the industry in autarky programmes (Chap. 18). Between 1925 and the early 1930s, three increasingly militaristic oligarchies were each associated strongly with one of the high-pressure nitrogen processes: Italy with Fauser, Japan with Casale and Germany with Haber–Bosch. The Soviet Union, after first adopting the Casale process, worked both the Fauser and the Nitrogen Engineering Corporation processes. Nitrogen capture enabled Italy, Japan and the USSR to make great strides in modern processes of chemical manufacture (Chap. 19).

Undoubtedly, the introduction of the Haber–Bosch process was an event in its time as big as the harnessing of nuclear energy or the space programme, though it rarely came into the public eye—apart from during debates over essential wartime supplies. While farmers could just get by without synthetic nitrogen compounds, the military could not. At least that was the perceived view after supplies of imported nitrates were cut off, as was the case in Germany from late 1914, or were threatened, as in Britain following the sinking of merchant ships by German U-boats. Notwithstanding the essential role of the nitrogen industry in warfare and strategic planning, this foray into the world of industrial nitrogen products mainly celebrates the development of an advanced technology that to this day contributes so much to feeding the population of our world.

For permission to reproduce illustrations I thank: BASF Unternehmensarchiv; Casale SA; Chemical Heritage Foundation; the Albert Einstein Archives at The Hebrew University of Jerusalem; the Leo Baeck Institute, New York; Evonik Industries AG, Corporate Archives, Hanau/Marl; Yara International ASA; Nick Lera; Archive of the Max Planck Society, Berlin (Archiv der Max-Planck-Gesellschaft, Berlin-Dahlem); the Central State Archive of Saxony-Anhalt, Merseburg Division (Landeshauptarchiv Sachsen-Anhalt, Abteilung Merseburg); Ostchem/Group DF International; the Tamaru Family Collection; and ThyssenKrupp, Corporate Archives, Duisburg. Uncredited figures and photographs are by the author. Figures 3.4, 5.5, 6.4, 6.11, 9.3 and 16.3 were prepared by the author for teaching purposes and appear in A. S. Travis, *The High Pressure Chemists* (Wembley: Brent Schools and Industry Project, 1984). They are based on diagrams in R. Coles, *Chemistry Diagrams* (London: John Murray, 1960).

Finally, there is the issue of the maze of company names, including changes of names, that, in order not to try the patience of the reader, has been simplified in a number of cases by introducing more recent or present-day names, shortened names and abbreviations. Modern names are retained both because of their familiarity and as indicators through historical examples of the origins of a company or corporation. However, in most cases, full names, as well as previous and subsequent name (s) and abbreviations, are given when first appearing in the text, according to context. In cases where the same names appear far apart in the text, old and new names are repeated. Important examples are:

Belgium

Union Chimique Belge: SA des Fours à coke Semet-Solvay & Piette. From 1928 merged into Union Chimique Belge

ASED: SA Ammoniaque Synthétique et Dérivés

France

Pechiney: Cie d'Alais, Froges et Camargue. Successor in 1921 to Compagnie des Produits Chimiques d'Alais et de la Camargue; forerunner of Péchiney; since 1950, Pechiney

Germany

BASF: Badische Anilin- & Soda-Fabrik, also once known as the Badische

Degussa: Deutsche Gold- und Silber-Scheideanstalt, vormals Roessler (German Gold and Silver Refinery, formerly Roessler), also once known as Scheideanstalt

Hoechst: Hoechst Dyeworks (Farbwerke vorm. Meister, Lucius & Brüning)

IG Farben (1916): Interessengemeinschaft der deutschen Teerfarbenfabriken

IG Farben (1925): IG Farbenindustrie Aktiengesellschaft (the 1925 amalgamation of BASF, Hoechst, Chemische Fabrik Griesheim Elektron, Bayer, Agfa, Cassella, Weiler-ter Meer, and Kalle)

Bamag: Berlin-Anhaltische Maschinenbau AG. From 1924, Bamag-Meguin

Bayerische Stickstoff-Werke AG (here often referred to as Bayersiche) was originally Bayrische Stickstoff-Werke AG. In 1939, through merger with a state-owned firm, it became Süddeutsche Kalkstickstoff-Werke AG (SKW) and from 1978 SKW Trostberg AG. Today, the original manufacturing site at Trostberg is owned by AlzChem

Great Britain

ICI: Imperial Chemical Industries, Limited (the 1926 amalgamation of Brunner, Mond & Co. Ltd, Nobel Industries Ltd, United Alkali Co., Ltd, and British Dyestuffs Corporation Ltd)

Italy

Azogeno: Società per la Fabbricazione dell'Ammoniaca Sintetica e Prodotti Derivati

SAFFAT: Società degli Alti Forni Fonderie e Acciaiere di Terni. Also: Società Alti Forni Fonderie & Acciaiere; Società degli Alti Forni e Fonderie di Terni

Carburo: Società Italiana del Carburo di Calcio, Acetilene e altri Gas (Roma). Also: Società Industriale del Carburo di Calcio, Acetilene e Gas. In 1922, taken over by SAFFAT, at which time Terni was formed (see below)

SIPA: Società Italiana per la Fabbricazione di Prodotti Azotati e di altri sostanze per l'Agricoltura. Also: Società Italiana Prodotti Azotati

Terni: Terni-Società per Indùstria e l'Elettricità. Also: Società per l'Industrie e l'Elettricità 'Terni'

Japan

Denka: Electro-Chemical Co.; Electrochemical Industries, Inc. (Denki Kagaku Kōgyō Kabushiki Kaisha). Associated with Hokkai Carbide. From October 2015, Denka Co., Ltd

Nitchitsu: Japan Nitrogeneous Fertilizer, Inc. (Nippon Chisso Hiryō Kabushiki Kaisha). Successor from 1908 to Sogi Electric and Nippon Carbide

Units of measurement are as given in original sources.

1 atmosphere = 101.325 kPa (kiloPascals)
1 metric ton = 0.97 short tons (US)
1 Quintal = 100 kg

The Ammonia Converter

An ammonia converter, or pressure or synthesis tube, also called a bomb, is a long steel cylinder, or shell, with lids tightly closed at each end, in which a mixture of nitrogen and hydrogen gases, under high pressure and at a high temperature, is forced to react on the surface of a catalyst. Typically the catalyst is located in a removable basket, or cartridge, held within an inner tube, the catalyst or reaction chamber. The converter is also fitted with a tube for heat exchange, and electric heating coils. The tubes are arranged in a concentric manner. The converter is one component of an assembly of machines and devices, linked through interconnected pipework, that together bring about the formation of synthetic ammonia. The process requires the production of extremely pure hydrogen. Ammonia is transformed into fertilizers that sustain life; it is transformed into explosives for munitions that destroy life.

Nitrates

In earlier literature when dealing with nitrogen-containing compounds of commercial or military importance, what were called "nitrates" (sometimes in the singular) referred to compounds that were not necessarily nitrates. This usage arose from the great reliance on the nitrate mineral—Chile saltpetre, or Chilean nitrate—exported from South America.

Essential biographical information, including dates, for the more important individuals in this story will be found in appropriate sections, in many cases not when they first receive mention in the text.

Jerusalem, Israel Anthony S. Travis

Acknowledgement

Some sections of Chapters 14 and 15 are derived in part from my article "Globalising Synthetic Nitrogen: The Interwar Inauguration of a New Industry," published in *Ambix*, vol. 64 (1)(2017):1–28, copyright the Society for the History of Alchemy and Chemistry, available online: http://www.tandfonline.com/<https://doi.org/10.1080/00026980.2017.1325585>

Abstract

The most far-reaching development in industrial chemistry during the early twentieth century was the capture of atmospheric nitrogen by the Haber–Bosch process of BASF. It used unprecedented brute force conditions—high pressure, a novel catalyst, and elevated temperature—to bring about the combination, or fixation, of the unreactive gas nitrogen with hydrogen, as ammonia. The process, at the peak of pre-1914 chemical technology, followed from a vast industrial research effort based on emerging technologies that at first drew on hydro-electric power to capture nitrogen utilising electric arcs (as nitric acid) and electrothermal reactions (as calcium cyanamide). The driving force was the need for self-sufficiency in nitrogen products, as fertilizers, in particular to reduce the great reliance of European nations and the United States on Chile saltpetre, or nitrates, imported from South America. More ominously, nitrates were also required in the manufacture of modern explosives. During World War I, the nitrate shortage in Germany stimulated the technical improvement and the massive expansion of synthetic nitrogen processes, as well as major developments in the production from ammonia of nitric acid by catalytic oxidation. In the aftermath of the war, the Haber–Bosch synthetic ammonia process inaugurated a new era in industrial chemistry. This happened after BASF, in the belief that its process could not be readily imitated, refused to license the novel technology, thereby stimulating research outside Germany, based on wartime investigations, into rival high-pressure synthetic ammonia processes. By the early 1920s, the huge technical challenges had been overcome in Italy and France, while British investigators successfully imitated the German process. The new synthetic ammonia technologies were of immense significance for the growth of chemical industries in nations that had previously been left behind in the establishment of science-based industry. In the geopolitical arena, synthetic ammonia was at the forefront of large-scale strategic self-sufficiency and state-sponsored programmes in Italy, Russia and Japan—at the very time that the new processes became widely available. As a result, the chemical industries of these nations, under the influences of fascism, communism and colonial modernization projects, began moving into the top ranks. At the same time, the widespread availability of high-pressure synthetic ammonia processes brought about the development of new areas of industrial chemistry, including the conversion of coal to oil.

Keywords Synthetic ammonia • Calcium cyanamide • Nitric acid • Fritz Haber • Carl Bosch • Luigi Casale • Georges Claude • Giacomo Fauser • Adolph Frank • Nikodem Caro • Samuel Eyde • Kristian Birkeland • Noguchi Shitagau • Friedrich Uhde • Friedrich Bergius

Contents

Chapter 1
Introduction: Food or Famine

1.1 Sir William Crookes

In late Victorian times, the parliament of British science was the annual gathering of the British Association for the Advancement of Science. In 1898, the president of that august body was the recently knighted chemist Sir William Crookes (1832–1919), famous for his work on thallium, cathode rays, and the radiometer. Other activities prior to 1898 included a study of rare earth elements, and the identification of helium on earth (it had first been observed on the sun). Crookes had started out as a skilful analyst who learnt his trade in the 1850s as assistant to the German chemist A(ugust) Wilhelm Hofmann (1818–1892) at the erstwhile Royal College of Chemistry in London. Crookes did not hold an academic post, nor did he possess formal qualifications in chemistry. Instead of pursuing academic studies in an ivory tower, his career was spent as a publicist for science, and himself, mainly as editor and proprietor of the journal *Chemical News*, and as discoverer, inventor, consultant, and expert witness in cases of litigation concerning scientific matters. Crookes during 1871–1880, among his other business activities, had been a director of the Native Guano Company, founded in London in 1869 to convert "unspeakable" (human) waste into fertilizer. He had, in addition, dabbled in theosophy, but seems to have been mainly forgiven this and other foibles. In any case, spiritualism and occult science were fashionable if dubious topics around 1900.

The 1898 meeting of the association was held in Bristol, south-west England, a city located on the River Avon and an important maritime hub. Here were the homes of merchants who had made their fortunes from the shipping trade. The trade had once included slaves and emigrants, but by the mid-1850s it was dominated by a vast business dealing in imported South American guano, a nitrogen-containing fertilizer (bird excrement), the supply of which declined in the early 1870s, by which time Bristol's port was replaced by new large docks at nearby Avonmouth. One of the main firms trading in guano, through Bristol and London, was Antony Gibbs & Sons, which by the early 1880s had shifted from guano to the import, also from South

A.S. Travis, *Nitrogen Capture*, https://doi.org/10.1007/978-3-319-68963-0_1

America, of another nitrogen-containing fertilizer, later named Chile saltpetre (sodium nitrate); also known as Chilean nitrate. Bristol's connection with a leading natural source of nitrogen fertilizer serves to heighten the significance of what follows here.

Now, in early September 1898, as president of the association and at the cusp of his career, the scientific celebrity William Crookes stood before the assembled delegates to give the president's address. It was both an annual rite and a rare privilege. Bespectacled and immaculately dressed, a man of sartorial style (as befitted the son of a tailor), sporting a bushy beard, and distinctive moustache, heavily waxed and tapering off at its extremities, Crookes had all the airs of a leading military gentleman or minor emperor of some faraway place. Not that his appearance was greatly different from many scientific and other notables of the day, particularly foreigners. (The German chemists Max Le Blanc and Carl Engler, for example, sported moustaches that were worthy rivals of Crookes's.)

Crookes, in order to keep his audience awake for at least an hour, tackled what he perceived to be a contemporary problem with the declaration of a great peaceful campaign intended to remove the threat of famine, at least among all the wheat-eating peoples of the world—the "great Caucasian race," as he described it. His rallying call was in the form of a stark warning: unless science and technology were harnessed in the service of enhancing the supply of food, wheat in particular, this race of people would soon be faced with mass starvation. Crookes may have broken with tradition in his approach, particularly the absence of drily presented scientific facts, but he succeeded in rattling his audience, and many others, as no doubt was his intention (Fig. 1.1).

Fig. 1.1 Sir William Crookes in his laboratory at his home in Notting Hill, London. (Sidney M. Edelstein Library)

1.2 "The Wheat Problem"

Wheat, of which Britain was the world's largest importer, having doubled imports per head of population since 1870, made good copy. Crookes, a wily publicist, was out to cause a stir and project his name into the headlines. Months earlier he had thrown himself into the task of preparing his presidential speech, and an extended version for distribution as a pamphlet, with his usual unstinting enthusiasm. The manuscript for the pamphlet was completed in June, then set in type, corrected and altered, and held ready waiting to be rolled out on the presses of Crookes's London printing office [1]. Distribution of "The Wheat Problem" to journalists and editors had the desired effect, garnering tremendous publicity at a level not normally achieved by scientific proceedings. Crookes wrote not only of a "catastrophe little short of starvation for wheat-eaters, but indirectly scarcity for those who exist on inferior grains, together with a lower standard of living for meat-eaters, scarcity of mutton and beef, and even the extinction of gunpowder."

The problem was fertilizer, or at least a lack of the main imported source, nitrates, within a few decades. In the nineteenth century fertilizer was often referred to as manure. As Crookes explained to his audience, in a more scientific vein: "It is now recognized that all crops require what is called a 'dominant' manure. Some need nitrogen, some potash, others phosphates." Wheat, however, "preeminently demands nitrogen, fixed in the form of ammonia or nitric acid. . .nitrogen is mainly of atmospheric origin, and is rendered 'fixed' by a slow and precarious process which requires a combination of rare meteorological and geographical conditions to enable it to advance at a sufficiently rapid rate to become of commercial importance" [2]. Nitrogen, as he implied, was the principal source for encouraging the growth of food plants. But it was also critical to the manufacture of gunpowder and the even more devastating new explosives, the nitro compounds, mostly based on coal tar products.

Apart from the media attention, Crookes generated a fair amount of criticism and controversy, as well as responses that flatly rejected his statistical argumentation and claims. These would arouse interest in, and fuel debates over, the yields of wheat-growing areas of the world, mainly through the publication in 1899 (and a reprint in 1905), of his *The Wheat Problem. . .Revised, with an answer to various critics* [3]. His claims were supported in chapters authored by two Americans, agronomist John Wood Davis of Peotone, Kansas, and John Hyde, chief statistician at the Department of Agriculture, whose pronouncements also came in for criticism. However, nobody could fault Crookes's chemistry when he drew attention to a specific and seemingly for most people secondary problem: the technical fixation of nitrogen from the air and its conversion into a stable compound for use as a synthetic fertilizer.

For capturing attention among those beyond the world of science and academe, the story of a possible calamity following a wheat famine was a brilliant ploy. After all, then as now, a threat to daily bread was the stuff of headlines. It was understood by everyone. But who except for a handful of chemists would give much attention to the scientific challenge? "The fixation of atmospheric nitrogen," Crookes explained, "is one of the great discoveries, awaiting the genius of a chemist." He was of course referring to a discovery of industrial value. Crookes, in fact, anticipated what was soon to come about, the imitation in the laboratory and then on a manufacturing scale of a natural process, that of nitrogen capture by lightning. He had even dabbled with the basis of a new technological system when, six years earlier, in 1892, he had demonstrated before the Royal Society in London nitrogen capture by burning air in a flame that was concentrated by a strong induction current flowing between terminals, not unlike the beam that provided illumination from the arc lamp. Subsequently, Lord Rayleigh (1842–1919), in 1897, during experiments aimed at the isolation of argon from the atmosphere, employed a similar apparatus "and succeeded in effecting the union of 29.4 g of mixed nitrogen and oxygen at an expenditure of one horse-power." Crookes offered a way out "of the colossal dilemma. It is the chemist who must come to the rescue of the threatened communities. It is through the laboratory that starvation may ultimately be turned into plenty." As for the technology, particularly the need for vast amounts of energy, he turned to the role of inexpensive electrical power, as derived for example from the Niagara River, that would one day, conceivably, enable the production of "electric nitrates." They would replace the reliance on imported Chilean nitrate, then the main source of nitrogen fertilizer, the future supply of which was so uncertain. He had every reason to project confidence in success. Electrochemistry had come of age with astonishing successes. Alpine regions, including in France, Germany, Italy, and Japan, in addition to Scandinavia and the Niagara Falls, were hydroelectric power centres of the future. No less astonishing were the accomplishments of the chemical industry, foremost in Germany, where in 1897, natural indigo dye was replicated. It was just the technology of nitrogen capture that needed to be sorted out, hopefully, as Crookes believed, by around 1930.

Crookes's own panoramic vista and ruminations over wheat-bearing areas of the planet became the source of vigorous polemics that simmered on for around two decades, and were then forgotten. But the historical impact of his call for a nitrogen crusade before an elite of scientific society, on behalf of future civilization, persisted—and indeed persists. In part this was because of Crookes's high standing, the prestigious event and his publicity campaign; and in part because the keys to viable technologies lay in progress made during the following decade. Before his death in 1919, William Crookes would read about and describe in print successes in industrial nitrogen fixation, even though they were associated as much with guns as with grain, notably the wartime battles to deny Germany nitrates and Great Britain wheat [4].

Although Crookes's lecture has become the standard starting point for technical and historical monographs on nitrogen fixation and he, along with Raleigh, purportedly inspired a number of inventors, including the Norwegian Kristian Birkeland, he was not the first to publicly promote the need for an artificial nitrogen fertilizer. Previously, in addition to electrochemical studies, there had been much interest in capturing nitrogen by its reactions with metals and non-metals, or through novel coal-based processes—destructive distillation and coke production—both investigated by the German-born chemist Ludwig Mond, co-founder of the British alkali manufacturer Brunner, Mond & Co. From the coal processes ammonia was obtained, which by reaction with sulphuric acid was readily converted into the stable solid ammonium sulphate, a nitrogen fertilizer second in importance after the South American nitrate. In 1890, before the British Society of Chemical Industry, Mond had drawn attention to the great commercial profit to be had from novel ways of satisfying the nitrogen fertilizer market, which then relied on "an abundant supply of ammonia [Mond's term for South American guano and nitrate], more particularly for the Old World, with its teeming population and worn-out soil." After alluding to the vast sums paid for imported "ammonia in [the] shape of guano, and more recently...sodium nitrate [from Chile]," he highlighted the challenge "of making ourselves independent of a country so far away for the supply of a material upon which the prosperity of our agriculture—our most important industry—depends... [It] is certainly one of the most important which our science has to solve" [5].

Almost a decade later, in May 1899, the chemist Carl Engler (1842–1925), provost of Karlsruhe's Technische Hochschule, at the dedication of a new institute building, also drew attention to the need for novel nitrogen capture technologies that would enable the world's population to be fed [6]. In the audience was his colleague, the expert in electrochemistry Fritz Haber.

Progress was rapid. By 1900, nitrogen had been captured through its reaction with calcium carbide, thanks to the efforts of Adolph Frank, Nikodem Caro, and Fritz Rothe in Germany, to yield calcium cyanamide, which, in the presence of steam, released ammonia. The viability of the process relied on the prior development of electric ovens capable of continuously withstanding extremely high temperatures—conditions that by 1910 were readily attainable. Around the same time electric arc processes were developed. Then there was the growing interest in the kinetics and thermodynamics of gas reactions, including the direct combination of nitrogen with hydrogen to give ammonia, as attempted in 1900 by Henri Le Chatelier in France and Wilhelm Ostwald in Germany. Novel methods for producing pure gases, improved electric arcs, and studies on dissociation and equilibria involving gas reactions—particularly in Germany by Walther Nernst and Fritz Haber—bore directly and indirectly on the technical problem of nitrogen capture.

All those mentioned in the foregoing played a role—some far more so than others—in the story of nitrogen capture. The modern chemist and historian of chemistry are in the privileged position of knowing those who met with success and by which method. However, the writing of a comprehensive and balanced account requires much circumspection and the need to include and explain pitfalls and dead ends as well as the successes. This is particularly the case when drawing on the vast amount of available literature, such as the technical monographs that appeared during the first three decades of the twentieth century—when nitrogen capture was a leading high technology—and that include numerous interesting but wildly impractical ideas. Earlier reporters, writing for or on behalf of those with academic or industrial interests, often relied on simply repeating patent information, promotional material, news reports or claims put forward in correspondence by inventors and firms. Not all of this information was reliable, while some of it was greatly exaggerated; and invariably there were instances of conflicting and contradictory information. In this respect, it is pertinent to quote John A. Harker, a British reviewer with a wide knowledge of the nitrogen industry, of Bruno Waeser's comprehensive study *Die Luftstickstoff-Industrie*, published in 1922: "…from a general perusal it is evident that while the treatise is one which should in some respects be of considerable value, a certain amount of the information it contains is quite unreliable, and has evidently been derived in a number of instances from the claims made for projects which never came to fruition, rather than from records of actual fact" [7]. Of the English translation, published in 1926, with brief supplements that included the new synthetic ammonia processes, Major Harry A. Curtis, who in 1918 was in charge of the US government's Sheffield, Alabama, synthetic ammonia-nitric acid laboratory, and later at the Bureau of Foreign and Domestic Commerce, was even more scathing [8]. The second German edition, published in 1932, and with fewer such reservations (and the absence of grandiose schemes), serves as a most important encyclopedia of a mature industry [9]. Then there is Alcan Hirsch's account of events in the Soviet Union, written not long after and at a critical time in this story. Hirsch's perspective, and the fact that it is the single most important source on the impact of Western technologies on the Russian nitrogen industry at the turn of the 1930s, justifies the inclusion of relevant biographical background (Chap. 14). This and similar material was often a mixture of advertising and propaganda, with a fair amount of bias, but it should not be dismissed out of hand; it informs us on a time and place, and enhances our appreciation of developed and developing processes.

Where possible, use has been made of archival collections for expanding on and correcting earlier historical accounts. The nature of their contents varies, with the result that there is some inevitable imbalance. There are also gaps—even within important archival sources. Thus British engineer Herbert Alfred Humphrey, who was intimately concerned with development of the Billingham synthetic ammonia factory of Brunner, Mond & Co. (later Imperial Chemical Industries, ICI), in the 1920s, wrote in his retirement: "All my data and records, and the contents of box files occupying 36 feet of shelves, all my drawings and 100 volumes of my private diary, have been burnt in an incinerator. It took my gardener 3 weeks to do all the job" [10]. Humphrey was a prolific inventor, and probably not a great deal of the lost material related to the history of ammonia synthesis.

Certainly, however, Humphrey's surviving papers add tremendously to the excellent account by W. J. Reader, historian of Imperial Chemical Industries, of the early history of synthetic ammonia at Billingham [11]. In many other cases, we have to accept at face value what participants wrote about themselves, or what their colleagues wrote in hagiographic obituaries, often long after the events and generally in a "romance of chemistry" vein.

There is also the problem of evaluating statistical information, particularly since here the presentation of an international history is one of the prime objectives. Commerce reports, government communications, year books, reports on trade and shipping, and official statistics, in addition to data supplied by manufacturers and their associations and formal agreements over production, were sources of information that appeared regularly in trade and chemical journals, and have been used throughout the text. In several cases statistics related to fixed nitrogen, sometimes referred to as "chemical nitrogen" (including Chilean nitrate and by-product ammonium sulphate), are for the agricultural year June 30–July 1. In instances of greatly conflicting data, the information, where possible, has been checked against additional sources. Occasionally data given for nitrogen output in the literature is for products and not for the nitrogen content alone. Where this is apparent, corrections and estimates have been introduced.

Sometimes, the data was deliberately misleading, particularly when companies and official bodies, respectively, painted a more optimistic picture concerning a process or annual output than was actually the case. Inadequate information, unreliable estimates, statistical sleights of hand, and confusion between fixed nitrogen and its products, has made it impracticable to adopt the same method of arrangement of statistical information in every case. However, even if sources are not always reliable, in most cases they can be taken as rough guides of capacities and outputs, and are no less valuable for that in indicating major trends.

Most important, however, this is a story of achievement that many of us now take so completely for granted that we barely ponder it at all.

References

1. Brock WH (2008) William Crookes (1832–1919) and the commercialization of science. Ashgate, Aldershot, pp 366–388.
2. Crookes W (1898) The world's wheat supply. Reports of the British Association for the Advancement of Science, London, p 4.
3. Crookes W (1899) The wheat problem. Based on remarks made in the presidential address to the British Association at Bristol in 1898. Revised, with an answer to various critics, by Sir William Crookes, F.R.S., with chapters on the future wheat supply of the United States, by Mr C. Wood Davis, of Peotone, Kansas, and the Hon. John Hyde, chief statistician in the Department of Agriculture, Washington. GP Putnam's Sons, New York. John Murray, London.
4. Offer A (1991) The first world war: An agrarian interpretation. Clarendon Press, Oxford.
5. Mond L (1889) The president's address. J Soc Chem Ind 8(7)(31 July):505–510, on 505.
6. Einweihung der Neubauten des Technischen Hochschule zu Karlsruhe, 17–19 May 1899.
7. Harker JA (1923), review of Waeser B (1922) Die Luftstickstoff-Industrie, mit besonderer Berücksichtigung der Gewinnung von Ammoniak und Salpetersaure. Otto Spamer, Leipzig. In: J Soc Chem Ind 42(17)(27 April):435–436 (Chem Ind London).
8. Curtis HA (1927), review of Waeser B (1926) The atmospheric nitrogen industry: with special consideration of the production of ammonia and nitric acid (trans. Fyleman E). P Blakiston's Son & Co., Philadelphia. In: Ind Eng Chem 19(3)(March):434.
9. Waeser B (1932) Die Luftstickstoff-Industrie, mit Berücksichtigung der chilenischen Industrie und des Kokereistickstoffs. Springer, Berlin.
10. H. A. Humphrey to Professor Robert Salmon Hutton, 12 September 1948. In: Pingree J, Smith D, compilers (1971), Herbert Alfred Humphrey, ACGI, FCGI, M Inst CE, M I Mech E, MIEE. List of his papers. Imperial College of Science and Technology, University of London, quoting from p i.
11. Reader WJ (1970) Imperial Chemical Industries: A history, vol I. The forerunners, 1870–1926. Oxford University Press, London.

Chapter 2
Agricultural Chemistry

2.1 Justus Liebig

Justus Liebig (1803–1873) is closely associated with promotion of the scientific development of agricultural chemistry. The career of this famous German chemist falls into two periods: before 1840 as organic chemist; and after 1840, in connection with agriculture. In 1825 at the University of Giessen he established a laboratory that served as a pharmaceutical-chemical institute. In 1831 he developed a method for the elemental analysis of organic substances, for which he gained considerable fame. Until 1835, most of his students qualified in pharmacy. Between 1839 and 1842, during which time enrolments increased considerably, his students intended to take up posts as chemists in industry, or as teachers, or enter into government service (Fig. 2.1) [1]. It was at this time that Liebig turned to the application of science to agriculture and soon attracted to his laboratory those interested in food supply as well as medicine.

Liebig's first monograph describing his method of organic analysis was published in 1837. In the same year he visited Ireland, Scotland and England, at the invitation of former students, and attended the annual meeting of the British Association for the Advancement of Science. He promised to draw up two reports for the association, but only one appeared—and even it was not quite what was expected. *Organic Chemistry in Its Application to Agriculture and Physiology*, published in London in 1840, better known as *Agricultural Chemistry*, was a rework of earlier material already prepared, in some cases for publication in German and French, and reflected the direction of his new interests. It was significant for the application of Liebig's method of analysis to the study of plant growth. Liebig pointed out that carbon and nitrogen were present in all plants. The carbon came from carbon dioxide in the atmosphere. Nitrogen, he believed, was available only through natural agencies—mainly ammonia dissolved in rainwater—as well as from excrement that had long been used as an effective manure. A number of inorganic nutrients were absorbed by plants. These were—in addition to carbon

A.S. Travis, *Nitrogen Capture*, https://doi.org/10.1007/978-3-319-68963-0_2

Fig. 2.1 Justus Liebig's laboratory at Giessen, around 1840. Liebig was ennobled in 1845. (From J. Hofmann, *Acht Tafeln zur Beschreibung des chemischen Laboratoriums zu Giessen*. Heidelberg, 1842. Lithograph by Trautschold and Ritgen. Sidney M. Edelstein Library)

dioxide, water, and ammonia—potassium citrate, calcium oxide (lime), magnesium oxide, iron, phosphorus and silicate. To maintain the level of these nutrients it was necessary to create artificial manures, but not, however, those containing nitrogen, the unreactive gas that constitutes about 80% of the earth's atmosphere.[1]

Liebig believed that the manures, natural or otherwise, should be insoluble, to prevent them from soaking into the soil. He toured British farming regions in 1842 and thereafter continued to promote his style of agricultural science [2]. In 1845, Liebig published his *Address to the Agriculturalists of Great Britain*, noting that the "fitness for producing wheat, or any other kind of plant bears a direct proportion to certain elements contained in . . . soil, which are absorbed by the plants" [3]. In the same year, 1845, London's Royal College of Chemistry was opened, headed by Liebig's star pupil A. Wilhelm Hofmann, and backed by British agriculturalists, whose hopes of increased productivity, however, were not fulfilled.

Following Liebig's recommendations, phosphates were the first fertilizer products to be manufactured. In 1845, on Liebig's behalf, the alkali manufacturer James Muspratt (1793–1886) patented a process in Britain for an insoluble product, calcium phosphate.[2] Liebig took out a separate patent in Germany. The process

[1]The role of soil bacteria in fixing nitrogen was unknown until 1893, when Serge Winogradsky (1856–1953) demonstrated how bacteria brought about the formation of nitrates.

[2]The collaboration with James Muspratt came about from their discussions at the 1837 meeting of the British Association for the Advancement of Science in Liverpool. The intention was to trial the manufacture in Liverpool and then expand into Germany and other countries. The patent was BP 10,616 of 1845.

was taken up in Britain, but the product served no useful purpose [4]. The production of insoluble phosphate as fertilizer was a major error, all the more so since the soluble form had already been introduced—with various but certainly greater degrees of success. The earliest commercial development was undertaken in Ireland and England by the Irish physician Sir James Murray (1788–1871) [5]. Soluble phosphate was made by treating bones with sulphuric acid. However, Murray's business failed—in part because, like Liebig, he had not conducted proper field trials—and was taken over in 1846 by agronomist John Bennet Lawes (1841–1900; baronet 1882). From 1843, Lawes, jointly with Liebig's former student John Henry Gilbert (1817–1900; knighted 1893), had conducted field experiments with fertilizers on Lawes's estate at Rothamsted, Harpenden, north of St Albans, England. This enabled Lawes to take the main initiative in commercialization, after filing a patent for a superphosphate and then embarking on its manufacture, from bone meal, in London. By mid-century, superphosphate was made from phosphate rock, in Britain and Germany. Moreover, as shown in 1850 by another British agronomist, J. T. Way, artificial fertilizers worked only if they were soluble. Around 1857 there became available in Britain a "triple superphosphate" made by reacting phosphoric acid with bone meal at the Triple Phosphate of Ammonia Company, according to the recipe in Murray's patent of the same year. It was advertised as available in the form of a compound fertilizer, in admixture with the nitrogen-containing guano.

Liebig knew that bone phosphorus could be made soluble by reaction with sulphuric acid, but he was so convinced of the correctness of his theory—based on the application of insoluble products—that he ignored the potential, at least until he revised his ideas. This happened frequently, including in connection with nitrogen. Lawes and Gilbert proved Liebig incorrect when it came to nitrogen. It was necessary to add nitrogen from an external source. The English workers were bitterly opposed by Liebig, who changed his mind only in 1856, the year in which Lawes and Gilbert commenced their Park Grass Experiment to determine hay yields based on different fertilizers. Less contentious than Liebig's approaches to nitrogen and phosphorus was another inorganic element of great significance in plant growth: potassium, available as potash. Its introduction was associated with one of the leading pioneers in nitrogen capture.

2.2 Adolph Frank: Disciple of Liebig

The European potash fertilizer industry was first developed in Germany by the multi-talented Adolph Frank (1834–1916) (Fig. 2.2). In common with many young students of chemistry, Frank's original interest was in pharmacy, and he probably had ambitions to open his own practice. However, as with others who studied pharmacy he was diverted from being an apothecary by the lack of suitable opportunities and by his teachers, who wisely, as it turned out, suggested a post in the beet sugar industry. The outcome was that Frank early on in his career gained

Fig. 2.2 Adolph Frank. (Sidney M. Edelstein Library)

the first of his special places in the history of modern fertilizers. He is credited with the development of Stassfurt potash as a principal source of potassium-based fertilizer in the 1860s.

Frank lived through a period of remarkable economic growth and challenges, which, in addition to the national unification in Germany (1871), provided entry to new and ultimately profitable ventures. The main turning point in his fortunes occurred after he became engaged in nitrogen capture in the early 1890s, which is why a biographical account deserves a place in this volume. It not only illustrates the life of an inventive chemist working outside the academic world during the second half of the nineteenth century but, through his scientific apprenticeship in industry, shares a number of parallels with the careers of others met here. It also restores to the history of chemistry a life that was widely praised as being part of the scientific elite in Germany prior to 1933, but was then promptly erased.

Adolph Frank was born in Klötze (Saxony-Anhalt) to a family of Jewish merchants. His parents, Ulrike and Salomon Frank, had five children. At the age of fourteen Adolph embarked on an apprenticeship with a pharmacist in Osterburg, after which he joined a pharmacy in Habelschwerdt, in Prussian Silesia. His military service in 1854 took him to the Royal Veterinary Medical School in Berlin, where during 1855–1857 he continued his studies, including at the university. His teachers were Eilhard Mitscherlich (1794–1863, chemist), Gustav Magnus (1802–1870, chemist, with an interest in chemical technology), Gustav Rose (1798–1873, mineralogist), and Heinrich Wilhelm Dove (1803–1879, physicist). Frank was awarded the state diploma in apothecary in the summer of 1857. It was on the recommendations of Mitscherlich and Magnus that he took up the position of chemist at the beet sugar processing factory of Bennecke, Hecker & Co., in Mecklingen, close to Stassfurt, near Magdeburg, in north-central Saxony-Anhalt.

The European beet sugar industry emerged during the Napoleonic era in France, then spread to Germany, and subsequently to other European countries. It grew so fast that by around 1880 beet would equal the output of sugar cane. In the German states, the beet was cultivated in the Prussian provinces of Saxony and Silesia, both with abundant flat terrain and rich soil. The main growing regions were centred on Halle and Breslau. Breslau, the capital of Prussian Silesia, is present-day Wrocłow, in western Poland's Silesia region, some 270 km south-east of Krakov (Cracow). There was plenty of work for Prussian processors, who also received beet from France, a major producer of the crop in Europe, particularly in its north-eastern reaches. For Frank, this employment provided an opportunity to participate in a major growth industry, the extraction of sugar from beet, and the stimulus to study the cultivation of beet and from it the steps involved in the production of sugar. Notably, he introduced the use of bone charcoal for the clarification of sugar and built a molasses charcoal plant. In 1858 he received his first patent—for the separation and cleaning of beet juices with silicates.

The beet sugar industry of Stassfurt also introduced Frank to the newly emerging agricultural fertilizer business, the products of which were in great demand in beet cultivation. According to his own account, Frank was aware from the writings of Justus Liebig and others that plant growth was encouraged by the addition of certain minerals, including potassium salts. Molasses, the by-product of sugar extraction, was fermented for alcohol, leaving a residue that was a useful source of potassium fertilizer. This was important since potassium in the soil was readily absorbed by the beets, and over time cultivation tended to denude the soil of this essential element. Frank later suggested that this threatened the future of the beet sugar industry.

Whatever the inspiration, Frank was prompted to examine the waste salts available from the local rock salt (sodium chloride) industry. These waste salts (Abraumsalze)—that became available when borings for rock salt were made at Stassfurt—had to be removed before the rock salt could be exposed and extracted. They were already known to contain carnallite (hydrated potassium magnesium chloride), as established by mineralogist Heinrich Rose (1795–1864; brother of Gustav) and chemist Richard Marchand (1813–1850). Frank's experiments showed that a soluble component of the waste mineral encouraged plant growth. Frank developed a method for isolating the component, a potassium salt that acted as an efficient fertilizer and was not harmful to the soil. This was potassium chloride (potash; muriate of potash). In August 1860 he filed a patent for an extraction process that was granted on 23 March 1861. Frank's potash was found to be ideal as a fertilizer for beet sugar [6].

The interest in potash led Frank to leave the beet sugar processing industry and explore possibilities for opening a potash fertilizer factory to serve the beet growers. In October 1860, he requested support from the state and local bodies responsible for allocating mining licenses in order to open a factory for processing minerals [7].

However, and although the response was supportive, Frank had difficulty raising capital—apparently due to the depression following a major financial crisis in 1857. Since Frank was unable at first to provide evidence of sufficient financial support, the local government refused to provide a lease for working an old salt mine in Leopoldshall, near Stassfurt; the situation changed after he managed to gain access to limited funding from the banking house Philip & Speyer, of Hamburg in April 1861. Frank's potash factory in Stassfurt opened in October 1861. From the start, he promoted the role of potash as a fertilizer for beet sugar. He also investigated crop protection, with particular reference to beet sugar, paying attention to losses in the field caused by nematodes (a type of worm parasite). These losses were included in his doctoral dissertation on sugar production at the University of Göttingen (1862).

In 1862, Frank joined the Association of German Engineers (Verein Deutscher Ingenieure), at which time he became co-founder and chair of the local division. Even though he established the German potash industry, his process of using open fires for heating was not perfected due to lack of financial backing, and he began to suffer when others with greater resources entered into the business, in particular Chemische Fabrik Vorster & Grüneberg of Cologne, relying on the investigations of chemist and partner Hermann Julius Grüneberg (1827–1894) [8]. This firm, founded soon after Frank opened his factory, employed steam heating. An important development relevant to expansion of the potash business was the opening of a new salt mine at Leopoldshall in 1864.[3]

Finding the competition overwhelming, and financially drained, Frank joined with and advised his former rivals, who worked improved extraction processes. As a pioneer of the industry, this must have given him some considerable satisfaction, at least an opportunity to further the development of local manufacture. In 1865, he obtained bromine from the potassium bromide present in mother liquors of the potash process. This was taken up by Vorster & Grüneberg about a year later and greatly reduced the price of bromine, enabling Germany to become a leading producer of the element.

Frank also engaged in widespread promotion of the German potash industry, and expanded the market to include the United States, which was a leading exporter of potassium-containing wood ash. In 1866, Frank, aided by Irish-born chemist John William Mallet (1832–1912), at the University of Louisiana, demonstrated the importance of potash to cotton and tobacco cultivation. Stassfurt potash was imported into the United States from around 1870; in 1871, 400 tons arrived from Germany. From 1872, following tests at the Pennsylvania State College, the United States became the largest export market for Stassfurt potash [9]; the previously profitable wood ash export business quickly declined. German-born chemical consultant and agronomist (John Christopher) Augustus Voelcker (1822–1884)

[3]Chemische Fabrik Vorster & Grüneberg was later Chemische Fabrik Kalk GmbH (1892–1945).

introduced Stassfurt potash to British farmers around 1870. Another use of potash that Frank investigated in Germany was its application as a fertilizer to peat bogs (during 1866–1868). Ongoing interest in the use of peat would later lead to one of Frank's two engagements with nitrogen. This was through the collaboration of his colleague Nikodem Caro with the Power-Gas Corporation of London, founded in 1901 to work the Mond patents for coal gas (Mond producer gas) and ammonia production [10]. Compared with other coal distillation processes, the Mond process afforded increased ammonia, for conversion into the fertilizer ammonium sulphate (see Chaps. 3 and 4).

By the early 1870s, Frank had become general manager of a group of potash companies based in Leopoldshall. However, on 1 January 1876, Frank resigned this post, no doubt as a result of the growing competition and reduced opportunities in the potash industry, in addition to the ongoing impact of the 1873 stock-market crash. He moved to Charlottenburg, where he became the technical and scientific director of a glass factory. Among other activities, he worked on the production of coloured glasses, including brown for beer bottles, which protected beer from sunlight better than the green then in use.

In 1885, he left the glass industry and became an independent chemical consultant, with an interest in the cellulose industry. He collaborated with his former teacher Eilhard Mitscherlich in developing the sulphite pulping process, and for many years was engaged in the design and construction of paper mills in the German Empire, the United States, and Japan. In 1893, he visited the Chicago World's Fair, and was involved in the construction of a paper mill at Portland, Oregon. In the same year he received the John Scott Award of the City of Pennsylvania, administered by the Franklin Institute, for his application of kieselguhr (silica) filters in the improvement of potable water. Frank claimed that he refused an offer from Jeremiah McLain Rusk, US Secretary of Agriculture (1889–1893), to assist in starting production of potash from sea salt in the United States, since he did not wish to become involved in an enterprise that competed with the German potash industry (from which perhaps he was still deriving an income) [11].

From 1878, Frank was an "unpaid adviser" to the Charlottenburg gas works, though there is no record of what duties he performed around that time. However, by 1898, he was certainly very closely involved with the coal gas industry—at Charlottenburg in particular, but also elsewhere. Around 1900, Frank and his colleagues designed gas generation plant (Fig. 2.3) and investigated the calorific values of various processes. Surviving correspondence shows that during the first decade of the twentieth century he took a strong interest in the production of water gas, a heating gas produced on an industrial scale from the late 1880s, that contained a considerable proportion of hydrogen mixed with carbon monoxide. Among the firms that he contacted was the Wassergas-Syndikat System Dellwick-Fleischer, of Frankfurt am Main, Bamag (Berlin-Anhaltische Maschinenbau AG), and Julius Pintsch, the latter two of Berlin (Fig. 2.4) [12]. The significance here is that both the Bamag and Pintsch water gas processes were later adapted to the production of hydrogen required in the larger synthetic ammonia factories. Around

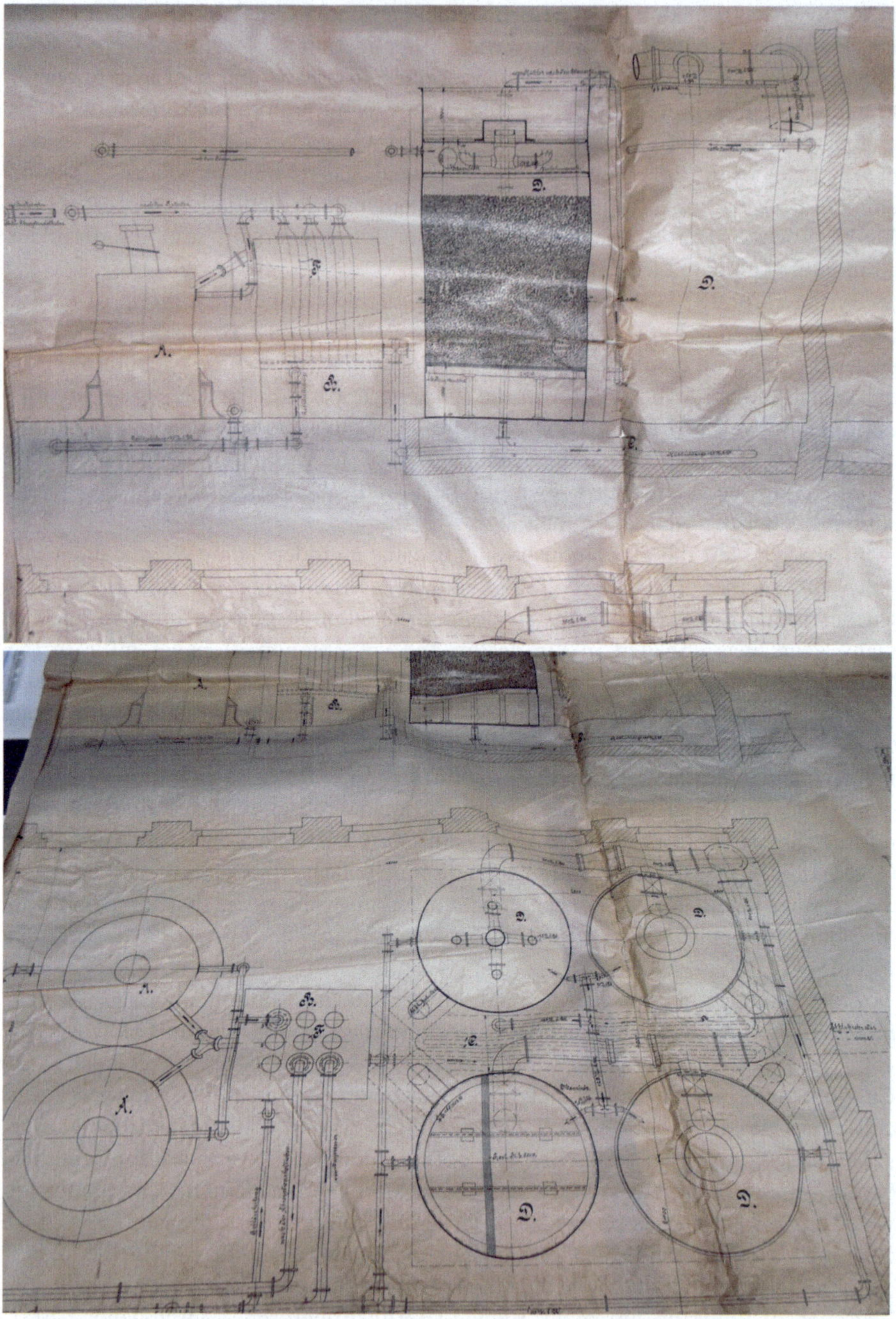

Fig. 2.3 Plan of commercial coal gas generating apparatus designed by Adolph Frank's chemical and engineering consultancy firm. Drawn on fabric, it probably dates from around 1903, when Frank was involved in the design of producer gas plant for use in generation of electricity. Key: A. Gas generator oven, B. Retort, C. Cooler, D. Gas washer, F. Retort tubes. (Adolph and Albert Frank Collection, Leo Baeck Institute, New York, AR 7176/MF722; Section XII, Folders TB16/TB17)

Fig. 2.4 Advertisement of Bamag (Berlin-Anhaltische Maschinenbau AG), around 1910. Bamag machinery and equipment were widely used in the coal gas industry, including for the manufacture of water gas (hydrogen and carbon monoxide) and the fertilizer ammonium sulphate from by-product ammonia. (Sidney M. Edelstein Library)

1904, jointly with Nikodem Caro, Frank investigated the use of peat as a source of fuel gas as well as ammonium sulphate fertilizer (Sect. 4.2). Soon after, Adolph Frank, his son Albert, and Caro collaborated with Carl von Linde in developing the Linde-Frank-Caro process for preparing pure hydrogen from water gas (Sect. 5.13).

As indicated earlier, Frank's career as inventor, scientist and entrepreneur deserves our attention because he was not untypical of many of those skilled in the art of chemistry and its application and who, through hands-on experience and their awareness of new needs, were enabled to recognize where opportunities lay. Frank was a leader in developing the chemistry-based agricultural industry in Germany. He contributed to fulfilling the need for a potassium based fertilizer.

By 1900, as a result of Frank's efforts, Stassfurt potash was widely available in Europe, and in ever greater demand from the United States.

Lawes and others in Britain had inaugurated the phosphate industry for the supply of phosphorus to plants and also proved the importance of added nitrogen-containing manures. The supplies of potash and phosphate, however, caused less concern than those of the nitrate imported from South America. Nitrogen, as in nitrates, as Crookes emphasized, was known to be the most important element contributing to crop yields—not only of wheat, but also of beet sugar, corn, potatoes, rice, and other important food crops. Nitrates were also essential in the production of explosives, which further increased concerns over the supply situation. This encouraged research in Europe into alternative sources, through nitrogen capture from the atmosphere. That challenge would in part be met through Adolph Frank's contribution to the development of a novel nitrogen fixation process (see Chap. 5).

References

1. Homburg E (1998) Two factions, one profession: the chemical profession in German society 1780–1870. In: Knight D, Kragh H (eds), The making of the chemist. The social history of chemistry in Europe, 1789–1914. Cambridge University Press, Cambridge, pp 39–76, on 66.
2. Brock WH (1997) Justus von Liebig: the chemical gatekeeper. Cambridge University Press, Cambridge, pp 150–168.
3. Reed P (2015) Entrepreneurial ventures in chemistry: the Muspratts of Liverpool, 1793–1934. Ashgate, Farnham, Surrey, quoting Liebig, on p 80.
4. Reed P (2015) Entrepreneurial ventures in chemistry: the Muspratts of Liverpool, 1793–1934. Ashgate, Farnham, Surrey, pp 79–82.
5. Alford WAL, Parkes JW (1953) Sir James Murray, M.D.: a pioneer in the making of superphosphate. Chem Ind London, no. 33 (15 August), pp 852–855.
6. The account of Frank's life is drawn from archival material held with the Adolph and Albert Frank Collection of papers and correspondence, Leo Baeck Institute, New York, AR 7176/MF772.
7. Adolph Frank, "Promenoria des Chemikers Adolph Frank betreffend die Anlage einer Chemischen Fabrik bei Stassfurt oder Leopoldshall," addressed to "Herrn Oberhauptmann Krug von Nidda in Berlin und an Herrn Staatsminister von Schaetzell in Bernberg mit Briefen von 20. Oktober 1860." The Adolph and Albert Frank Collection of papers and correspondence, Leo Baeck Institute, New York, AR 7176/MF772; series I, personal papers, II, box 1, folders 1–4.
8. Domheim A, Brügelman W (2006) Forschergeist und Unternehmermut: der Kölner Chemiker und Industrielle Hermann Julius Grüneberg (1827–1894). Böhlau Verlag, Cologne, esp. pp 117–120.
9. Lodge FS (1938) Potash in the fertilizer industry. Ind Eng Chem 30(8)(August):878–882, on 879.
10. The Times (London), 23 July 1901.
11. The Adolph and Albert Frank Collection of papers and correspondence, Leo Baeck Institute, New York, AR 7176/MF772.
12. The Adolph and Albert Frank Collection of papers and correspondence, Leo Baeck Institute, New York, AR 7176/MF772; series V, Charlottenburg gas works, box 2, folders 22–25.

Chapter 3
The Quest for Fixed Nitrogen

3.1 The Background to Fixed Nitrogen

By the 1870s, chemists and agronomists recognized the important roles of nitrogen, phosphorus, and potassium in plant growth. While the manufacture and supply of phosphorus and potassium fertilizers had been placed on a sound footing in Europe, this was not the case for products of nitrogen, the most important of the three elements essential to plant growth. Formerly the principal supply was animal excrement, from which was obtained potassium nitrate (saltpetre, or nitre), also employed in the manufacture of gunpowder and nitric acid. In Europe, where demand for nitrogen fertilizer was greatest, there were few products, apart from "manures," in which the unreactive element was found in a combined state, for direct or indirect application.

The two main natural sources, as mentioned earlier, were located in South America: guano (containing uric acid and other nitrogen compounds), available until the late 1870s; and, from then on, sodium nitrate, which replaced guano and would remain the leading source of nitrogen fertilizer until well into the twentieth century. These plant nutrients brought about greatly enhanced crop yields in industrial and industrializing nations alike. Their importance was such that the three-month journey to Europe by sea was not an impediment. Guano was the basis of a vast enterprise from the 1840s, at a time when agricultural science was developing, with an emphasis on the role of fertilizers, nitrogen compounds in particular—despite Liebig's objections. Agricultural experiment stations were established, managed by individuals trained in chemistry, notably Lawes and Gilbert.

In the 1890s, debates over how long the South American nitrate would last drove laboratory and industrial efforts towards direct fixation of the unreactive atmospheric nitrogen in compounds that could be used as fertilizer. For this purpose, the nitrogen had to be made available as a stable product, for storage and shipping. Concerns of a strategic nature arose, since it was nitrate that was converted into the nitric acid used in the manufacture of explosives, such as Alfred Nobel's dynamite.

A.S. Travis, *Nitrogen Capture*, https://doi.org/10.1007/978-3-319-68963-0_3

Europeans had come up with a new nitrogen fertilizer: ammonium sulphate—made from by-product ammonia available from coal gas and, later, from coke works. Coal gas was the principal source of illumination in cities and towns, prior to the introduction of acetylene and electric lighting. The tremendous scale of gas production encouraged increasing recovery of the by-product ammonia by reaction with sulphuric acid. The output of ammonium sulphate fertilizer, while considerable, did not bear comparison with the import of Chilean nitrate.

The growing availability of electrical power during the 1890s directed the main research effort into fixing atmospheric nitrogen through its combination with oxygen in the flame generated by an electric arc. This became a new source of nitric acid, as required by the explosives industry, as well as of nitrogen-containing fertilizer. Electrical power was also used to produce calcium cyanamide, by the high-temperature reaction between nitrogen and calcium carbide. This process became closely associated with Adolph Frank. The product, a solid generally referred to as cyanamide, or lime nitrogen, released ammonia on application to the soil. These two processes were introduced in 1905 and offered some independence from the naturally occurring nitrate.

Uncertainty over how long it would take before the South American nitrate became exhausted was one thing. There was also the fact that the supply was controlled by commercial monopolies in which British entrepreneurs, merchants and banks were greatly involved. This stimulated research in Germany, where natural resources apart from coal and potash were limited and there was almost total dependence on imported nitrate for supporting Germany's extensive agricultural sector, as well as enabling explosives manufacture. The research effort was taken up at the large German chemical corporations founded in the 1860s and 1870s to engage in the manufacture of synthetic or coal tar dyes. During the 1890s and early 1900s, they had begun to diversify, including with research into inorganic nitrogen products. Attempts at nitrogen capture—particularly by its combination with hydrogen to give ammonia—were abandoned mainly as a result of poor, almost insignificant yields, explosions, and flaws in the apparatus. However, in 1909, physical chemist Fritz Haber and his English assistant Robert Le Rossignol, at the Karlsruhe Technische Hochschule, assembled a metre-high steel apparatus in which nitrogen and hydrogen were combined as ammonia at a high pressure and temperature, in the presence of a catalyst. This became the basis of an industrial process that, from September 1913, produced the fertilizer ammonium sulphate. The process—subsequently called the Haber-Bosch process—was developed by the German dye-making firm Badische Anilin- & Soda-Fabrik (BASF) of Ludwigshafen, almost opposite Mannheim, on the west bank of the River Rhine (Fig. 3.1).

World War I created an unprecedented demand for nitrogen compounds, as well as for coal tar toluene and phenol (carbolic acid), the two aromatic organic chemicals that, on nitration, gave the explosives TNT (trinitrotoluene) and picric acid, respectively. From December 1914, German manufacturers of nitric acid and explosives were denied access to Chilean nitrate. Priority was given to the capture of atmospheric nitrogen for both feeding the population and fighting the Entente

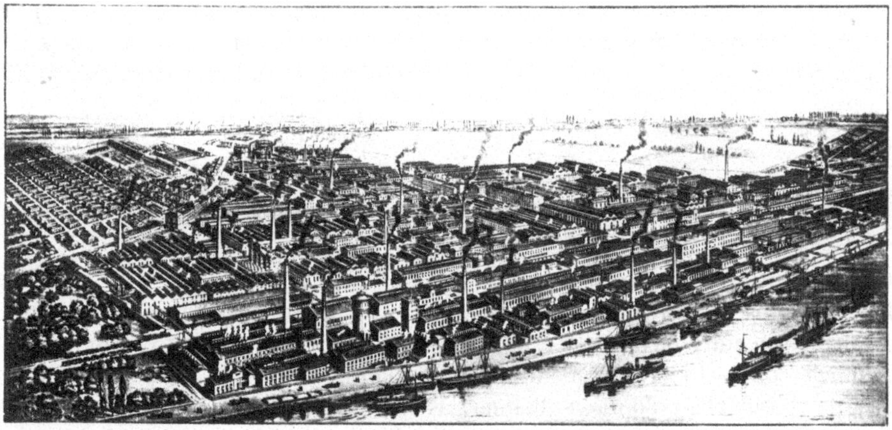

Fig. 3.1 BASF factory, Ludwigshafen, 1890s. (Edelstein collection)

Powers (the United Kingdom, France and Russia, joined by Italy in 1915, and assisted by Japan). Vast works for the manufacture of nitrogen products were erected in Germany; what was achieved by German entrepreneurs, scientists, and technologists could not be matched elsewhere.

From 1918, Germany's former enemies fought to catch up in nitrogen fixation processes and the development of high-pressure chemical technology. For every industrial nation the synthetic nitrogen industry became a key strategic sector for the production of food and munitions.

There are many leading personalities in the nitrogen capture story, though none more so than Fritz Haber. Through his invention of the synthetic ammonia process, he has been accorded the dubious honour of enabling Germany to conduct war after September 1914 in addition to introducing large-scale gas warfare in April 1915. Notwithstanding his ingenious method of capturing nitrogen, the former was mainly the result of rhetoric and propaganda put out by military men, industrialists and scientists during and after World War I. In later years, Haber's role in these two chemistry-related sectors of modern warfare became somewhat sensationalized. In response, I examine here certain purported facts and myths concerning Haber's roles, and reinterpret the foundations of stories about the "Haber factories" that some still believe enabled Germany to conduct war from the autumn of 1914, or even earlier, right through to the Armistice on 11 November 1918.

It is appropriate to note that in 1904 BASF and the other leading German dye firms formed two groups of loose alliances: BASF, AGFA (Actien-Gesellschaft für Anilin-Fabrikation), and Bayer & Co. (Farbenfabriken vorm. Friedrich Bayer), together known as the Dreibund or Triple Confederation; and Leopold Cassella & Co. GmbH and Hoechst (Farbwerke vorm. Meister, Lucius & Brüning), the Double Alliance, in 1907 joined by Kalle & Co. Aktiengesellschaft, and from then on known as the Tripartite Association, or Dreiverband. In 1916, at the height of World War I, the two groups created a community of interests, the Interessengemeinschaft der deutschen Teerfarbenfabriken, often called the IG. In 1925, they merged to create the behemoth IG Farbenindustrie Aktiengesellschaft, better known as IG Farbenindustrie AG, or IG Farben.

3.2 Early Studies on Nitrogen

Investigations of nitrogen-containing gases go back to the last quarter of the eighteenth century. Joseph Priestley (1733–1804) was the first to examine the chemical composition of ammonia (NH_3), which in 1774 he decomposed with an electric spark. The colourless alkaline gas was known long before. Ammonia is named after the Temple of Jupiter Ammon in the oasis of Siwa in Egypt's western desert [1]. It was there that, so the story goes, sal ammoniac (ammonium chloride) was found in camel dung and the decomposing remains of animal sacrifices left out in the searing desert heat.

In 1772, Priestley established the composition of nitric oxide (NO, nitrogen monoxide), and also reported nitrous oxide (N_2O). In 1774 he collected gaseous ammonia ("alkaline air"), and in 1777 he prepared nitrogen dioxide (NO_2) [2]. Later, Henry Cavendish (1731–1810) made nitric oxide using frictional electricity to spark together nitrogen and oxygen, and obtained what he called nitrous acid (nitric acid) [3]. However, Sir William Ramsay (1852–1916), while giving these two famous chemists due credit for their pioneering work, opined that "It was true that neither of those experimenters knew what they were doing. They were dealing with 'phlogisticated' and 'dephlogisticated air' [oxygen]" [4]. By a group decision in France in 1787, the unreactive, nonflammable gas in phlogisticated air that we now call nitrogen was called azote. In 1790, Jean-Antoine Chaptal (1756–1832) called it nitrogène, from nitre, or potassium nitrate. (In 1823 Chaptal published *Chimie appliquée à l'agriculture*.) In German it was Stickstoff.

Around the turn of the century Humphry Davy (1778–1829) prepared and analysed oxides of nitrogen [5]. He also reported the capture of nitrogen as ammonia and nitric acid by electrolysis—a method that was repeated in 1922 [6, 7]. The French chemist Frédéric Kuhlmann (1803–1881) in 1838 reacted nitric oxide with air to form nitrogen dioxide, which with water afforded nitric acid [8]. This became the basis of a French patent filed in 1859 by Madame Louise Lefêbvre. Unfortunately theories of physical chemistry were not advanced sufficiently to assist her, nor was there an economical source of electricity to bring about the continuous combination of oxygen with nitrogen. Nitric acid had been made centuries before from saltpetre (potassium nitrate) [9].

3.3 Natural Nitrogen Fertilizers

3.3.1 Guano and Saltpetre

From the 1840s until the late 1870s, guano from the west coast of South America was the main European source of fixed nitrogen for use in agriculture. Guano had accumulated over thousands of years, principally on the three Chincha Islands, off the coast of Peru, close to the town of Pisco (south of Lima). The foul-smelling material was investigated by the geographer Alexander Humboldt (1769–1859) and botanist Aimé Bonpland (1773–1858) while exploring Spain's American colonies during 1799–1804. Humboldt was told by the Spanish, who learned it from the Indians, that guano aided the growth of plants. On return to Europe, Humboldt asked the French chemists Antoine-François Fourcroy (1755–1809) and Louis-Nicolas Vauquelin (1763–1829) to undertake an analysis of guano. In 1805 they demonstrated its nitrogen content: it contained fixed nitrogen (mainly uric acid) in the form of bird droppings; and it showed potential as a concentrated form of nitrogen fertilizer.

Concerns that food supplies could not continue to keep pace with the growth of populations had been expressed not long before by the English cleric and economist Thomas Robert Malthus (1766–1834) in *An Essay on the Principle of Population* (1798)—the outcome of which would be competition for available resources. By the 1840s, this concern had turned into enthusiasm for the study of fertilizers. The central role of excrement—human and animal—and waste from crop production as sources of nitrogen fertilizer on arable land was recognized by various individuals, including John Mechi (1802–1880), in England, and Liebig. Liebig was originally against adding guano to the soil, claiming that Nature, including lightning and excrement, provided sufficient nitrogen to satisfy plant growth. However, he changed his mind after the phosphorus content of guano was established [10].

Guano was first exported to England, from where the European market was mainly financed and controlled. Between 1841, when 2000 tons were imported, and 1850, there was a hundredfold increase in imported guano. Such was the importance of guano that in 1856 the US Congress passed the Guano Islands Act, empowering US citizens to take possession of unoccupied islands containing guano.

The other significant source of nitrogen, as already mentioned, was saltpetre (potassium nitrate), used in the manufacture of gunpowder in Europe. It was made from nitrogen-containing waste by the action of lime, mortar and wood ash on animal excrement and urine. Antoine Lavoisier (1743–1794) had investigated its manufacture and use—both in agriculture as fertilizer and in gunpowder—and supervised the French saltepetre beds, known as nitrières. Saltpetre was exported on a large scale from India.

3.4 Fertilizers and Agricultural Experiment Stations

The scientific approach to agriculture, with a major input from chemistry, relied on agricultural experiment or research stations, dedicated colleges, and associations of agriculturalists. Their histories in Europe went back to the 1830s and early 1840s. In France during 1836, Jean-Baptiste Boussingault (1802–1887) opened a research station some 60 km north of Strasbourg, where his greenhouse and field tests included pioneering studies on the role of nitrogen in plant growth—work that would be brought to an abrupt halt at the outbreak of the Franco-Prussian War of 1870–1871. In 1843, James F. W. Johnston (1796–1855) was appointed chemist to the new Agricultural Chemistry Association of Scotland. This was absorbed by the Highland and Agricultural Society of Scotland.

In 1843, Lawes and Gilbert commenced their field experiments and soon after emphasized the role of nitrogen as fertilizer. Around 1850, the first German agricultural experiment station was established at Möckern, near Leipzig; in the 1860s a Union of Experiment Stations existed in Bavaria. The rapid growth in agricultural chemistry, with its focus on fertilizers, was demonstrated in 1862 at the International Exhibition in London. There, the "industry of manures" occupied a prominent place in "Section A, Chemical Products and Processes," with a "Tribute to Messers. Lawes and Gilbert," and a "Homage to Liebig" [11].

The academic study of agriculture and the soil expanded during the last four decades of the nineteenth century. Notable were two experiment stations in the Province of Posen, Prussia, opened in 1861 and 1873; in 1877 they were merged to create the Posen station. By 1890 there were over 70 such stations in the German Empire [12–15]. In Holland an experiment station opened at Wageningen in 1877; other important Dutch stations were established at Groningen, Hoorn and Breda. Notable also were Belgium's State Agricultural College in Gembloux (1860), and agricultural colleges and universities in Italy, Denmark, and the union of Sweden and Norway. In Sweden, agricultural institutes opened in 1848 at Ultuna and in 1862 at Alnarp. Norway's first agricultural college opened in 1859 at Ås, near Kristiania (present-day Oslo). By the 1880s these institutions and their allied associations and unions had facilitated the emergence of agricultural science as a profession in its own right. The interest in fertilizers encouraged the transmittal of information on analyses, use, and application through specialist technical and chemical journals. Agricultural research stations published monographs, and analysed and tested novel fertilizers, under laboratory and field conditions.

The first two US state agricultural experiment stations were opened in 1876, in Connecticut and California. The American Association of Official Agricultural Chemists was founded in 1884 under the auspices of the Department of Agriculture. Its members included chemists charged with setting standards for analysis of fertilizers and enforcement of state fertilizer laws. Other important activities included sugar analysis, to detect adulteration as well as to protect Federal coffers, which relied heavily on a system of sugar taxation.

The agricultural sector in Germany was highly dependent on imported nitrogen products as a result of the limited natural resources and a growing population. It provides a useful example of how the demand for and application of nitrogen fertilizer increased productivity of the soil, notably in potato and beet sugar cultivation [16]. By 1890, this enabled Germany to become, as it had been decades earlier, a net exporter of agricultural products.

3.5 Expansion of the Nitrogen Industry

3.5.1 Decline of Guano

Guano created huge fortunes not only in Peru but also in England for William Gibbs of Antony Gibbs & Sons, which controlled much of the European trade. By 1857, 2.4 million tons of guano had arrived in Bristol and London, whence considerable amounts were sent to France, Prussia, and later the Netherlands [17, 18].

The demand for guano from just before mid-century, as well as phosphates, was driven by the massive growth in European populations, contributions of chemistry to agricultural science, and the expansion of steam-driven transport networks on both sea and land (sail was still important for long distance cargo transport). While trade in the export of guano brought about a boom in the Peruvian economy, the emergence of the nation as a new economic power encouraged corruption and created acute political tensions with neighbouring countries, in addition to international rivalries. The outcome was the Chincha Islands War of 1864–1866, involving intervention by Spain, which hoped to regain control of its former colony.

The peak of guano production was reached in 1870, when 280,000 tons were imported into Britain. In 1872, the Peruvian government placed restrictions on export from the Chincha Islands because the quality of the remaining guano was less suited to use as a fertilizer. By 1880, just 20,000 tons arrived in Britain [19]. This marked almost the end of the export trade. Guano from other sources never matched that of the Chincha Islands and Peru, which contained as much as 14% nitrogen in terms of value as a fertilizer [20].

3.5.2 South American Nitrate

The other local source of nitrogen, that of nitrate, became far more important than guano. This nitrate industry was based on sodium nitrate—variously called Chile (an) saltpetre, Chile nitre, Chilean nitrate, nitrate of soda, "nitrates," and *salitre*; until early in the twentieth century also Chili saltpeter—obtained from the mineral caliche. Caliche was found near the coast, in the provinces of Tarapacá (in Peru) and Antofagasta (in Bolivia), and later some distance inland in the Atacama Desert, a plateau, approximately 700 km long, lying between the South Pacific Ocean and the Andes (mainly in Peru). In July 1835, Charles Darwin had observed caliche in the north of Tarapacá, and suggested that it might once have formed the coastline of the sea. The main source was the dry, desolate Atacama Desert. The richest deposits of caliche, once containing 50% or more of sodium nitrate, were in what is today the northern part of Chile (the provinces of Antofagasta and Tarapacá).

From the 1830s, and for almost a century, the South American saltpetre (sodium nitrate) industry satisfied much of the need for gunpowder production, as well as for nitric acid. The central significance of this saltpetre, in addition to its later use in agriculture, arose from major changes in the technology of destruction with the introduction of organic nitro compounds as explosives during the second half of the nineteenth century.

In the manufacture of gunpowder, the moderately deliquescent sodium nitrate was converted into potassium nitrate, known as artificial saltpetre. In the United States, during 1857, Lamont du Pont (1831–1884) substituted South American sodium nitrate for potassium nitrate in blasting powder. Sodium nitrate was used directly by manufacturers of gunpowder in the South during the Civil War years, when the Union blockade cut off supplies of Indian potassium nitrate from Confederate forces. South American saltpetre, apart from replacing guano and the Indian nitrate, through its use in manufacture of gunpowder facilitated the mining industries and railway construction. The nitrate industry expanded initially under ownership of the Antofagasta Nitrate and Railway Company (Compañía de Salitres y Ferrocarriles de Antofagasta), founded in 1872. Chilean, British, and other European interests participated in infrastructure investment, including an extensive narrow gauge railway driven inland into the desert from the coast.

To isolate saltpetre from the caliche, a mechanical means, known as the Shanks process, was used until the 1920s. The crushed mineral was placed in dissolving tanks, the aqueous mixture was steam heated, and after cooling run into settlers. The liquid saturated with nitrate was drawn off, and crystalline sodium nitrate was allowed to form, removed and dried. Due to the scarcity of water supplies in the desert the valuable mother liquor was reused. Mining the caliche and extracting the nitrate were labour-intensive operations and wasteful in energy. The high levels of profit did not encourage improvement [21]. The nitrate served the expanding European agricultural sectors and nitric acid industry [22]. In Britain alone, the production of acid from imported sodium nitrate in the mid-1870s was around 10,000 tons annually [23].

During the 1870s Chile emerged as a new economic power, including through investments in the nitrate industries of its northern neighbours, Bolivia and Peru. The great profits that accrued as a result of the trade in South American nitrate, much of it controlled by British and Chilean interests, led to taxes being imposed by Bolivia, in 1878, and the nationalization of mines in Peru. This contributed to rivalry and eventually warfare (in April 1879) between Chile and Bolivia, the latter aided by Peru. This War of the Pacific ended when Bolivia and Peru capitulated in 1883 (though Bolivia did not sign a peace treaty until 1903). The borders between Chile and Bolivia, and between Chile and Peru, were changed to the benefit of Chile, which now became the principal world supplier of sodium nitrate for the agricultural and explosives industries. Chile saltpetre, as it was now called, was exported to London, Liverpool, and Hamburg for distribution throughout Europe; and it was shipped to both the east and the west coasts of the United States. Exports from the Tarapacá and Antofagasta regions quadrupled in the 1880s, from 226,090 tons in 1880 to 1,065,277 tons in 1890 [24]; over 60% went to Europe and the remainder to the United States [25–27]. The commercial product, around 96% nitrate, consisted of irregular crystals. It was analysed in Chile by European chemists, and again at the various destinations by inspectors and analysts representing the interests of merchants and consumers [28].

3.6 Ludwig Mond

From the 1860s, another source of nitrogen fertilizer was the ammonia obtained during the destructive distillation of coal at gas and, later, coke works (Figs. 3.2 and 3.3). At gas works the ammonia liquor obtained by scrubbing the coal gas with water was absorbed in sulphuric acid; the resulting ammonium sulphate was used directly as fertilizer. By 1890, 140,000 tons of ammonium sulphate were produced from European gas works ammonia each year. Ludwig Mond (1839–1909)—who studied under Hermann Kolbe (1818–1884) at Marburg and Robert Wilhelm

Fig. 3.2 Manufacture of coal gas by destructive distillation of coal. (From Frederick Accum, *Description of the Process of Manufacturing Coal-Gas. For the lighting of streets, houses, and public buildings, with elevations, sections, and plans of the most improved sorts of apparatus. Now employed at the gas works in London*. London, 1819. Sidney M. Edelstein Library)

Fig. 3.3 Movements of coal and coke, as bulk commodities, were mainly reliant on rail transport, for long distances over main lines and within factories on industrial railways. This is part of Granton coal gas works, Edinburgh, Scotland, in 1961. Opened in 1902, for Edinburgh and Leith Corporation, the site was a major producer of coke and by-products from the distillation of coal, including ammonia liquor, and incorporated a chemical works where ammonia was converted into ammonium sulphate. The view includes part of the internal railway system with, in the foreground, a narrow gauge locomotive of reduced dimensions for working in the restricted confines of the gas retort house (at right). The locomotive was used to haul trucks loaded with ash from the retorts. At left is a standard gauge train of the type used for delivering coal to gas works and for removing coke. Coal was stored in bunkers and delivered to the retort house by a system of enclosed overhead conveyors, as seen in the background. For some large gas and coke works on mainland Europe overhead electric tramway systems, with suspended skips, were in use by the early twentieth century. Similar overhead tramways were used for moving sacks of Chilean nitrate and ammonium sulphate between warehouses and ships and railway wagons

Eberhard Bunsen (1811–1899) at Heidelberg, prior to entering industry, first in Germany, then in the Netherlands, before settling in England—gained practical experience in ammonia production from 1872 in connection with his working of the Solvay ammonia-soda process for sodium carbonate (soda ash, or carbonate of soda), used in the manufacture of textiles, soap, paper, and glass. This was the first continuous process introduced into chemical manufacture, as developed successfully by Ernest Solvay (1838–1922), founder of Solvay & Cie, at Charleroi in Belgium during 1863–1865. Overall, the reaction involved treating sodium chloride (salt) with calcium carbonate to give sodium carbonate and calcium chloride, the latter as waste. However, the process had to be carried out in steps. Brine-soaked ammonia was treated with carbon dioxide, released from heated limestone (calcium carbonate) in tall towers; the resulting sodium bicarbonate was heated to give the carbonate, soda ash. Ammonia was regenerated by heating the by-product ammonium chloride with the lime obtained from limestone; this left the calcium chloride. Solvay's process eventually replaced the Leblanc alkali process that was associated with severe and unhealthful environmental releases.

From 1873, Mond, in partnership with the Unitarian John Tomlinson Brunner (1842–1919; made a baronet in 1895) erected a factory employing Solvay's process next to Winnington Hall, near Northwich, Cheshire, north-west England (in later years often referred to as the Northwich works). It was close to the River Weaver, and with good railway connections. This represented the foundation period of the prominent British alkali manufacturer Brunner, Mond & Company [29]. In 1881, William Cogswell and Rowland Hazard co-founded in the United States the Solvay Process Company of Syracuse, New York, in which from 1887 Brunner, Mond had an interest. In 1895 the first by-product coke ovens in America were introduced by the Semet-Solvay Company, based on the design of Solvay's brother-in-law, Louis Semet, associated with the Brussels gas works, founded by Florimond Semet. It was at the gas works in the early 1860s that Solvay's founder developed his technological know-how from tinkering with chemicals and studying the by-products, including ammonia. This was the first step in his rediscovery of the ammonia-soda process, as taken up at Charleroi. Solvay & Cie's licensing agreements stipulated the free exchange of technical developments, which would prove of considerable value during the early years of Brunner, Mond's research into synthetic nitrogen.

Though Mond's main interest during the 1880s was in ensuring a reliable supply of ammonia for his alkali factory, he also realised the great potential for satisfying the artificial nitrogen fertilizer market.

According to Mond, in 1889, 750,000 tons of Chile saltpetre were exported, of which 650,000 tons arrived in Europe. Mond at that time was promoting his new process for coal gas production—an alternative to the Semet-Solvay coke ovens—in order to provide a cheap supply of ammonia. In Mond's process, the combustion of low-quality coal with steam in a current of air gave what became known as Mond, or Mond producer, gas, a mixture high in hydrogen that also contained nitrogen, methane, carbon dioxide, and carbon monoxide [30]. Producer gases were used as fuels. A form rich in nitrogen was obtained by blowing air over coke; it was important in the early synthetic ammonia industry (Fig. 3.4).

Mond's process, for which a more detailed description is given in Chapter 4, was worked at Brunner, Mond's Winnington works from 1902 and at various other sites in Britain, as well as in Argentina, Spain and the United States by 1903. In 1902, of the total British production of 220,000 tons of ammonium sulphate, some 148,500 tons came from gas works and 19,000 tons from coke, carbonizing and Mond gas works. Two-thirds of this ammonium sulphate was exported for use as fertilizer [31]. In Europe around 1900, firms such as Semet and Evence Coppée worked on increasing by-product ammonia recovery from coke ovens. Another leader was Heinrich Koppers, of Essen, who established Heinrich Koppers AG in 1904 and introduced his coke oven into the United States in 1907 [32, 33]. In 1910 the British production of ammonium sulphate from gas works, coke ovens and Mond producer gas works amounted to 367,587 tons and supplemented the supplies of fixed nitrogen from Chile [34]. Japan was an important market for ammonium

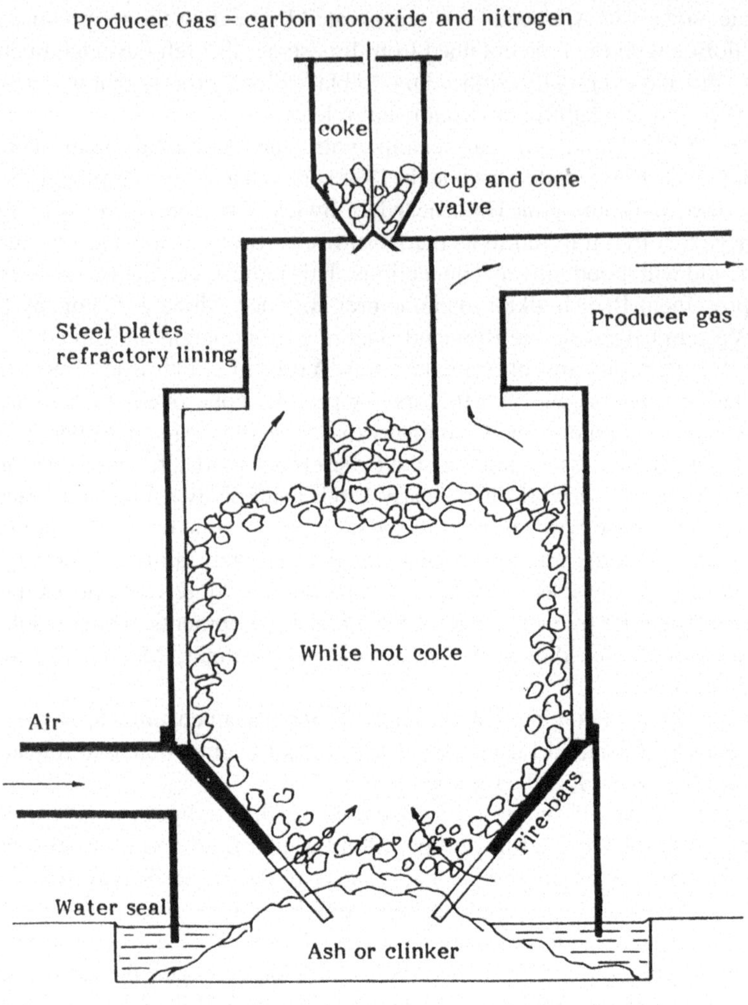

Fig. 3.4 Manufacture of producer gas (mainly carbon monoxide and nitrogen)

sulphate, supplied until 1914 mainly by Great Britain. In wetland agriculture, ammonium sulphate was preferred to sodium nitrate, which ensured its popularity in the Far East.

3.7 Nitro Compounds

From the 1880s, when demands on agricultural output accelerated, Chile saltpetre was the nitrogen fertilizer of choice, particularly in Belgium, France, Germany, Holland, Hungary, and Italy. Until 1914, it represented two-thirds of the world's supply of fixed nitrogen fertilizer and was the principal raw material employed in the manufacture of nitric acid. The acid was used to nitrate such aromatic compounds as benzene, toluene, and phenol. The large-scale industrial manufacture of nitro compounds began at the end of the 1850s, pioneered by William Henry Perkin (1838–1907; knighted 1907) at his dyestuff manufacturing factory, located at Greenford Green, north-west of London. Perkin's nitro compounds (and those of other dye-making firms that emerged from around 1860), after reduction to such aromatic amines as aniline, were reacted with various reagents to afford synthetic dyes [35]. The German chemist-colorist Heinrich Caro (1834–1910) designed the first modern nitration and reduction equipment while working in Manchester around 1860 (Fig. 3.5). Later he would spearhead the massive scientific–technical endeavours that enabled BASF to become the leading science-based chemical corporation by 1900.

In the manufacture of nitro compounds, a mixture of both nitric and sulphuric acids was required. Some of the nitro compounds were used as dyes, such as picric acid (trinitrophenol). There was also trinitrotoluene (TNT), discovered during 1863 in Germany by Julius Wilbrand (1838–1906) while investigating synthetic dyes. This was around the same time that Alfred Nobel (1833–1896) developed nitroglycerine as an explosive, followed in 1866 by dynamite [36]. The first artificial silk process, that of Chardonnet, involved the nitration of cellulose. The nitrated cellulose, also a potentially explosive compound, was the basis of guncotton.

Fig. 3.5 Heinrich Caro. (Edelstein collection)

Picric acid's use as a dye was no more than a few years; at the end of the century it became an important explosive. Trinitrotoluene, despite its explosive properties, was not adopted for military purposes until 1902, when it was used to fill artillery shells in Germany. The aromatic nitro compounds used in explosives were manufactured in factories similar to those in which dye intermediates were synthesized. This meant that any nation with a well-established synthetic dye industry was well prepared for the contingencies brought on by modern armed conflict—as was the case in Germany during 1914, though not in Britain, which had lost most of its dye industry to German rivals by around 1880. However, German production of explosive nitro compounds relied on adequate supplies of Chilean nitrate for the manufacture of nitric acid. This would have important implications from late 1914.

3.8 Dead Ends: Nitrides and Cyanides

3.8.1 Nitrides

Serpek ingeniously combines the production of aluminium nitride with the manufacture of pure alumina, thus associating the aluminium and nitrogen fixation industries.

Bruno Waeser, *The Atmospheric Nitrogen Industry. With Special Consideration of the Production of Ammonia and Nitric Acid* (trans. E. Fyleman). Philadelphia: P. Blakiston's Son & Co., 1926, vol. II, p. 421.

In 1876, American chemist John W. Mallet, a former lieutenant general in the Confederate Army, while attempting to prepare a carbide of aluminium, was struck by an unexpected odour, that of ammonia. His experiment had led not to aluminium carbide but to aluminium nitride, which released the pungent gas. A similar observation had been made in France in 1859 by Charles Tellier (1828–1913), who in 1880 formed a short-lived company to license a manufacturing process. Around that time, Ludwig Mond was also investigating the synthesis of nitrides as a source of ammonia, but without success.

This remained a laboratory curiosity until shortly after 1900, when Czech-born chemist Ottokar Serpek, then residing in Madulain, Switzerland, investigated formation of aluminium nitride from bauxite (alumina, aluminium oxide), as a route to both ammonia and pure aluminium oxide, from which aluminium was prepared. Around 1906, a pilot plant based on Serpek's electrothermal process was set up near Mulhouse. Serpek's interests were merged with the Société Générale des Nitrures, of Paris, which was established by Adrien Badin (1863–1917) of the Compagnie des Produits Chimiques d'Alais et de la Camargue, forerunner of Péchiney (hereafter Pechiney). Serpek soon after moved to Paris, where he vigorously promoted his novel process. In 1909, his nitride work attracted the notice of Japanese chemists, particularly Shiraishi Genjirō of Japan Steel Pipe (Nippon Kōkan), who was also interested in the production of aluminium. However, while in Europe, he was dissuaded from further consideration of the process by physical chemist Tamaru Setsuro, at that time working with Fritz Haber at Karlsruhe's Technische Hochschule [37].

From 1902, Carl Bosch at BASF investigated metal nitrides as possible sources of stable nitrogen products. All the BASF nitride patents were transferred to Société Générale des Nitrures to avoid future patent litigation; in return, the French firm granted licenses to BASF for the Serpek process. In 1911, BASF increased the yield of the aluminium nitride reaction by adding certain oxides [38, 39]. However, by then BASF had made considerable progress with scaling up Haber's synthetic ammonia method, and lost interest. Also interested in aluminium nitride and other nitrides was Gebrüder Giulini, a leading German manufacturer of alumina, Du Pont in the United States, and Norsk Hydro in Norway.

Since there was a clear understanding of the reaction, there was much optimism that it could be commercialized. In August 1912, Société Générale des Nitrures, jointly with the Norwegian Elektrokemisk AS (Det Norske Aktieselskab for Elektrokemisk Industri; Elkem since 1969)—mainly organized in 1904 by Samuel Eyde of the Birkeland-Eyde electric arc nitrogen capture process (see Chap. 5)—formed Det Norske Nitridaktieselskab, which constructed a factory to work the Serpek nitride process near Arendal (north of Kristiansand), close to the extreme south of Norway. Prior to 1914, great hopes were placed in the Serpek process, which during 1912 included at the Bussi aluminium works (Società Italiana per la Fabbricazione dell'Alluminio) in Italy, near Pescara, Abruzzo region, between the Appenine Mountains and Adriatic coast. While some funds were raised for the Italian works, achieving the required balance relied on proving that the process was workable, which was not the case [40]. According to a French patent filed by Société Générale des Nitrures on 24 March 1914, the process employed a three-phase electric arc type furnace in which a mixture of carbon and alumina or some other aluminous substance was carried into the zone of the arc flame by nitrogen. The process required pure nitrogen.

By 1914, however, the Serpek process proved to be unworkable on a commercial scale in Norway (though it was not the end for the company name Société Générale des Nitrures) [41–44]. What was probably a last attempt at improvement, at least in Europe, was made in 1919 by the Internationale Nitridgesellschaft, of Zurich, which set up a Serpek pilot plant at Niedermorschweiler, near Mulhouse [45]. (The company was absorbed by Société Générale des Nitrures.) Well before that, the Serpek process had not lacked critics—for several reasons. There were disagreements over the optimum reaction temperature and the quality of the carbon, although problems with the furnaces were considered to be the greatest impediment.

Nevertheless, for some years after the process was abandoned in Europe, efforts were made to bring about improvements in the United States. There, Serpek had filed one patent in 1907 and two in 1911. On 20 December 1913, he filed an application for fixing nitrogen by means of ferro-aluminium or other ferro alloys. Again, the major problem was the design of a suitable furnace. Paul Hershmann, of Milwaukee, originally from Vienna, took up the Serpek process and developed a carbon-alumina refractory material capable of withstanding temperatures of around 2400 °C. It was used to line the interior of an aluminium nitride furnace operating at 1700 °C [46].

Among critics of the Serpek process was French chemist Camille Matignon (1867–1934) at the Collège de France, specialist in thermochemistry and electrometallurgy, who had made a close study of the process, and E. Herre. In 1919, Hans Goldschmidt (1861–1923), associated with the aluminothermic (thermite) reaction, stated that before the war [47]: "Professor Matignon spoke favourably of the process in general, but said openly that practical success had not been obtained. [Paul L. S.] Héroult and [Theodore] Richards were of the same opinion. The process has been greatly advertised. The shares of the Société Générale des Nitrures, of Paris, were initially dealt with at fabulous prices. At the last International Chemical Congress at New York [Eighth International Congress of Applied Chemistry, 1912]... much misleading information was published by interested parties, so that it appears necessary to state that the process is still in the experimental stage, unless practical results have been obtained quite recently of which something would certainly have been heard."

Following a thorough laboratory investigation of the Serpek nitride process in the United States during the early 1920s, the conclusion was that while it suited the manufacture of pure alumina, it was not a viable process for nitrogen fixation. In other words, the process was not worthwhile commercially [48].

3.8.2 Cyanides

The preparation of metal cyanides was another method of fixing atmospheric nitrogen. Cyanides were required for making pigments as well as for other purposes, and their synthesis attracted investigators as early as the 1830s. In 1838, British chemist Lewis Thomson prepared sodium cyanide by reacting together potash, coke and iron turnings. The cyanide was then converted into the pigment Prussian blue [49].

The manufacture of potassium cyanide was taken up near Paris in 1845 and then transferred to Newcastle; but the high temperature requirement of the process led to its demise in 1847. In 1860, French chemists L. T. F. Margueritte and A. L. de Sourdeval produced barium cyanide in the laboratory by heating a mixture of barium carbonate, coke (carbon) and nitrogen at 1200 °C. On cooling, in the presence of water vapour, the cyanide released ammonia [50, 51].

Further investigations were carried out during the latter part of the nineteenth century, including by Ludwig Mond as part of his studies into the fixation of atmospheric nitrogen. In 1879, Mond commissioned the Austrian chemist Josef Hawliczek to study first the production of ammonia from coke, and then the formation of nitrides and cyanides. While the nitride work did not meet with success, cyanide held out some promise. Barium carbonate (as briquettes) was heated with coke bonded with pitch, to form barium cyanide. Treatment of the cyanide with superheated steam gave ammonia. In 1882, Mond filed a patent for the process, but it proved unworkable on a commercial scale [52]. As with other thermochemical fixation processes prior to the early 1900s the high temperature requirements of the reaction (over 1400 °C), for which no suitable materials for furnace construction were available, led to abandonment. Though improvements were made to the process of Margueritte and de Sourdeval, their process was not viable. Mention should also be made of the mid-1890s laboratory work of Frank and Caro on the reaction between nitrogen and carbides, that at first focused on cyanides (see Chap. 5), but resulted in the preparation of calcium cyanamide, and a revival of interest in cyanides during World War I (see Chap. 8).

References

1. Flood WE (1963) The origins of chemical names. Oldbourne, London, pp 34–35.
2. Partington JR (1946) General and inorganic chemistry for university students. Macmillan & Co., Ltd, London, pp 545, 570, 581.
3. Jungnickel C, McCormmach R (1999) Cavendish: the experimental life, rev. ed. Bucknell University Press, Lewisburg, PA, pp 366–367.
4. Guye PA (1906) The electro-chemical problem of the fixation of nitrogen. J Soc Chem Ind 25(12)(30 June):567–578, quoting Ramsay, on 575.
5. Knight D (1992) Humphry Davy: science & power. Blackwell Publishers, Oxford, pp 34–35.
6. Davy H (1807) The Bakerian Lecture. On some chemical agencies of electricity. Phil Trans R Soc London 97:1–56
7. Jewess M, Crabtree RH (2016) Electrocatalytic nitrogen fixation for distributed fertilizer production. ACS Sustainable Chem Eng 4:5855–5858, on 5856.
8. Kuhlmann F (1887) Mémoire sur nitrification, 1838. In: Recherches Scientifiques et Publications Diverses. Victor Masson, Paris, pp 240–247.
9. Karpenko V (2009) Some notes on the early history of nitric acid: 1300–1700. Bull Hist Chem 34(2):105–116.
10. Brock WH (1997) Justus von Liebig: the chemical gatekeeper. Cambridge University Press, Cambridge.
11. Iselin JS, Le Neve Foster P (eds) (1863) Reports by the juries, International Exhibition, London, 1862. Class II, section A. Chemical and pharmaceutical products and processes. Society for the Encouragement of Arts, Manufactures and Commerce, London, p 7.
12. Rossiter M (1975) The emergence of agricultural science: Justus Liebig and the Americans, 1840–1880. Yale University Press, New Haven, CT.
13. Finlay MR (1991) Science and practice in German agriculture: Justus von Liebig, Hermann von Liebig, and the agricultural stations. In: Woodward WR, Cohen RS (eds), Scientific discipline formation: science studies in the German Democratic Republic. Kluwer, Dordrecht, pp 309–320.
14. Brassley P (1995) Agricultural research in Britain, 1850–1914: failure, success, and development. Ann Sc 52:465–480.
15. Harwood J (2005) Technology's dilemma: agricultural colleges between practice and science in Germany, 1860–1934. Peter Lang, Bern.
16. Perkins JA (1981) The agricultural revolution in Germany, 1850–1914. J Eur Econ Hist 10:71–74.
17. Leigh GJ (2004). The world's greatest fix: a history of nitrogen and agriculture. Oxford University Press, Oxford, pp 78–81.
18. Hoffmann R (ed) (1862) Jahresbericht über der Fortschritte der Agriculturchemie, Dritter Jahrgang 1860–1861. Springer, Berlin.
19. Melillo ED (2012) The first green revolution: debt peonage and the making of the nitrogen fertilizer trade, 1840–1930. Am Hist Rev 117(4):1028–1060.

20. Smil V (2001) Enriching the earth: Fritz Haber, Carl Bosch, and the transformation of world food production. MIT Press, Cambridge, MA, pp 41–48.
21. Guttmann O (1895) The manufacture of explosives: a theoretical and practical treatise on the history, physical and chemical properties, and the manufacture of explosives, vol. I. Whittaker, London, pp 41–42.
22. Hoffmann R, Peters E (eds) (1868) Jahresbericht über der Fortschritte auf dem Gesammtgebiete der Agricultur-Chemie, Zehnter Jahrgang: Das Jahr 1867. Springer, Berlin.
23. Hardie DWF, Pratt JD (1966) A history of the modern British chemical industry. For the Society of Chemical Industry. Pergamon, Oxford, p 48.
24. Blakemore H (1974) British nitrates and Chilean politics, 1886–1896. Athlone Press, London.
25. Gustav Müller (1909) Die Chemische Industrie. BG Teubner, Leipzig, pp 201–203.
26. Loeber K (2010) Der Niedergang des Chilesalpeters: Chemische Forschung, militärische Interessen, ökonomische Auswirkungen. Wissenschaftlicher Verlag, Berlin.
27. Wiley HW (1908) Principles and practice of agricultural analysis: a manual for the study of soil, fertilizers, and agricultural products; for the use of analysts, teachers, and students of agricultural chemistry, 2nd ed, vol. I. Fertilizers and insecticides. Chemical Publishing Company, Easton, PA, p 294.
28. Travis AS (2014) The emerging role of titrimetry in late nineteenth-century industrial problem solving: the example of trace analysis for perchlorate in Chile saltpetre. Ambix 61(1):67–94.
29. Morris PJT (1989) The legacy of Ludwig Mond. Endeavour 13(1):34–40.
30. Mond L (1889) The president's address. J Soc Chem Ind 8(7)(31 July):505–510.
31. Wood RD & Co. (1903) Mond gas. R.D. Wood & Co., Philadelphia, p 82.
32. van Rooij A, Homburg E (2002) Building the plant: a history of engineering contracting in the Netherlands. Walburg Pers, Eindhoven, pp 23–25.
33. Koppers Company (1943) Koppers-Beckers coke ovens, 1893–1943. Koppers Company.
34. Kilburn Scott E (1912) The manufacture of nitrates from the atmosphere. Nature 89(2227) (4 July):463–465, on 463.
35. Travis AS (1993) The rainbow makers: the origins of the synthetic dyestuffs industry in western Europe. Lehigh University Press, Lehigh, PA.
36. Fant K (1993) Alfred Nobel: a biography (trans. Ruuth M). Arcade, New York.
37. Molony B (1990) Technology and investment: the prewar Japanese chemical industry. Council on East Asian Studies, Harvard University. Distributed by Harvard University Press, Cambridge, MA, pp 117–118.
38. Farber E (1966) From chemistry to philosophy: the way of Alwin Mittasch (1869–1953). Chymia 11 (1966):157–178, on 160.
39. Waeser B (1926) The atmospheric nitrogen industry: with special consideration of the production of ammonia and nitric acid (trans. Fyleman E), vol II. P Blakiston's Son & Co., Philadelphia, p 423.
40. Benegiamo M (2013) Bussi e la grande chimica in Abruzzo. Un'ambizione fallita. Textus Edizioni, L'Aquila.
41. Cerruti L, La fissazione dell'azoto, ovvero l'ambiguità della scienza, 5–6. http://www.minerva.unito.it/storia/chimicaclassica/azoto/azoto4.htm (accessed 9 September 2015).
42. On Pechiney's engagement in the aluminium industry, see Le Roux M (1998) L'Enterprise et la Recherche. Une siècle de recherche industrielle à Pechiney. Editions Rive Droite, Paris.
43. Le Roux M (2015) From science to industry: the sites of aluminium in France from the nineteenth to the twentieth century. Ambix 62(2):114–137.
44. Périères M (1955) Une grande compagnie industrielle français: Péchiney. Revue de géographie alpine 43(1):151–212, on 163.
45. Waeser B (1926) The atmospheric nitrogen industry: with special consideration of the production of ammonia and nitric acid (trans. Fyleman E), vol II. P Blakiston's Son & Co., Philadelphia, pp 415–433.

46. CH MacDowell to Williams Haynes, 16 October 1943. In: Haynes W (1945) American chemical industry: a history, vol. II. The World War I period, 1912–1922. D. Van Nostrand, New York, pp 370–372, on 371.
47. Waeser B (1926) The atmospheric nitrogen industry: with special consideration of the production of ammonia and nitric acid (trans. Fyleman E), vol II. P Blakiston's Son & Co., Philadelphia, p 432.
48. Krase HJ, Thompson JG, Lee JY (1926) The fixation of nitrogen as aluminum nitride. Ind Eng Chem 18(12)(December):1287–1290.
49. Thomson L (1839) Improvements in the manufacture of Prussian blue. Dinglers Polytechnisches Journal 73:281.
50. Farrer WV (1971) Some early ventures in the fixation of atmospheric nitrogen. Ambix 18(2):123–138.
51. Martin G, Barbour W (1915) Industrial nitrogen compounds and explosives. Crosby, Lockwood and Son, London, p 78.
52. Mond L (1889) The president's address. J Soc Chem Ind 8(7)(31 July):505–510, on 505–507.

Chapter 4
Ammonium Sulphate

4.1 Coke Oven and Mond Gases

Throughout the period under review here, ammonium sulphate, because of its several sources, was the most important nitrogen fertilizer. The sulphate's industrial origins ranged from coal-based processes—namely the coal gas and coke oven works—to production from calcium cyanamide and then synthetic ammonia. By the mid-1920s, ammonium sulphate made from the nitrogen capture processes had considerably reduced demand for Chilean nitrate.

Pure ammonium sulphate contains 21.2% of nitrogen (compared with 16.5% of nitrogen in sodium nitrate). It was available as a white powder or in crystalline form, of low hygroscopicity. Ammonium sulphate was a source of both nitrogen and sulphur; its chemical stability was well suited to long distance transportation and storage. It was acid forming when applied to the soil and thus useful for treating alkaline conditions.

The destructive distillation of coal, which contains around one per cent of nitrogen, was the first source of ammonia for the large-scale production of ammonium sulphate. By 1850, coal gas works provided by-product ammonia that was converted into ammonium sulphate; for every 100 kg of coal destructively distilled, around 1 kg of ammonium sulphate was obtained. From the late 1890s, ammonium sulphate was increasingly made from coke oven ammonia, which had previously gone to waste. An important difference between the coal gas and coke oven gas processes was that in the former an ammonia liquor was obtained that required further treatment, and concentration, whereas in the latter the coke oven gases were passed directly through sulphuric acid, with the formation of a more concentrated ammonium sulphate. Ludwig Mond modified the distillation process, in order to increas the yield of ammonium sulphate.

Mond found that if distillation was conducted in a stream of superheated steam, the amount of ammonium sulphate obtained from 100 kg of coal was increased to 3 kg. Fine and waste coal was charged into generators where gasification took place

A.S. Travis, *Nitrogen Capture*, https://doi.org/10.1007/978-3-319-68963-0_4

through the action of air and steam. By keeping the temperature relatively low and using a large excess of steam, most nitrogen present in the coal formed ammonia. The mixture of gaseous products was directed through heat exchangers, where incoming air was preheated. The cooled gas was passed through a washer to remove tar, then up a tower against a current of descending sulphuric acid, with which the ammonia formed ammonium sulphate. The acidic solution was then passed up another tower in the presence of more acid until it was saturated with ammonium sulphate. The liquid was run off and evaporated to afford the solid, stable ammonium sulphate product. The sulphate was separated out, dried and stored in large sheds. This was the basis of the Mond producer gas process, put into practice from 1901 and promoted by the British Power-Gas Corporation, of Stockton-on-Tees, directed by Emile Mond (1865–1938), the French-born nephew of Ludwig Mond. The process was further developed by Brunner, Mond at its Winnington works from 1902.

Mond producer gas was used in gas engines, which were more efficient than steam engines for driving dynamos, pumps, and other devices, in addition to its use as a heating fuel. The Power-Gas subsidiary, the South Staffordshire Mond Gas (Power and Heating) Company, was established in 1901 to work the Mond process in the Midlands of England. The gas was piped under pressure to industrial consumers through an extensive distribution network. By 1918, Mond gas plants, following extensive promotion by Power-Gas, were in use in Japan and Manchuria.

The South Staffordshire Mond Gas consulting engineer was Herbert A. Humphrey (1868–1951), who had studied engineering at Finsbury Technical College (1883–1885) and the Central Institution (1885–1887; the present-day Imperial College London). By all accounts, Humphrey was an outstanding student and at Finsbury, according to his chemistry lecturer, Henry E. Armstrong (1848–1937), displayed an interest in chemistry such that it might be of use to an engineer, something that was not common. In 1887 he joined the engineering firm Heanan & Froude, of Manchester, where after a short time he was appointed manager of the engine and electrical department at the Newton Heath Iron Works, in Lancashire, and later became manager of the general engineering works in Birmingham. In December 1890, Humphrey was offered a post at Brunner, Mond's works at Winnington, as engineering manager of the refined bicarbonate and crystal department where crude sodium bicarbonate was processed into a purer product and washing soda (1891–1901). In 1892, Ludwig Mond, who was greatly impressed with the newcomer's efforts, supported his application for fellowship of the City and Guilds of London Institute. Through his close association with the Mond enterprises, Humphrey would have a long and important association with the nitrogen story in Britain (see Chap. 9).

Humphrey had made a careful study of the Winnington ammonia-soda process, in his own time, and by 1897 was investigating the manufacture of Mond producer gas. Humphrey's initiatives, and his ambition, were recognized and appreciated by Ludwig Mond, who made every effort to provide support. Humphrey in early 1899 had written to Mond in Rome (Mond's residence) regarding greater independence in his work, and advancement. A meeting with Mond was agreed in the billiard room at Winnington Hall on June 13, "about 11.15 pm." There, "after talking over utilizing exhaust gases," the discussion turned to a post in the Brunner, Mond engineering department under a Mr Johnson, something that Humphrey was not prepared to accept. The meeting continued [1]:

> *Dr. Mond*: Well I did not expect you would (with a smile) and I think it would be more suitable and acceptable work if you were to become associated with me in the important work of getting twice the amount of mechanical energy out of coal than can be obtained by steam engines... Also I wish to feel myself free to call upon your services in the future. So I propose to make up your salary to what you have asked for, out of my own pocket, until such time as I find I have enough work in connection with the gas producer and power plants to occupy all your time.
> Does this meet your wishes?
> *HAH*: I think I have every reason to feel satisfied with your proposal and....
> *Dr Mond (breaking in):* Well I am glad as we will regard that matter as settled.
> *HAH:* Thank you Dr.

This was Humphrey's introduction into the gas industry. In 1901, with the approval of Ludwig Mond, he set up as a consulting engineer in London. Almost from the start Humphrey became closely associated with South Staffordshire Mond Gas, and served as part-time consultant to both Power-Gas and Brunner, Mond. From 1903 other clients included British Westinghouse Electric Manufacturing Ltd, and Salt Union Ltd, a firm that had connections with Brunner, Mond.

From 1904, South Staffordshire Mond Gas provided facilities for experimental work on the uses of Mond producer gas, including a gas turbine, driven by the gas, intended to replace steam turbines in electrical power stations. This led to what became known as the Humphrey gas pump, a direct-acting gas-powered water pump patented in 1906, described to great acclaim in 1909, and installed by the Metropolitan Water Board at Chingford reservoir in 1913 [2].

4.2 Peat and Ammonium Sulphate

Around 1900, Adolph Frank became associated with Power-Gas following his renewed interest in the uses of peat. This included its combustion when partially dried for the generation of electricity, and gasification for general use as a source of power, as suggested by Frank in 1897 [3]. By 1904, Frank and his colleague Nikodem Caro became aware that the Mond gas process offered a use for peat in power generation in Germany and elsewhere. This happened after Caro had worked with Power Gas at its Stockton-on-Tees facility on the utilization of waste colliery products, and also of German Seeschlick, a lake slime containing considerable nitrogen. Not long before, peat had been successfully gasified in England by the Mond process. Peat was now further investigated by Caro. Power-Gas erected a plant in Germany suited to peat, and around 1906 two further plants were built by the company, one in Italy that worked until the 1930s. Frank reported that in May 1908 at the works of the Deutsche Mondgas- und Nebenprodukten-Gesellschaft mbH, at Sodingen, Westphalia, 40 tons of wet peat containing 50–55% water were gasified without difficulty to provide power for a Deutz gas engine. In addition, ammonium sulphate corresponding to 75–80% of nitrogen in peat could be obtained [4]. Arthur Henry Lymn modified the Mond process for the improved extraction of ammonia.

4.3 Marketing Ammonium Sulphate

Chile saltpetre was poorly suited to the wet soil cultivation of rice. As a result, during 1908–1912, Japan was the largest export market for British coal-derived ammonium sulphate. This provides an example of the enormous international trade in the sulphate after 1900. In Japan from around 1910 ammonium sulphate was produced from ammonia that, by the action of steam, was released from calcium cyanamide. This was important, since farmers disliked working with the newly introduced cyanamide, which among other things was unpleasant to handle. The ammonium sulphate business elsewhere was no less significant. Spain—with a substantial agricultural sector focused on citrus fruits and olives—and the United States followed Japan as leading importers. In Germany during 1913, ammonium sulphate accounted for 111,000 tons of fixed nitrogen consumed as fertilizer.

The large-scale production of ammonium sulphate from coal gas and coke works led to the formation of associations of manufacturers, particularly in Britain and Germany, the principal producers, for promoting both home markets and exports. In 1895, Deutsche Ammoniak Verkaufs-Vereinigung GmbH was founded in Bochum by the German coke-oven producers. There was also the Verkaufs Vereinigung für Stickstoffdunger. In 1897, the British Sulphate of Ammonia Committee was established, representing the coal-based sulphate industry. It became the British Sulphate of Ammonia Association just before the outbreak of World War I. In 1920, one year after a British wartime export embargo was lifted, its assets were transferred to the British Sulphate of Ammonia Federation Ltd. In 1921, there were 396 members. Sales were co-ordinated by the federation, which in 1923 appointed Brunner, Mond & Co. as the sole agent in Japan, then still an important and extremely robust export market. Brunner, Mond's "Crescent" brand ammonium sulphate, dark brown with white crystals, was a success in China and Japan, in China outselling the cheaper but equally good product available from the Sulphate of Ammonia Federation, which created a conflict of interests [5]. British exports of the sulphate to Japan fell off in the mid-1920s but was compensated for by considerably increased export to Spain [6].[1]

In 1925, the Brunner, Mond subsidiary Synthetic Ammonia and Nitrates Ltd, then recently engaged in the manufacture of synthetic ammonia at Billingham, in

[1]In 1925, the United States exported 123,000 tons of ammonium sulphate to Japan, the Philippines, and Dutch East Indies (present-day Indonesia), out of home production of 544,000 metric tons.

the north-east of England, joined the federation. In December 1926, Brunner, Mond became part of ICI (Imperial Chemical Industries). This was a response to competition from the newly formed German conglomerate IG Farben, the merger of BASF and other firms in 1925, in which year the German production of ammonium sulphate (mainly from BASF-made synthetic ammonia) was estimated at 1,200,000 metric tons [7].

Synthetic Ammonia and Nitrates and the federation in 1926 established Nitram Ltd to market and control, at first, sales in Britain of nitrogen products other than ammonium sulphate, particularly urea, then supplied by IG Farben, and, should it be produced there, from the Billingham factory of ICI. Later Nitram promoted sales of ammonium sulphate. One of the directors was Lieutenant-Colonel George P. Pollitt, who had overseen development of Brunner, Mond's synthetic ammonia factory at Billingham (Chap. 9). Nitram was in effect a Brunner, Mond company. It was dissolved in 1930, when there was a glut of nitrogen products, and demand for the federation's coal-based ammonium sulphate had declined [8].

Ammonium sulphate was generally shipped in strong bags. In the 1920s at the Clairton By-Product Coke Company Works, of the American Carnegie Steel Company, "the largest coke producing plant in the world," the sulphate was packed in "double bags using sewing machines which sew first the inside bag and then the outside bag independently." Twelve men weighed and sewed 3800 bags in ten hours. Some 250 tons were shipped every day—sold to US consumers in the growing season, and exported out of season [9].

British chemist James West developed a method whereby ammonium sulphate was made by converting ammonia into ammonium carbonate, using by-product carbon dioxide from the synthetic ammonia process, and then reacting the carbonate with gypsum (anhydrite). This did away with the need for sulphuric acid, and was adopted at the Billingham ammonia facility, below which a seam of the mineral anhydrite was discovered. Until the Billingham anhydrite mine was complete, in the late 1920s, this mineral came from Cumberland and Nottingham, and later from nearby West Hartlepool. A similar process, often called the Merseburg method, was used by BASF/IG Farben, which mined gyspum in the Harz Mountains.

One attractive and important feature of the fertilizer business, at least for manufacturers, was the frequency of repeat orders, since fertilizers had to be reapplied to the soil. Sales to farmers in Britain of ammonium sulphate was mainly in the hands of merchants and corn chandlers. As elsewhere, farmers were in debt to these suppliers, which restricted sales. In order to enhance sales of the Billingham product, during the second half of the 1920s ICI seriously considered direct sales to farmers by introducing favourable credit terms and reduced prices [10].

4.4 The Far Eastern and South East Asian Markets

For the first three decades of the twentieth century, Japan was the principal importer of ammonium sulphate from, successively, Britain, Germany, and the United States. Japanese home production of sulphate from coal began on a small scale in 1901 at Tokyo Gas. All ammonium sulphate and one-third of the Chilean nitrate imported was used for manurial purposes, the balance of nitrate being employed in the manufacture of explosives. In 1914, Japanese demand for ammonium sulphate amounted to 120,000 tons [11]. World War I had a major impact on the British supply, which was cut off for the duration of hostilities. This led to a tripling of Japanese home production (Tables 4.1 and 4.2).

From 1916, the Electro-Chemical Co. (Denki Kagaku Kōgyō Kabushiki Kaisha, Denka) produced ammonium sulphate from calcium cyanamide at Fushun, Manchuria, the site of Asia's largest opencast coal mine [12]. Fushun coal contained a relatively high percentage of nitrogen, 1.6%, which made it suitable for processing in the Mond gas facility built by the South Manchuria Railway at the mine in 1914 and enlarged in 1917 and again in the early 1920s. By 1918, ammonium sulphate was obtained from Mond gas plants in Japan installed by Ida Coal Mining, Fuji Paper, and the government railways. These were the main producers of coal gas-derived sulphate, after Tokyo Gas.

After 1918, imports of ammonium sulphate grew rapidly, exceeding previous levels and reaching 203,500 tons in 1925. Home production was 130,000 tons, including from the recently introduced synthetic ammonia process of Japan Nitrogeneous Fertilizer, Inc. (Nippon Chisso Hiryō KK; Nitchitsu) on Kyūshū, Japan's third largest and most southerly island [13, 14].

Until the early twentieth century, bean cake imported from China had been an important fertilizer in Japan. Periods of shortages and during the 1920s of decrees that supported the home nitrogen industry encouraged the growing use of ammonium sulphate. By around 1930, Nitchitsu and Denka accounted for 40% of the total supply of sulphate, and 75% of home production.

In 1932, during a period of overproduction brought on by the global depression, the Japanese Ammonium Sulphate Distribution Agency was established to control

Table 4.1 Imports of nitrogen products to Japan, in tons

	Ammonium sulphate	Chile saltpetre
1896	None	4539
1900	1730	6440
1904	11,139	5633
1908	65,850	5887
1912	83,928	19,037
1913	110,634	26,539
1915	19,790	26,976
1917	14,980	54,649

Source: "Japan: The Nitrogen Industry," *Journal of the Society of Chemical Industry*, vol. 37, no. 21 (15 November 1918):416R

Table 4.2 Home production of ammonium sulphate in Japan, in tons

1913	7402
1914	15,904
1915	31,571
1916	40,620
1917	52,917

Source: "Japan: The Nitrogen Industry," *Journal of the Society of Chemical Industry*, vol. 37, no. 21 (15 November 1918):416R. The above figures do not include Manchuria, where ammonium sulphate production in 1916 was 3462 tons; and in 1917, 10,178 tons (5400 tons from the Mond gas process). Japanese cyanamide production (mainly converted into ammonium sulphate) in 1914 was 10,990 tons; in 1915, 29,600 tons; in 1916, 32,934 tons; and in 1917, c. 50,000 tons

prices and set production quotas. By the mid-1930s, partly as a result of Nitchitsu's new synthetic ammonia factory at Hûngnam in northern Korea, Japan was self-sufficient. Over 95% of Korean exports of sulphate from Hûngnam went to Japan. Geographical and transport factors, particularly the high cost of rail transport, favoured shipping sulphate from Hûngnam to Hokkaido, and from Kyūshū to southern Korea. For the first nine months of 1935, ammonium sulphate production was 684,000 tons, and calcium cyanamide production (also converted into sulphate) was 147,000 tons [15]. At the close of the 1930s Nitchitsu and Showa Fertilizer were the leading producers of ammonium sulphate made from synthetic ammonia (see Chap. 15).

4.5 The International Market in the Mid-1920s

German exports of ammonium sulphate grew from 11,600 metric tons in 1921 to 119,200 metric tons in 1923 and 344,400 metric tons in 1925. Japan was the main customer for German sulphate, of which 142,609 tons arrived in 1925 [16]. The second chief purchaser of German sulphate was the Netherlands (88,260 tons). Belgium was also an important market, importing 27,400 metric tons in 1925, supplementing home production of 64,000 metric tons [17]. These figures reflected the high usage per acre on arable land of nitrogen fertilizer: the Netherlands, 70.6 lb; Belgium 45.9 lb, and Japan 16.8 lb (compared with Germany at 22.3 lb, and the United Kingdom at 10.4 lb) [18].

Other destinations for German ammonium sulphate were Italy, Denmark, Spain, Czechoslovakia, the United States, and Cuba. Two-thirds of German nitrogen products, mainly ammonium sulphate, came from two Haber-Bosch plants, one-quarter from ammonium sulphate produced in coke and gas plants, and the balance (around 10%) from cyanamide.

Germany imported 24,428 tons (possibly considerably more) of Chilean nitrate, of which 15,428 tons were re-exported, chiefly to the Soviet Union, which required refined nitrate for explosives [19].

Italy, an important consumer of ammonium sulphate, was from the mid-1920s a major producer. Domestic output in 1925 was around 30,000 tons and in 1926, 61,000 tons, during which year 65,000 tons were consumed (21,700 tons were imported). Some 44,000 tons of sulphate were derived from synthetic ammonia; 5570 tons came from gas works; 5460 tons from coke ovens; and 4570 tons from gasification of peat and lignite (sub-bituminous coal) in Mond gas plants [20]. Just 1200 tons were made from cyanamide. Of 47,000 tons of cyanamide manufactured in Italy, most was for direct application to the soil. In France, another major producer and consumer of ammonium sulphate, production grew from 52,000 metric tons in 1921, to 117,000 metric tons in 1925. Over the same period, imports grew from 68,000 to 133,000 metric tons [21]. Polish production of ammonium sulphate from coal gas and coke works increased from 15,400 tons in 1922 to 37,800 tons in 1929 [22]. After 1931, the output of gas and coke works sulphate in industrialized countries declined (Fig. 4.1).

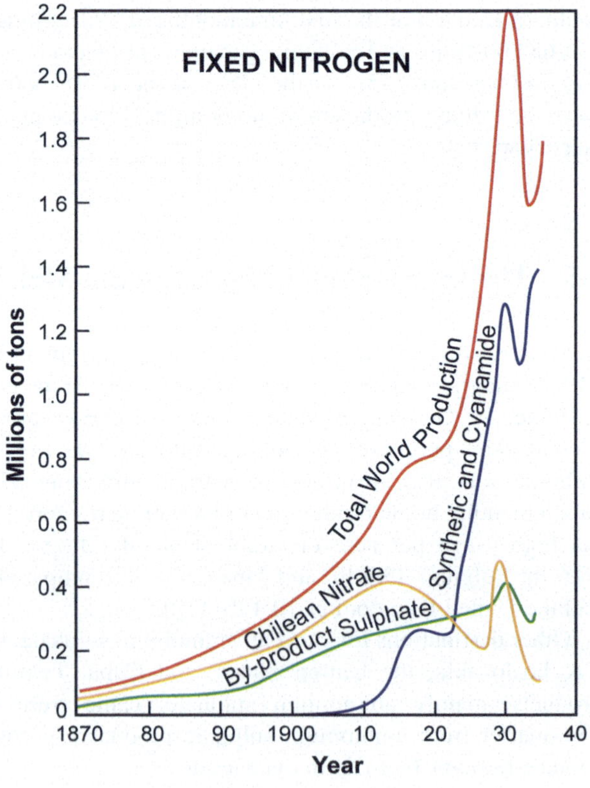

Fig. 4.1 Production of fixed nitrogen (from synthetic ammonia and calcium cyanamide), Chilean nitrate, and by-product coal gas and coke works ammonium sulphate, until 1932–1933. (Redrawn from report of the British Sulphate of Ammonia Federation appearing in *Chemistry and Industry*, no. 2, 18 January 1935, p. 65)

4.5.1 Improving Ammonium Sulphate

From around 1930, Dutch investigators at Royal Dutch-Shell's MEKOG (Maatschappij tot Exploitatie van Kokesoven-Gassen) synthetic ammonia facility at IJmuiden tackled problems related to the production of ammonium sulphate, including the corrosive action of dilute sulphuric acid on hard lead equipment, and caking of the sulphate. The corrosion problem was overcome during 1931–1932 through the introduction of recently developed stainless steel alloys. The caking problem, which also affected transportation and application, was taken up in collaboration with the Dutch States Mines (DSM), and solved in part by production of large ammonium sulphate crystals through the addition of double superphosphate (October 1934). The crystals were better suited for shipping and application to the soil. Further work at the state mines during 1935–1936 resulted in an improvement based on oiling with spindle oil. The process was licensed to firms in England, Canada, Belgium, and Japan. Reflecting the decline in demand for ammonium sulphate and a growing interest in ammonium phosphates and then calcium nitrate during the late 1930s, MEKOG ceased manufacture of the sulphate in 1939 [23, 24]. Calcium nitrate, which contains 17% nitrogen, was useful on acidic soils. Its manufacture for use as a fertilizer was carried out mainly in the Netherlands and Norway.

Compound nitrogen-phosphorus fertilizers were investigated before 1914 by American Cyanamid, through combination of cyanamide with phosphate. Interest in the solid crystalline or granular fertilizers, based on the combination of nitrogen products and superphosphate or phosphoric acid, as well as potash, had emerged in the 1920s particularly in Germany, Norway, the United States, Britain and France. Agricultural research stations, agricultural unions, and firms were involved in bringing about improvements. Investigators included Giacomo Fauser, for Montecatini, in Italy, Georges Claude, in France, and Erling Johnson, at Odda Smeltverk in Norway. Labour-saving methods of application, including machines, for spreading fertilizer, and harvesting, were of interest due to the loss of farmworkers as a result of the war and the migration away from rural areas. In 1926 the French firm Compagnie de Saint-Gobain created the Bureau Central de Reseignment Agricole et de Propagande to undertake research into improved forms of fertilizers [25]. At the end of the 1920s, Imperial Chemical Industries (ICI) in Britain commenced field trials aimed at improving its nitrogen fertilizers and educating farmers in their use. The first major commercial success with a three-component nitrogen-phosphorus-potassium fertilizer was IG Farben's product, given the brand name Nitrophoska, introduced in 1927. The granular mixed products had advantages in transport, storage and application.

References

1. Interview with Dr Mond on the evening of June 13th Tuesday. Mr R Mond present. Transcript dated 14 June 1899. Papers of Herbert Alfred Humphrey, Archives, Imperial College London, item C4–C5.
2. Smith D (1970) The Humphrey pump and its inventor. Trans Newcomen Soc 43(1):67–92, on 67–68.
3. Frank A (1897) Verwerthung der nordeutschen Moore insbesondere für elektrische Kraftstationen. Verein zur Beförderung des Gewerbefleisses, Berlin (4 October).
4. Frank A (1908) Chemical industry in relation to agriculture. J Soc Chem Ind 27(22) (30 November):1093–1100.
5. Reader WJ (1970) Imperial Chemical Industries: a history, vol I. The forerunners, 1870–1926. Oxford University Press, London, p 341.
6. Reader WJ (1975) Imperial Chemical Industries: a history, vol. II. The first quarter-century, 1926–1952. Oxford University Press, London, pp 103–112.
7. League of Nations (1927) Armaments year-book: general and statistical information. Third year. 1926–1927. League of Nations, Geneva, p 487.
8. Reader WJ (1970) Imperial Chemical Industries: a history, vol I. The forerunners, 1870–1926. Oxford University Press, London, pp 369–370.
9. Humphrey HA (1927) Preliminary report on visit to United States of America. 9th April to 28th May 1927. Papers of Herbert Alfred Humphrey, Archives, Imperial College London, item M6, pp 45–51, on 50.
10. Reader WJ (1975) Imperial Chemical Industries: a history, vol. II. The first quarter-century, 1926–1952. Oxford University Press, London, pp 98–115.
11. Ammonium sulphate in Japan (1927) J Soc Chem Ind 46(6)(11 February):123 (Chem Ind London).
12. Japan. The nitrogen industry (1918) J Soc Chem Ind 37(21)(15 November):416R.
13. League of Nations (1927) Armaments year-book: general and statistical information. Third year. 1926–1927. League of Nations, Geneva, p 595.
14. Khosla A (2006) Trade and industrial organisation: Japanese ammonium sulphate industry in the interwar period. Austr Econ Hist Rev 46(1) (March):1–22.
15. Increased Japanese production of chemicals (1936) J Soc Chem Ind 55(13)(27 March):263 (Chem Ind London).
16. Wilson O (1926) Fertilizer trade developments. Ind Eng Chem 18(4)(April):401–404, on 402.
17. League of Nations (1927) Armaments year-book: general and statistical information. Third year. 1926–1927. League of Nations, Geneva, p 59.
18. Wilson O (1926) Fertilizer trade developments. Ind Eng Chem 18(4)(April):401–404.
19. Slade RE (1926) The nitrogen industry and our food supply. J Soc Chem Ind 49(37)(12 September):760–766 (Chem Ind London).
20. Ammonium sulphate in Italy (1928) J Soc Chem Ind 47(36)(7 September):934 (Chem Ind London).
21. League of Nations (1927) Armaments year-book: general and statistical information. Third year. 1926–1927. League of Nations, Geneva, p 470.
22. Zamoyski T (1930) Chemical products. The Times Trade and Engineering Supplement, Poland and Free City of Danzig Industrial Number (18 January), p 22.
23. Forbes RJ, O'Beirne DR (1957) The technical development of the Royal Dutch/Shell, 1890–1940. EJ Brill, Leiden, pp 504–509.
24. Homburg E, Small JS, Vincken PFG (2000) Van carbo- naar petrochemie, 1910–1940. In: Schot JW, et al. (eds), Techniek in Nederland in de twintigste eeuw, vol. II. Walburg Pers, Zutphen, pp 332–357.
25. Martin P (2017) Twentieth century fertilizers in France from natural mixing to artificial making (1890–1970). In: Tessier P, Mody CCM, Van Tiggelen B (eds), From bench to brand and back: the co-shaping of materials and chemists in the twentieth century, series III, no. 2. Cahiers François Viète, Université de Nantes, pp 69–91.

Chapter 5
Electricity and the Chemical Industry

5.1 Electric Arcs

Investigations into technologies for the direct capture of atmospheric nitrogen after 1900 were increasingly motivated by concerns over the depletion of caliche and the Chilean hold on the saltpetre monopoly. Estimates of how long reserves would last varied, from two decades to half a century, and in the case of lobbyists for the nitrate industry to over a century. Electrical methods for forming "nitrates," which came to mean fixed nitrogen in any of its forms, as Crookes had emphasized, appeared to offer the way forward [1, 2].

Electrochemistry had around 1890 become an established industrial field. The first processes developed, for aluminium and inorganic chemicals, were based on electrolysis. In 1886, Charles Martin Hall (1863–1914) in the United States and, independently, Paul Louis-Toussaint Héroult (1863–1914) in France reduced alumina, the mineral containing aluminium, to the free metal [3]. To achieve this, the alumina was first dissolved in fused cryolite. Next came a process for preparing magnesium. In Germany, Chemische Fabrik Griesheim from 1884 developed electrochemical processes, in association with two other firms. The outcome was the foundation of Chemische Fabrik Elektron in 1892 to work a chlor-alkali (chlorine-alkali hydroxide) process based on the electrolysis of a solution of potassium chloride. The Elektron factory, opened in 1895 at Bitterfeld, north-east of Halle, used inexpensive local brown coal, or lignite, for the generation of electricity. An identical electrolysis process at BASF, Ludwigshafen, produced the potassium hydroxide required in its synthetic indigo process. In the United States, Herbert H. Dow (1866–1930), in Midland, Michigan, during 1889–1890 produced bromine by electrolysis and adapted his process to the electrolysis of aqueous sodium chloride to give sodium hydroxide (caustic soda) and chlorine.

Other processes relied on electrothermal reactions, conducted at high temperatures in electrically heated furnaces. In 1891, the American Edward Goodrich Acheson (1856–1931), by heating a mixture of coke and quartz, created the

A.S. Travis, *Nitrogen Capture*, https://doi.org/10.1007/978-3-319-68963-0_5

refractory, abrasive material silicon carbide. From 1894 he marketed it as Carborundum. In May 1892, Canadian inventor Thomas Lovell Willson (1860–1915), then in Spray (later part of Eden), North Carolina, attempted to produce calcium metal from lime by heating with tar in an electric arc furnace. This failed. Instead he obtained calcium carbide, which with water produced the highly reactive gas acetylene. Willson collaborated with James Turner Moorhead, who started carbide manufacture in 1894. By 1900, acetylene was an important source of lighting—its flame was far brighter than that from coal gas—for home and industrial purposes. Train lighting with acetylene, or carbide, lamps was introduced in Germany in 1896. Carbide lamps were also used on bicycles. The carbide and acetylene industry was of interest to the young German engineer Walther Rathenau (1867–1922), son of Emil Rathenau (1838–1915), founder of the Allgemeine Elektrizitäts-Gesellschaft (AEG) electrical-engineering corporation [4]. In 1899, Walther Rathenau joined the board of AEG.

AEG recognized the extreme brilliancy of acetylene lighting, which was seen as a competitor to electrical lighting, a mainstay of its business. Not wishing to lose out to the rival form of illumination, AEG in 1893 established the Elektrochemische Werke GmbH at Bitterfeld for the manufacture of carbide and electrochemicals. Carbide, however, was not an important product. Instead the factory produced chlorine and sodium hydroxide by electrolysis. Five years later, AEG built a second electrochemical factory, this time equipped with a carbide furnace designed by Walther Rathenau, at Rheinfelden, on the Lower Rhine, near Basel, Switzerland. Technical difficulties led to the facility being leased to Elektron, which in 1898 merged with Griesheim to create Chemische Fabrik Griesheim-Elektron [5].

The Schuckertwerke electrical firm, Elektrizitäts-AG, of Nuremburg, equipped and held interests in carbide factories in Norway, under the ownership of a Swiss subsidiary; in Bosnia at Jajce (Bosnische Elektrizitäts-AG, Elektrobosna, 1897); and in Switzerland at Gampel (Lonza AG, Basel, 1898). It also supplied equipment and plant for the first major carbide factory in Germany, at Lechbruck (1901), later partly owned by Lonza AG. Schuckert's agent in Norway was the engineer Knud Bruyn (1855–1941), who had completed his engineering education in Munich (1877) before studying electrical engineering, also in Germany. He then spent some time on railway projects. In 1898 Bruyn founded the Hafslund company, which, using Schuckert equipment, built the first large hydro-electric power station in Norway and was a major supplier of power, including to Kristiania (present-day Oslo). Soon after, Bruyn was engaged in the manufacture of calcium carbide. Cheap hydro-electricity became a critical factor in the siting of many carbide factories in Norway, located at first in the south-east. Overproduction encouraged moves into other electrochemical processes, notably ferro-alloys and aluminium from 1906. By this time carbide had become an important starting point for one of the main nitrogen capture processes, that of calcium cyanamide (Sect. 5.6).

On the west coast of Scotland, the Acetylene Illumination Company manufactured carbide for acetylene from 1896 until 1903; and in England, the United Alkali Company took up limited carbide manufacture during 1902–1904. In both cases, high production costs and limited demand, partly as a result of the widespread use of electrical lighting, brought this manufacture to an end. For similar reasons, Schuckert in 1903 sold out to Siemens & Halske, apart from the electrochemicals division. (The Siemens & Halske power division was named Siemens-Schuckert.) However, new and lasting uses for acetylene emerged, in welding, and, through Wacker Chemie, founded on 13 October 1914 as successor to Schuckert's electrochemicals division, synthesis of chlorinated solvents and acetic acid. Wacker originally planned to draw hydro-electric power from the River Alz in Upper Bavaria [6].

5.2 The Burning of Air

Electrothermal processes were investigated by most inventors interested in fixing atmospheric nitrogen. However thermal reactions carried out continuously at elevated temperatures required the development of suitable furnaces. In addition, lack of reliable high-voltage transmission systems restricted the geographical distribution of technically viable processes to localities that could offer cheap hydro-electric power, or access to sources of low-cost coal, such as brown coal (lignite). Lord Rayleigh, taking the basic idea from Priestley and Henry Cavendish, and as a result of his argon experiments, in 1897 suggested the technical combination of nitrogen with oxygen in an electric arc, which directed studies into gaseous electrothermics. This was not only a stimulus to further investigations in Europe and the United States, but also an indication of the frantic efforts being made to gain fame and fortune through nitrogen capture. Several electric arc methods were devised, though few were adaptable to industrial use. The general idea was to turn discharges from the application of, mainly, single-phase alternating current into flames when subjected to blasts of air, encouraging the formation of nitric oxide.

It was daunting work and it invariably ended in failure when scaled up. This was the case for the electric arc of Arthur McDougall and Frederick Howles in Manchester, England (1899). The similar arc process of Charles Schenk Bradley (1853–1929) and Robert D. Lovejoy in the United States, introduced in 1902 at Niagara Falls, New York State, by the Atmospheric Products Company, was not viable commercially due to the complexity of the furnaces; the factory was closed within two years [7].

Another electric arc method was devised during the first years of the twentieth century by the Polish chemist and inventor Ignacy Mościcki (1867–1945), who studied chemistry at Riga Polytechnic, graduating in 1891 under organic chemist Karol A. Bischoff. He then resided in Warsaw, in the western part of the Russian Empire. However, he did not remain there long. A Polish patriot and socialist, he schemed with others to assassinate the governor of the city. In 1892, sought by the tsarist police, he quickly moved to England, where he followed chemistry courses at London's Finsbury Technical College. In 1896 he was appointed assistant to physicist and fellow Pole Jósef Wierusz-Kowalski at the University of Fribourg,

Switzerland. There he put together an electric arc for fixing atmospheric nitrogen, and with the collaboration of Wierusz-Kowalski set up an experimental plant in the city around 1903. Trading as Société de l'Acide Nitrique, they soon after produced nitric acid, and Mościcki continued to work on modifications and improvements [8].

5.3 Birkeland and Eyde

As indicated in the foregoing, the natural resources of a given location frequently play a role in determining the development of appropriate new technologies. In the case of Norway, the most mountainous country in Scandinavia, water power was the principal resource. At first, the low waterfalls in the south-east were exploited. Later, from around 1900, advantage was taken of high generating heads and fast flowing rivers in the west for generating hydro-electricity. From substantial dams, water was directed through tunnels and along steel pipes, falling hundreds of metres to turbines, designed and supplied by Swiss and German companies, as were generators and other electrical devices. The supply of water was inexhaustible and the cost of electricity in many cases was the lowest in Europe. (This was unlike the situation in Italy, where hydro-electric power was also important, but which relied on mountain lakes that on occasion dried out, at least until new and larger facilities were constructed in the 1920s.)

Norwegian developments in chemical industry were closely associated with the engineer and entrepreneur Samuel Eyde (1866–1940), who played the most prominent role in the production of aluminium, ferro-alloys, and one of the two main electrothermal processes used in nitrogen capture. Here we review the early years of the process for which he is mainly known, the Birkeland-Eyde electric arc process. This began in 1903, after Eyde met Norwegian physicist and inventor Kristian Olaf Bernhard Birkeland (1867–1917), during a dinner party at the home of Norwegian cabinet minister and Secretary of Agriculture Gunnar Knudsen (1848–1928), who later served twice as prime minister of Norway. Birkeland, born in Kristiania, was based at the city's university. He had participated in high-latitude Norwegian polar explorations to gather magnetic field data, from which global patterns of electric currents in the region were established. He developed a theory of the Aurora Borealis that was ridiculed, particularly by British scientists, though proved correct in the 1960s; and he designed electromagnetic devices, including a cannon, though none were technically viable (Fig. 5.1) [9]. Birkeland, invariably short of funding for his research, received some assistance towards his expeditions from Knudsen, who had a financial interest in the cannon. During a demonstration at the Domus Academica (Gamle Festsal) of the University of Kristiania on 3 February 1903, a short circuit caused the cannon to fail. Massive flames appeared at the muzzle, but Birkeland noticed that the flames were deflected by the electromagnetic field. This was just a week or so before Knudsen's dinner party.

Fig. 5.1 Kristian Birkeland. (Yara International ASA)

Eyde was born in Arendal and studied engineering at the Technische Hochschule Charlottenburg, Berlin, prior to working on railway infrastructure projects in Hamburg and then forming the engineering firm of Gleim & Eyde with his former employer. Through offices in Kristiania and Stockholm, this firm enjoyed considerable success [10]. By 1903, Eyde had taken up an interest in exploiting hydro-electric power and acquired the rights to waterfalls in Telemark. He was also aware of the recent work of the Americans Bradley and Lovejoy in capturing nitrogen with an electric arc. This is why at Knudsen's party, aware of Birkeland's interest in electromagnetism, Eyde suggested the capture of atmospheric nitrogen by a powerful flash of artificial lightning—in other words an electric arc. Birkeland, according to one account, drew on his observation of the deflection of flames from the mouth of the cannon to assemble within a week an electric arc incorporating an electromagnetic field to spread out and concentrate the discharge. This enabled the capture of nitrogen by what appeared to be a technically viable method. The first patent application was filed on 20 February 1903.

Birkeland, with Eyde's engineering expertise and backing, developed what was to become known as the Birkeland-Eyde electric arc process. In 1903, an experimental installation was tested at the Frognerkilens Fabrik, of Kristiania (Figs. 5.2 and 5.3). In October, a pilot plant was assembled at Ankerløkken, close to the city, where a one-metre-diameter furnace was tested. As in the other arc processes, atmospheric nitrogen, introduced near the electrodes, was captured as nitric

Fig. 5.2 Birkeland-Eyde experimental set up, Frognerkilens Fabrik, Kristiania, 1903. At right is Kristian Birkeland; at left is probably Eivind Bodtker-Næss, later head of Elektrokemisk (according to Knut Sogner, *Creative Power. Elkem 110 Years, 1904–2014*. Oslo: Elkem, 2014, p. 15). (Yara International ASA)

oxide. Birkeland's flame, also called an electric sun, was a circular sheet of fire, created when the arc between the two electrodes (supplied with alternating current), at right angles to a magnetic field, was rapidly deflected from one side to the other. This arrangement concentrated the discharge, expanded the arc and permitted the working of large furnaces (Figs. 5.4 and 5.5).

By the end of 1903 a prototype two-metre-diameter iron electric arc apparatus was installed in a new test facility at Vassmoen, near Arendal. The product, nitric oxide, was converted into nitric acid which was then reacted with limestone to give

Fig. 5.3 The most prominent feature of the Birkeland-Eyde apparatus was the large electromagnet, here seen on the experimental version at Frognerkilens Fabrik, Kristiania, 1903. (Yara International ASA)

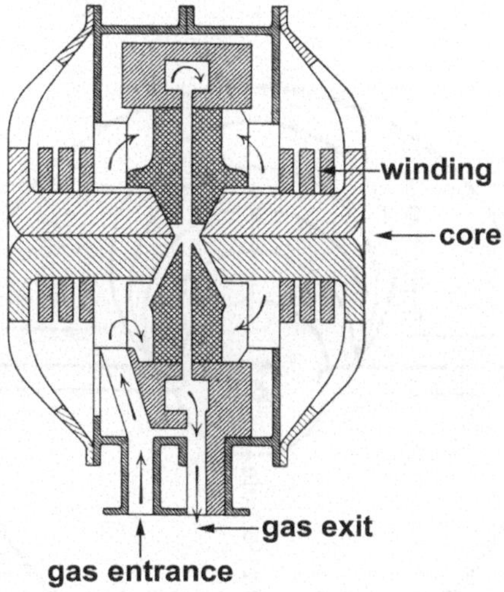

Fig. 5.4 Sectional elevation of the Birkeland-Eyde electric arc apparatus for capture of atmospheric nitrogen, at right angles to the arc disc. (Adapted from Georg Lunge, *Handbuch der Schwefelsäurefabrikation und ihrer Nebenzweige*. Braunschweig: Friedr. Vieweg & Sohn, 1916, vol. I, p. 259)

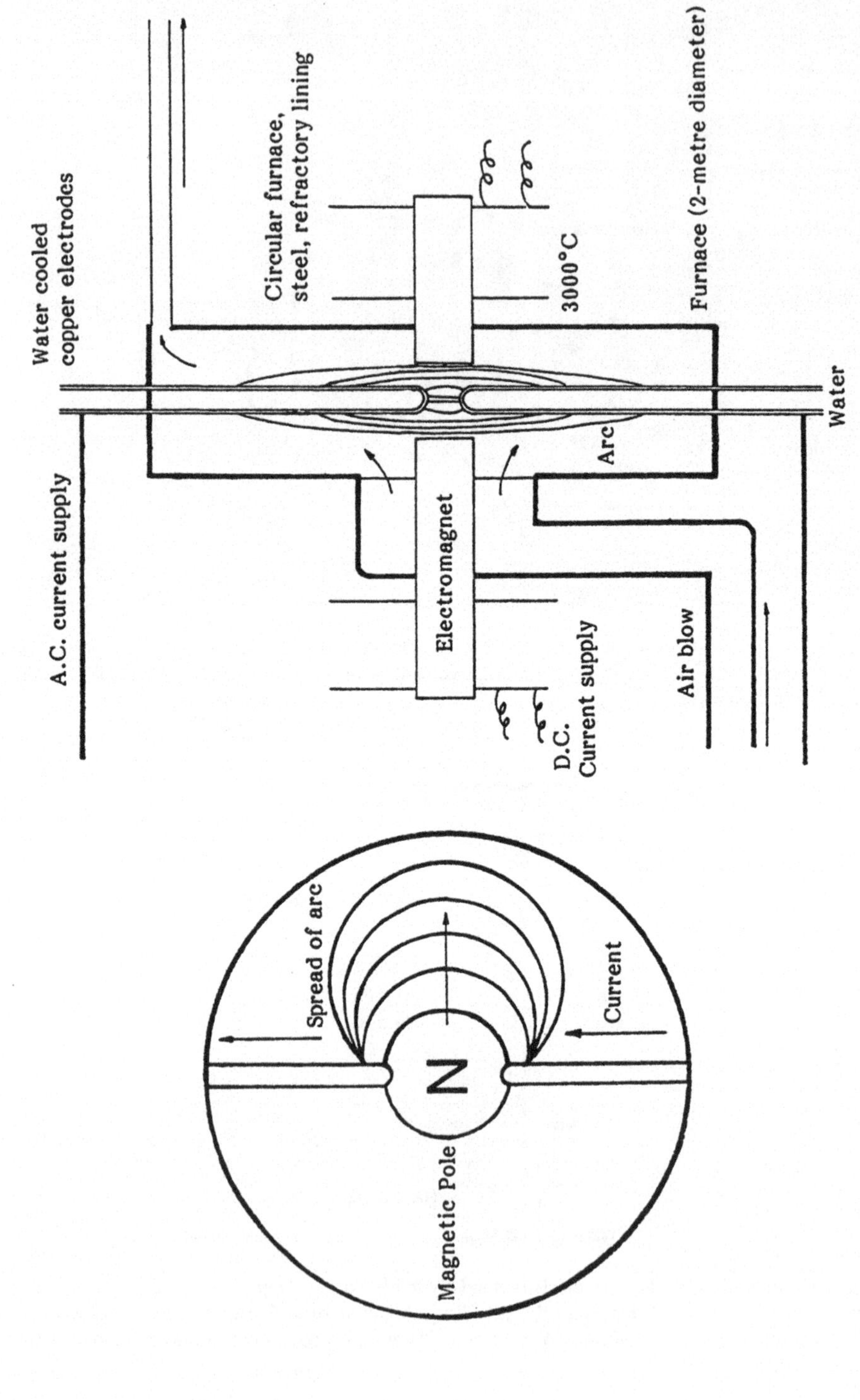

Fig. 5.5 General arrangement of the Birkeland-Eyde electric arc apparatus. The rapid to-and-fro movement of the semi-circular arc created an apparent spherical shape

calcium nitrate. Trials as fertilizer were undertaken at the Ås agricultural college and indicated that calcium nitrate was as good a fertilizer as sodium nitrate, especially on sandy soil. Eyde and Birkeland, jointly with Swedish politician and investor Knut Tillberg (1867–1946), and the Swedish bankers Knut Agathon Wallenberg (1853–1938) and Marcus Wallenberg (1864–1943), of Stockholms Enskilda Bank, in 1904 set up the electrochemical corporation Elektrokemisk to develop the electric arc process [11]. From this time Eyde was in partnership with the Norwegian Admiral Urban Jacob Rasmus Børresen (1857–1943), a prolific inventor and writer, who in 1905—the year he almost started a naval battle with Sweden—joined the Norwegian admiralty staff. As a gentleman of some standing, in both politics and naval affairs, Børresen smoothed the way through beaurocratic channels in trade negotiations with foreign countries, a strategy often adopted elsewhere by co-opting prominent military and political figures on to company boards.

Elektrokemisk was followed by the founding of Notodden Saltpeterfabriker, to undertake nitrogen fixation at Notodden, in Telemark. On 2 May 1905 a small factory equipped with three 500-kilowatt furnaces, using power supplied by the Tinfos hydro-electric station (opened in 1901), began operating [12]. The annual capacity was 2000 tons of calcium nitrate.

As with other arc processes, the yield of nitric oxide was low, which justified locating the process only at a site where cheap electrical power was available. The gas leaving the furnace contained 2–3% of oxide, which was rapidly quenched in an aluminium vessel, lowering the temperature sufficiently to prevent decomposition. The cooled nitric oxide gas was then passed through oxidation chambers, vertical iron tanks lined with acid-proof stone, to form dioxide and other oxides, particularly nitrogen tetroxide, N_2O_4 (the main reacting species in the liquid phase; in the past it was often called, incorrectly, nitrogen peroxide). In the presence of water these oxides turned into dilute nitric acid. The dilute acid was then directed through a series of four separate stone absorption towers, each one 20 metres tall and filled with broken quartz, over which the acid trickled. The nitric acid was concentrated in the first tower to 50%, and to progressively lower concentrations of acid in succeeding towers; in the fourth tower it was just 5%. More dilute acids were progressively recirculated to provide a greater volume of 50% acid. In a fifth stone tower, filled with bricks, residual acid and limestone gave calcium nitrate. Following concentration, the calcium nitrate was run into 200-litre iron drums, ready for shipping. Later, wooden barrels of 100 kg capacity were introduced. The facility also produced nitrous acid, which was reacted with sodium carbonate (soda) in wooden towers to give sodium nitrite. The nitrite was exported to German dye-making firms for use in the manufacture of azo dyes.

The new Birkeland-Eyde process was described by chemist Otto Nicolaus Witt (1853–1915) on 25 November 1905 at the inauguration of the new technical chemistry institute at the Royal Technical College Berlin (Königliche Technische Hochschule zu Berlin, also known as the Technische Hochschule Charlottenburg). Witt, originally a dye chemist, had made his name while working in England, in particular with azo dyes, from which he developed a theory of colour and constitution. Now he was a promoter of electrothermics, acquainted with the Birkeland-Eyde process after serving as one of twelve consultants and investors who arrived at Notodden by continental train in the early hours of 14 July 1905 to evaluate the commercial viability of the electric arc process [13]. The technical experts were headed by British physicist Silvanus Phillipps Thompson (1851–1916) from London's Finsbury Technical College. Also included was French chemist-agronomist Alphonse-Théophile Schoesing (1856–1930). The investors included Marcus Wallenberg and representatives of the Banque de Paris et des Pays-Bas (Paribas). Scientific experts and investors alike were duly impressed with what appeared to be the realization of Crookes's great challenge: a technical process for producing "electric nitrates" from cheap electric power, though not at Niagara Falls.

Following Crookes, and Carl Engler, Witt at the Berlin event also warned of future shortages of Chile saltpetre, suggesting that the caliche would become exhausted by the mid-1920s. Witt's solution was nitrogen capture with the electric arc [14]. The audience included Birkeland and Eyde, as well as Heinrich von Brunck (1847–1911; ennobled 1905), managing director of BASF, and his colleague the chemist Rudolf Kneitsch (1854–1906). BASF was also at that time committed to the capture of nitrogen through electric arcs. Witt invited von Brunck and Eyde to his home for an informal business discussion on electric arcs.

On 2 December 1905, the Norsk Hydroelectric Nitrogen Company (Norsk Hydro-Elektrisk Kvælstofaktieselskab), better known as Norsk Hydro, was established, backed by Swedish capital, from Enskilda, which held 50% of the stock, and Paribas and three other French banking houses which together held 30% of the stock. Birkeland and Eyde together held around eight per cent. The managing director was Eyde. The eight other directors on the board included J. Edmond Moret, of Paribas, the Wallenbergs, and Børresen. The first chairman was Marcus Wallenberg [15]. The technical director was Birkeland. The French investors insisted that concessions for water rights in Telemark be obtained from the state for future expansion. Work began on enlarging the Notodden factory. German and Swiss turbines and generators were ordered for a new power station, dedicated to the Birkeland-Eyde electric arc process. Meantime trials with applications of Norsk Hydro's nitrate were conducted in Sweden, France, Germany, Austria-Hungary, Italy, and Great Britain. Patents were filed in 25 countries. Norsk Hydro was at the heart of what was now to become Norway's first multinational corporation. Six months earlier, in June, Norway had gained its independence from Sweden.

Thompson described Norsk Hydro's electrothermal capture of nitrogen as the oxide at London's Royal Institution on 2 February 1906. Nitrogen also captured the public's interest. Ramsay reviewed progress in the engineering supplement of the London *Times* on February 21 [16]. It was certainly a heady time for those following the latest developments in electrothermal nitrogen fixation.

Promoters of new electrothermal nitrogen capture technologies tended to descend on London, the world's pre-eminent financial capital, and a leading centre for studies in technical electricity. Inventors and entrepreneurs had to convince academic audiences of the viability of their processes in order to raise capital and find investors. Thus on 21 May 1906, Swiss chemist Philippe-August Guye (1862–1922), at the University of Geneva and founder in 1903 of the *Journal de Chimie* (Paris), described the calcium cyanamide and electric arc processes as well as promoting his own arc process, which was still under development, before the London Section of the Society of Chemical Industry, at Burlington House (the Chemical Society's offices). The lecture, on "The Electro-chemical Problem of the Fixation of Nitrogen," was attended by chemists and industrialists, including, from Germany, Nikodem Caro, co-inventor with Adolph Frank of the calcium cyanamide process. With an eye to economy, Philippe-August Guye proposed that the pure oxygen remaining after the isolation of pure nitrogen from the air by liquefaction, as required in the cyanamide process, could be used in the electric arc process, and that both processes could be worked together [17].

On July 2, Birkeland addressed the Faraday Society in London on nitrogen capture with electric arcs, with special reference to the original, small factory at Notodden. The discussion revealed that Thompson and Witt had also visited the test facility near Arendal. There, as Thompson pointed out [18]:

> . . . all manner of small variations in the conditions of manufacture were being carefully and thoroughly investigated, using the most modern scientific methods of recording, [and] making frequent analyses. . ..Not satisfied with a few days' working of the process, Professor Otto Witt had the whole apparatus cleared out and a new test run started with perfectly fresh raw materials. This test lasted for two weeks. . .The various results agreed sufficiently closely to enable the Commission to assert that an output of 500 kilogrammes of nitric acid, per kilowatt year, could be looked for with confidence in any new plant that might be laid down.

Thompson also suggested that it was Birkeland's work on the aurora that led to his studies of the deflection of "electric flames in magnetic fields." This was in contrast to the story in which inspiration came from the effect of electromagnetism on the flame exiting from Birkeland's misfiring cannon. Another version was given by Norsk Hydro officials in 1916 to British investigator John Allen Harker (1870–1923), a physicist who was closely involved with British wartime munitions production. Harker was told that the inspiration was provided by Crookes's 1898 address and Lord Rayleigh's experiments [19].

In October 1907, Notodden began the full-scale production of calcium nitrate (Norges-salpeter, Norwegian saltpetre, lime nitrate, nitrate of lime, air-saltpetre). Electricity came from the new Svælgfoss power station, 5 km away. It was the largest in Europe and second largest in the world after that of the Ontario Power Company at Niagara Falls (opened in 1905) [20]. Eyde commissioned the artist Theodor Severin Kittelsen (1857–1914) to prepare a series of paintings celebrating this great taming of nature to provide mineral fertilizers that nourished seeds.

By 1909, there were 36 furnaces, 32 powered by Norsk Hydro's Svælgfoss power station and four by the Tinfos power station. The total investment was equivalent to six million US dollars, and the annual production capacity was then claimed to be of the order of 20,000 tons of calcium nitrate. Further expansion took place at Notodden in 1910.

Fig. 5.6 Birkeland-Eyde furnaces, Rjukan, 1912. (Yara International ASA)

In 1907 work had started on the large Rjukan facility near Såheim in Vestfjorddalen, Upper Telemark, north-west of Notodden (Fig. 5.6). Electricity was generated nearby, at the Vemork power station, next to the powerful Rjukanfossen waterfall on the Måne (Maan) River. The Vemork station, also known as Rjukan I, was the largest power station in the world when built (1907–1911). It generated 140,000 hp from ten dynamos. The Rjukan electric arc facility was equipped with 120 furnaces. The first consignment of nitrate fertilizer left the still incomplete works on 8 December 1911.

At the Eighth International Congress of Applied Chemistry held in New York in 1912, Eyde announced that as of November 1911, Norsk Hydro employed 1340 workmen and 143 officers [21]. In 1912 a new factory was built at Glomfjord, in northern Norway. One year later, at the Rjukan facility alone, 2000 thousand workmen and 400 "engineers and officers" were employed. During 1912–1913 demand for Norwegian nitrogen doubled. An even larger power station at Såheim (Rjukan II) began operation in 1916, with nine generators supplying 167,000 hp to 40 furnaces, all within the same building. Rjukan lay between Såheim and Vemork.

Concentration of the dilute acid was achieved at Rjukan in quartz-filled granite absorption towers, 70 feet high and 20 feet in diameter. Later, earthenware filling enabled operations in smaller and less expensive towers. The maximum concentration of acid from these and similar towers was 50–60% by mass. With extensions, there were in total 36 towers.

Fig. 5.7 Southern Norway, early 1900s. Norsk Hydro (Birkeland-Eyde) factories were located at Notodden and Rjukan. The North-Western Cyanamide Company's factory was located at Odda, south-east of Bergen

The remote location and the large number of workers required for the factory and power plant led to the founding of a company town, Rjukan, and the construction of a railway line during 1908–1909 by a new company, Norsk Transportaktieselskab, initially for carriage of supplies used in building the factory and then for the transport of nitrogen products to Mael, on Lake Tinnsjø. At Mael, the railway wagons were loaded on to a train ferry, which crossed the lake to Tinnoset, where they continued by another railway to Notodden (Figs. 5.7 and 5.8). The nitrogen products were then transshipped to canal barges and taken to Skien for export. Returning wagons carried limestone to Rjukan. Concentrated acid was safely transported in aluminium tanks. By 1914, designs had been drawn up for steamships to carry 600–700 tons of acid to Britain [22]. An extension of the railway south from Notodden replaced the canal to Skien in 1919. The output of Norgessalpeter in 1906 was 1200 tons; in 1911, 15,000 tons; in 1912, 71,000 tons; and in 1916, 110,000 tons. The other products were sodium nitrite and refined sodium nitrate. The main export customer in 1908 was Germany, which took 3841 tons out of a total production of 7082 tons; two years later Germany and Britain each took just under 5000 tons out of a total of 13,530 tons, according to the Central Bureau for Statistics at Kristiania [23, 24].

The supply of ammonium nitrate to the British explosives industry by Norsk Hydro is not without interest. Ammonia liquor from British gas works was sent by

Fig. 5.8 Nitrogen products from Norsk Hydro's remote Rjukan facility were transported by both rail and water to the sea port at Skien. The journey involved the railway from Rjukan to Mael (16 km), a railway ferry from Mael to Tinnoset across Lake Tinnsjø (34 km), another railway to Notodden (34 km), and, until 1919, carriage by barge on a branch of the Telemark Canal to the main canal, and then to Skien (54 km). In 1919, a railway was opened from Notodden to Skien. This is the train ferry terminal at Tinnoset in 2009. (Photo by Nick Lera)

sea and canal to Notodden, where it was reacted with nitric acid to yield ammonium nitrate. The nitrate was then sent to Britain, mainly to the Nobel's factory at Ardeer, Ayrshire, on the west coast of Scotland. This procedure, and the fact that in one direction the liquid sent from gas works was mainly water, caused a British observer to comment somewhat wryly: "It is this kind of thing that has caused scientific and engineering circles in Germany to get the impression of us which they have" [25].

Birkeland resigned as technical director of Norsk Hydro in 1906; thereafter he acted as consultant. He developed related processes, including the production of concentrated nitric acid, until around 1910. He sold some of his shares in Norsk Hydro in 1906 and 1907, and the remainder in 1910, which gave him the means to conduct further expeditions and academic work of his own choice (Fig. 5.9). He died while on a visit to Tokyo in 1917, apparently after taking an overdose of Veronal, the first synthetic barbiturate, introduced by Bayer.

Fig. 5.9 Portrait of Kristian Birkeland, from a painting by Asta Nørregaard, 1906. (Yara International ASA)

5.4 Otto Schönherr

On 11 September 1906, Norsk Hydro and its Swedish and French investors had signed an agreement with BASF, in partnership with Bayer and AGFA—the Dreibund—for joint developments at Rjukan. The BASF connection followed meetings between Birkeland, Eyde, Heinrich von Brunck, and Rudolf Kneitsch in 1905 on the occasion of Witt's Berlin lecture. Norsk Hydro took care of the hydroelectric plant, and BASF was responsible for the chemical side, including the production of nitric acid, and work on absorption towers.

BASF intended to work its electric arc process, designed by Otto Schönherr, at Rjukan. Otto Schönherr (1861–1926), born in Chemnitz, studied chemistry at the Technische Hochschule Dresden, then spent a few years in industry before joining Karl Elbs (1858–1933) at the University of Giessen in 1894 to undertake research in electrochemistry [26]. Schönherr received his doctorate in 1895, and in June 1896 joined BASF. From 1899, working under Kneitsch and encouraged by Brunck, he investigated electrothermal nitrogen capture [27]. In 1905, a long, slender discharge furnace—an iron tube incorporating within concentric iron pipes that surrounded an inner electrode—was developed in collaboration with the electrical engineer Johannes Hessberger (1871–1934).

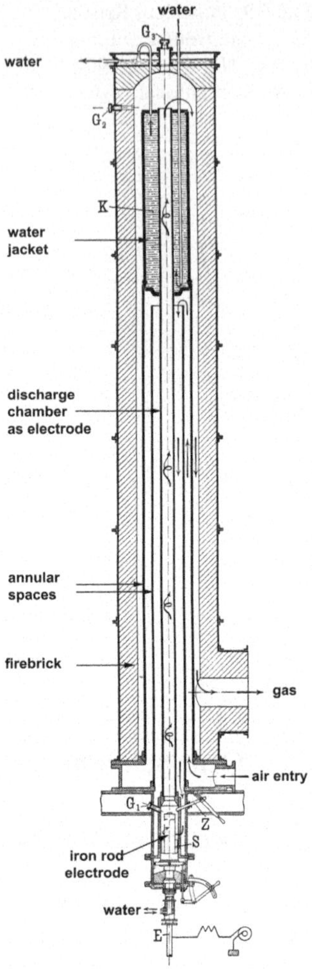

Fig. 5.10 Schönherr (BASF) electric arc apparatus for the capture of atmospheric nitrogen. (Adapted from Georg Lunge, *Handbuch der Schwefelsäurefabrikation und ihrer Nebenzweige*. Braunschweig: Friedr. Vieweg & Sohn, 1916, vol. I, p. 266)

The operation of this furnace relied on a long, steady column of discharge rather than the spread of an arc in a magnetic field. The lower end incorporated an insulated electrode, comprising an iron rod inserted into a water-cooled copper block. This was used to start the arc in the main, inner electrode chamber. At the base, air was introduced with a rotary motion; it moved up an intermediate pipe and then down a second pipe that surrounded the inner electrode chamber, where it was preheated. The hot air then entered the discharge region, moving up in a whirling motion, and again down, this time within the outer annular ring next to the firebrick lining of the furnace (Fig. 5.10). This arrangement ensured long contact with the arc and heated area. The gas mixture left the furnace near its lower end. The process consumed 14,800 kWh per metric ton of nitric acid, equivalent to just under 67,000 kWh per metric ton of nitrogen [28].

In the autumn of 1907, the new BASF electric arc furnace was tested at a special laboratory erected in Kristiansand. In 1909 ten such furnaces, each one 23 feet in height, were installed at Notodden [29]. In 1911, taller furnaces were installed at the new Rjukan facility. They were worked alongside Birkeland-Eyde furnaces for comparative purposes. Eighty per cent of the furnaces were of the BASF type.

The yield of nitric oxide from the BASF process was higher than from the Birkeland-Eyde process; also the cost of the BASF furnace was lower and the repair and maintenance costs were not great. However, the taller BASF furnaces were problematic in operation. BASF commissioned the academic physicist Jonathan Zenneck of the Technische Hochschule Braunschweig to take up residence in Rjukan for over a year in an effort to bring about improvements. In 1910, BASF consultant Fritz Haber was also called upon to investigate. One of his assistants performed experiments with a laboratory-size Schönherr furnace. They found that an increase in the pressure of the air blown into the tube had no advantage [30]. Little came out of these efforts. The Birkeland-Eyde furnaces were found to be more efficient than the newer BASF furnaces.

Relations between the various partners and investors became strained. There were considerable polemics over the most suitable type of electric arc furnace. Eyde, who naturally favoured the Birkeland-Eyde arc, was opposed by the French and Germans for his management, and was forced to step down from involvement in Norsk Hydro. He went off to develop Elektrokemisk which no longer had access to nitrogen capture technology with an electric arc.

Meantime BASF was losing interest in electric arc capture. BASF and Norsk Hydro parted company on 28 September 1911 after Eyde returned, backed by Marcus Wallenberg and other investors [31]. Rjukan was enlarged, this time with Birkeland-Eyde furnaces alone.

5.5 Other Arc Processes

Among other electric arc processes developed on an industrial scale was that of chemical consultant Harry Pauling (1875–1956), citizen of Brandau, Bohemia, Austria-Hungary. In Pauling's process, invented in 1902, a stream of fast moving air was blown against an arc struck between two diverging electrodes. This horn-arrester, imitating the horn-break lightning arresting device, was comprised of two hollow iron electrodes arranged to form a V. Charge was applied to thin horizontal knives at the base of the furnace to start the arc; air blown up through a nozzle into the lower part of the arc was deformed into a flame (Fig. 5.11) [32]. The process was used at the Cologne-based Salpetersäure-Industriegesellschaft Gelsenkirchen (later named Luftverwertungs Gesellschaft), at Patsch, near Innsbruck, where a pilot plant was assembled in 1906 (with 24 furnaces installed by around 1910); at La Nitrogène Cie, La Roche-de-Rame, Hautes-Alpes, France, from 1908 (with 18 furnaces installed by the end of 1912); and at Legnano, near Milan, from 1911 (Società

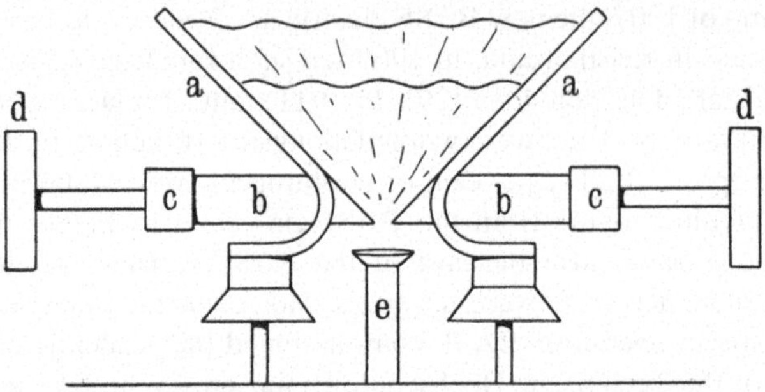

Fig. 5.11 Pauling electric arc apparatus for the capture of atmospheric nitrogen. Key: (**a**) electrodes; (**b**) iron knives; (**c**) insulation; (**d**) distance regulators; (**e**) air nozzle. (Georg Lunge, *Handbuch der Schwefelsäurefabrikation und ihrer Nebenzweige*. Braunschweig: Friedr. Vieweg & Sohn, 1916, vol. I, p. 254)

Elettrochimica Dr Rossi). The latter facility was established by Italian architect-chemist Carlo Rossi (1877–1924).

Nitric acid was obtained, as elsewhere, from absorption towers. Harry Pauling guaranteed "60 g of 100%. HNO_3 per kw.-hour of electrical energy, measured at the entrance of the electric transmission line into the factory" [33]. Pauling worked in partnership with his brother, G. Pauling, and they employed the British research chemist Glyn William Arnold Foster, who shortly after 1900 had undertaken research with Fritz Haber. The Pauling process was investigated in the United States by James B. Duke (1856–1925), and also by the Du Pont company. Duke had made his fortune in tobacco, invested in hydro-electric power from 1904 and obtained US rights to the Pauling process. He set up the Southern Electrochemical Company, of Nitrolee, South Carolina, using hydro-electric power to work the process from 1913. However, the high cost of the acid, 19,500 kWh per ton, led to closure of the plant in 1916. Until the 1930s, Harry Pauling's consultancy business in Berlin offered services in the design and construction of equipment employed in nitrogen fixation and the manufacture of concentrated nitric acid.

Philippe-August Guye, who had lectured in London on nitrogen capture, developed his electric arc process during 1905–1908 jointly with his brother, the physicist and electrical engineer Charles-Eugene Guye (1866–1942), also at the University of Geneva, and chemist and agronomist Aloys Adrien Naville (1853–1912). They claimed to have begun investigations in the mid-1890s. Their process was similar to the Pauling process, in that a horn arc, with cylindrical electrodes, was used. Though they undertook experiments for some time in cooperation with the Société d'Etudes Electrochimiques at Geneva there is no record of successful technical application [34].

Ignacy Mościcki improved his electric arc process based on what was described as a rotating circular arc. In 1905, Albert Einstein, as a patent examiner in Bern, reviewed

Mościcki's Swiss patent application, and his great interest in the method led to meetings between the two men in Bern and at the University of Fribourg [35]. Mościcki's process was adopted in Switzerland by Aluminium-Industriegesellschaft Neuhausen at Chippis, on the River Rhône [36]. A concentrated nitric acid was produced there from 1910. In 1912, Mościcki moved to Lwów (Lvov), where he served as head of the chemical technology department (Physical Chemistry and Technical Electrochemistry) at the Polytechnic National University until 1925.

Interest in electrochemistry in Russia led to development by Gorbov and Mitkievitch of an electric arc process that was claimed to require less energy than the Birkeland-Eyde process, and in 1908 it was stated that the process was worked on a large scale [37].

On the occasion of the 1906 jubilee celebrations for the foundation of the synthetic dye industry by William Henry Perkin there was considerable emphasis on electrical methods for capturing nitrogen. At that time, it was believed, foremost among the international companies and inventors engaged in research towards their exploitation was BASF, based on its Schönherr electric arc furnace. Celebrants at the New York event were informed by Hugo Schweitzer that [38]:

> The experience gained in electric methods is being applied to the problem of using the nitrogen of the air, and to judge from past successes it will not be many years before the Badische Anilin- und Soda-Fabrik [BASF], in whose laboratories experiments in this direction are being conducted, will bring into the market nitrates, nitrites and nitric acid made from atmospheric nitrogen instead of Chile saltpetre, the supply of which is calculated to last no longer than about twenty years.

In November 1907, BASF's von Brunck advised the Prussian Ministry of Trade and Business that the firm's "epoch-making" electric arc process offered an increase in yield over the Birkeland-Eyde process. Also, that the cyanamide process (see next section) was not suited to the manufacture of nitric acid required for explosives and for use in chemical industry. BASF planned to build a factory to work the Schönherr process in Bavaria, drawing power from the River Alz, with the steepest gradient of any river in Germany. The proposed BASF site was seven kilometres downstream of the rural market village of Trostberg (population 174). However, the Bavarian government was not keen on the project, and as a result BASF cancelled its plans in 1910. By then, BASF was testing its Schönherr electric arc furnace in Norway. Meantime the Bavarian authorities had sequestered the water rights of the Alz in order to control developments on the river, including regulation of the flow by excavation of a canal near Trostberg, and construction of hydro-electric power stations.

5.6 Calcium Cyanamide

Another approach that relied on cheap electrical power involved the reaction in an electrically heated oven between calcium carbide, made from coal and lime, and atmospheric nitrogen to yield the calcium salt of the compound cyanamide,

NC–NH_2. The early commercial product (sometimes called lime nitrogen) reportedly contained 12–15% nitrogen, later increased to 23% nitrogen.

In the early 1890s Ferdinand-Frédéric-Henri Moissan (1852–1907), professor of toxicology at the School of Pharmacy in Paris, and an expert in carbides, reacted coke and limestone (calcium carbonate) in an electric furnace in his laboratory to form, in two steps, calcium carbide.

$$\underset{\substack{\text{calcium carbonate}\\ \text{(limestone)}}}{CaCO_3} \rightarrow CaO + CO_2$$

$$CaO + 3\,C \rightarrow \underset{\text{calcium carbide}}{CaC_2} + CO$$

Moissan conducted his experiments under more careful conditions than those used by Willson and claimed to have obtained crystalline calcium carbide. Intriguingly even in the 1950s, it was stated that "no one has been able to repeat his preparation of 'pure' calcium carbide" [39]. Moissan reported that the carbide did not react with nitrogen even at an elevated temperature. He was soon proved wrong. It is here that Adolph Frank re-enters the nitrogen capture story.

In February 1895, Frank gave a lecture before the Berlin manufacturers' association (Verein zur Beförderung des Gewerbefleisses) on the potential of carbide in both the lighting and the chemical industries. It was probably at this time that he decided to investigate the reactions of carbides that might have commercial potential. Frank was in his early sixties, and probably needed a more youthful man to assist with experimental work. He found such an individual in recently qualified Nikodem Caro, some thirty years Frank's junior, who had recently opened a private laboratory offering research and consulting services in Berlin.

5.7 Nikodem Caro

Nikodem Caro (1871–1935) (Fig. 5.12), a relative of Heinrich Caro, head of research and development at BASF, grew up in Lodz (Łódź), the industrial centre of Russian-occupied Poland. His father, Albert, was a son of Rabbi Josef Hayim Isaak Selig Caro (1800–1895), a prominent figure in the German-Jewish reform movement; his mother was Rosa Rubinstein (Stein). Nikodem, fluent in Russian and German, arrived in Berlin in 1888, where he studied organic chemistry at the Technische Hochschule Charlottenburg under Carl Liebermann (1842–1914). Four years later, he was awarded his doctorate from the University of Rostock (at that time the Technische Hochschule Charlottenburg could not confer doctorates) [40]. Caro also studied with organic chemist Adolf Pinner (1842–1909) at Berlin's Friedrich-Wilhelms-Universität, later Humboldt Universität. (Pinner in 1885 had joined the Prussian Department of Commerce.) Caro, probably with little

Fig. 5.12 Nikodem Caro. (Evonik Industries AG, Corporate Archives, Hanau/Marl)

hope of an academic post, then acted as a chemical consultant, with his own laboratory, which is how the association with Adolph Frank came about. Over the next three decades or so Caro published widely on a range of topics related to chemistry and chemical industry. His wife, Else, was a daughter of brewery director Heibrich Friedman.

Early in 1895, Frank and Caro heated barium carbide and calcium carbide in streams of nitrogen and found that both of these carbides absorbed nitrogen. They had, unlike Moissan, established that carbides of alkaline earths fixed nitrogen. Their first patent, filed on 31 March 1895, was for barium cyanide, obtained in around a 30% yield. Although it mentioned in passing that calcium carbide also absorbed nitrogen, the main product was not identified. Probably at this time they drew the attention of Dynamit AG (formerly Alfred Nobel & Company), of Hamburg, to their work. Frank and Caro undertook research for Dynamit into a method for making sodium cyanide, used in the recovery of precious metals by the process of the MacArthur-Forrest Company [41]. The main firm interested in the production of cyanide for the extraction of gold and rare metals was German Gold and Silver Refinery, formerly Roessler (Deutsche Gold- und Silber-Scheideanstalt, vormals Roessler; Degussa).

Chemist Fritz Rothe (1867–1958) at A. Beringer & Söhne of Charlottenburg undertook similar work around 1895, but his employer did not complete Rothe's patent application. Rothe resigned and joined Elektrochemische Werke of Bitterfeld, from where in March 1896 he was in touch with Frank and Caro in connection with Caro's patent for preparing cyanides from carbides. Within a year he had moved to Dynamit, where he collaborated with Frank and Caro. Following Rothe's earlier studies, around 1898 they determined that dry nitrogen was required in the reaction and that the product from calcium carbide was not calcium cyanide but calcium cyanamide. The calcium carbide had to be heated at around 1100 °C in order to take up the nitrogen. Patents were filed for the cyanamide process.

$$CaC_2 + N_2 \rightarrow \underset{\text{calcium cyanamide}}{CaCN_2} + C$$

However, the main interest was still in cyanides. On 29 July 1899, Adolph Frank, jointly with Deutsche Bank, Siemens & Halske, and Degussa formed Cyanid-Gesellschaft mbH, in Berlin. The capital was half a million marks, of which 63.76% represented the Degussa interest, and 10.62% each the Siemens & Halske and Deutsche Bank stakes in the enterprise. Adolf Frank's share was 15%, representing the value of the Frank and Caro cyanide patents in Germany and elsewhere. Apart from Caro and Rothe, chemists engaged in further research and development were Hermann Freudenberg, Johannes Pfleger (1867–1957), directors Hector and Heinrich Rössler at Degussa, and Georg Erlwein (1863–1945) head chemist at Siemens.

A factory was erected in Frankfurt and the first product there was barium cyanide. Sodium cyanide was manufactured soon after, starting from calcium cyanamide, which was fused with calcium carbide and sodium chloride in an arc furnace, followed by rapid chilling of the melt; the presence of carbide prevented loss of nitrogen at the high temperature required. The resulting sodium cyanide found some use in gold extraction.

In 1901, Pfleger improved on the Hamilton Young Castner cyanide process. At the same time the price of cyanide, mainly used in gold extraction in South Africa, fell sharply as a result of the slowdown in mining during the Boer War (1899–1902). Degussa withdrew from the partnership since the Frank-Caro cyanide product could no longer compete in the rare metals cyanide market.

The reorganized Cyanid-Gesellschaft was directed by Nikodem Caro and managed by Frank's son, Albert Rudolph Frank (1872–1965), jointly with Hermann Freudenberg. Albert Frank had studied chemistry at the University of Munich before moving on to the Technische Hochschule Charlottenburg where he received his PhD in 1900, just one year after the institute was permitted to award the doctorate degree. In 1900, Caro found that calcium cyanamide when heated with steam gave ammonia.

$$CaCN_2 + 3\,H_2O \rightarrow CaCO_3 + 2\,NH_3$$

Albert Frank then demonstrated that cyanamide on contact with soil released ammonia. Thus it appeared to offer potential as a fertilizer. On 14 January 1901 this information was sent by Adolph Frank to Siemens & Halske, in a letter marked "Strictly confidential." Adolph Frank had just returned from a meeting in Frankfurt with Degussa, which was not informed of the experimental work. He advised Siemens that calcium carbide, in an autoclave under pressure of 11 atmospheres, took up nitrogen, which could "almost theoretically be converted into ammonia and becomes water soluble." He believed that "Our cyanamide could excel ammonium sulphate and Chili nitrates, with the advantage of lime content." However "It still needs to be determined whether cyanamide in soil is equivalent for fertilizing purposes." After a short trip he would consult with two agricultural chemists regarding field tests. Adolph Frank urged the "strictest discretion toward our Frankfurt friends [Degussa] with regard to this subject." Siemens & Halske replied on January 16, agreeing to maintain confidentiality regarding "our Frankfurt friends" [42].

In 1901, Albert Frank and Freudenberg filed patents in Germany, Britain, the United States and elsewhere for the application of cyanamide as a fertilizer [43]. Later, details appeared in the scientific literature [44]. Caro established the exothermic nature of the reaction, a factor of considerable value in reducing the cost of production. During the early 1900s, while field tests and technical developments were in progress, Frank and Caro pursued their various consultancy services, including for fertilizer chemicals, and the coal gas and paper industries.

In 1903, interest in cyanide and cyanamide was taken up in the United States by Charles Schenk Bradley and Charles B. Jacobs, of the Ampère Electro-Chemical Company, at Niagara Falls, with backing from Edward Dean Adams (1846–1931) who in 1895 had opened the first large-scale alternating current power station, at Niagara Falls. Bradley, whose electric arc process was facing difficulties at the time, made claim to discoveries that rivalled those of Frank and Caro. He suggested an arrangement over patents, one that was not taken up, probably because the American claims did not represent a danger to the strong patent position of the Frank-Caro process [45].

As for calcium cyanamide, one problem encountered by all who come into contact with it was the unpleasant effect on the skin. Farmers were advised to apply the new product with caution. Concerns over methods of application had earlier been raised following problems with the fertilizers potash and Chilean nitrate, notably with beet sugar. Manufacturers and users of novel agricultural products were alerted to the need for careful testing on the soil following studies of the efficacy of Chilean nitrate in the 1890s. These revealed the adverse impact of perchlorate contamination [46]. Exhaustive tests with cyanamide were undertaken at agricultural experiment stations during 1901 and 1902 by chemist Paul Wagner, head of Darmstadt Agricultural Experiment Station, and M. Gerlach of the Posen station. These were reported to be successful, and further tests were carried out in the following year at numerous stations. Wagner found that cyanamide was good for winter fruits, though not in abnormal circumstances. Experiments were undertaken with spring and winter wheat, winter rye, oats, corn, potatoes, and beet sugar in Germany, Italy (by R. Perotti, of Rome), and Vienna [47].

There were, however, objections that the cyanamide contained cyanide. Adolph Frank, around 1905, asked Wagner to identify any toxic action arising from the use of calcium cyanamide. Wagner's report was favourable, suggesting conditions for optimal application, though drawing attention to problems arising from certain soil types and concentrations [48]. In early practice, the crude cyanamide was generally applied directly to the soil as fertilizer; however, not every farmer was made familiar with the product and methods of dispersal. There were also problems in the United States and Japan. When first placed on the market, free lime present in the cyanamide was converted into the hydrate, which due to expansion caused the jute sacks to burst. Cyanamide was then packed, free of moisture, in double paper and jute sacks.

In 1905 a pilot plant for the manufacture of cyanamide was set up at Westeregeln, near Magdeburg. It operated until 1912. The earliest full-scale cyanamide production, however, was undertaken in Italy.

5.8 The First Cyanamide Factory, Piano d'Orta

Before 1914, the Italian chemical industry was poorly developed. There were few reserves of coal, apart from lignite. Some 90% of imported coal came from Great Britain and the importers had to contend with price fluctuations. However, the limited extent of tillable land stimulated interest in the production and efficient application of fertilizers. Thus there had existed since the 1870s a low-technology superphosphate industry, which was dependent on locally produced sulphuric acid. Manufacture of the acid was increasingly based on pyrites mined in Tuscany by the Montecatini Mining Company, founded in 1888. It moved into other chemicals after engineer and businessman Guido Donegani (1877–1947) became president, chief administrator, and chairman of the board in 1918. Montecatini in 1920 first broadened its activities by taking control of the leading manufacturers of superphosphate fertilizer, and was renamed Montecatini Mining and Agriculture Company (Società Generale per l'Industria Mineraria e Agricola Montecatini).

The lack of carboniferous products had earlier hindered the entry of Italy into the science-based chemical industry arising from the discovery of synthetic dyes [49]. While that may have been the case until around 1900, Italian entrepreneurs soon after adopted emerging new chemical technologies well suited to the country's natural resources. This was brought about by the engagement with electrochemistry and thermochemistry, drawing on abundant hydro-electric power. The growth of the Italian hydro-electric industry was encouraged by the same factor that held back chemical industry, the great scarcity of coal reserves. Many business leaders believed that hydro-electricity would not only herald in a new era of industrialization, nothing less than a technology-based economy, but also remove the great dependence of the energy sector on mainly imported coal. This suited well the leading German manufacturers of electrical equipment, who were eager to exploit the Italian market through the export of machinery and participation in joint ventures [50].[1] Widespread electrification in transportation and manufacture was

[1]As a result, the Italian energy industry was dominated by a trust formed by, among others, Siemens & Halske, and Imprese Elettriche Conti, with support from Deutsche Bank.

rapid. Gas lighting for homes was replaced by electric lighting. Hydro-electricity enabled innovations that spawned new industries. The large-scale generation of electricity provided sufficient surplus power for its use in chemical manufacture, including of calcium carbide (1898) for acetylene, the important source of lighting. These circumstances made Italy, with its great need for fertilizer, a suitable location for inauguration of the Frank-Caro calcium cyanamide process in the early 1900s.

Following extensive field tests, agricultural chemist Angelo Menozzi (1854–1947) at the University of Milan joined with Cyanid-Gesellschaft to promote in Italy the electrothermal Frank-Caro process. Menozzi emphasized the access to Italian and foreign capital, abundant hydro-electricity, well-established carbide manufacture and the role of experts in educating farmers in the use of the new nitrogen product. On 9 September 1904, the Società Italiana per la Fabbricazione di Prodotti Azotati e di altri sostanze per l'Agricoltura (SIPA) was established in Rome to manufacture and promote the use of nitrogen and other fertilizers. The Frank-Caro process was also promoted by Fausto Morani, a founder of the Italian carbide industry and head of Società Italiana del Carburo di Calcio, Acetilene e altri Gas (Carburo), of Rome, established in May 1896.

In July 1905, Menozzi, Cyanid-Gesellschaft, Carburo, and SIPA, backed by German and Italian investors, formed the Rome-based La Società Generale per la Cianamide, with a capital of three million lira, as assignee of the Frank-Caro patents [51, 52]. This company was headed by Morani, whose Carburo operated carbide factories at Collestate and Papigno, near Terni, in Umbria, north-east of Rome. The principal German investors were Siemens & Halske and Deutsche Bank [53].

A factory with an intended initial capacity of 500 tons of cyanamide per year began operating in 1905 at Piano d'Orta, close to Pescara. It made use of hydro-electric power available from a nearby generating station opened in 1904 [54]. However, there were technical difficulties because of the limited lifetime of the externally heated horizontal ovens in which nitrogen was passed through powdered carbide. This external heating led to an accumulation of carbide on the inner wall of the ovens—which was removed by hammering out—causing excessive wear on the walls. In addition, the heating was not efficient.

Pure nitrogen was necessary. The Brins process for isolating atmospheric oxygen, as developed in the 1880s in France by Quentin and Arthur Leon Brin, was tested as a source of nitrogen and may initially have been used at the factory. Later, nitrogen was obtained by the new Linde air liquefaction process, which gave nitrogen as well as oxygen.

Usefully, Adolph Frank suggested that since the conversion of carbide to cyanamide was exothermic, the heat given out by the reaction, once started, could be used to help maintain the carbide at the reaction temperature. A vertical oven in the form of a drum with internal heating was designed. This incorporated a carbon "pencil" electrode. Frank's thin electrode provided the heat to start the reaction (Fig. 5.13). The horizontal ovens were abandoned. In 1907 the Italian factory, now fitted with batteries of self-heating ovens, commenced production at the rate of 4000 tons of calcium cyanamide per year. The carbide was held within paper sections inside the oven. Centrally placed was a vertical paper tube for the carbon "pencil" electrode. After the oven was closed and made air tight, the pencil

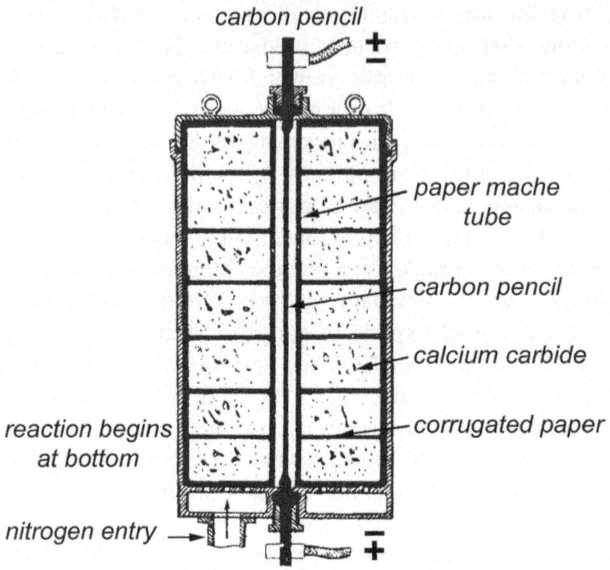

Fig. 5.13 Early Frank-Caro oven for production of calcium cyanamide. (Redrawn from Ernest Kilburn Scott, "The Manufacture of Nitrates from the Atmosphere," *Nature*, vol. 89, no. 2228 [11 July 1912]:492)

electrode was inserted into the paper tube and the alternating current switched on. Nitrogen under pressure was fed in at the bottom of the oven. The heat destroyed the paper tube and partitions. The absorption of nitrogen took place at the point of contact of the pencil with carbide and continued gradually outward [55]. After around 25 hours the current was switched off and the oven left a further 10 hours, by which time the reaction was complete. The cyanamide, appearing as a black-grey clinker, was turned out, crushed, ground, and packed in bags for shipping.

Challenges in manufacture and application were still receiving attention when in May 1906, Philippe-August Guye spoke on technical fixation of atmospheric nitrogen in London. He and respondents, including Nikodem Caro, discussed the costs of energy, nitrogen, and calcium carbide for the cyanamide process, and technical and other issues. At that time, the price and quality of carbide had only just become suited to cyanamide production; and Carl von Linde's air liquefaction process was facing difficulties in operation. J. A. Voelcker drew attention to the fact that tests with calcium cyanamide in England at Rothamsted and Woburn were not satisfactory. Over the following year or so, the technical difficulties were mainly overcome. Though application remained problematic there was a general feeling of optimism and an understanding that chemical engineering was taking over from chemistry in the novel nitrogen capture processes.

An exclusive cyanamide license for France was granted to Le Société Française des Produits Azotés, associated with Henri Gall's Société d'Electrochimie (established in Paris in 1889), which undertook production at Notre-Dame-de-Briançon, Haute Savoie, Hautes-Alpes, the site of a carbide factory opened in 1897. As at Piano d'Orta, equipment for an annual capacity of around 4000 tons was installed. The factory opened in 1908 and used the new vertical ovens. Carbide, and later cyanamide, were also manufactured at Villers-Saint-Sépulcre (Oise).

Several innovations were later introduced, including spraying the cyanamide with water to remove free carbide that might otherwise form explosive acetylene. Despite improvements, the high operating temperature kept the power cost high. Annual production of the commercial product, Kalkstickstoff, lime nitrogen, or nitrolime, by 1909 was 10,000 tons.

The choice of Italy as a centre for local operations, drawing on abundant hydro-electric power, and the need to improve crop yields, as well as for international licensing, was pragmatic. Subsequently, Menozzi studied mixtures of cyanamide and phosphate for use as a compound fertilizer. In addition, the Italian license holder intended to appropriate water rights it owned in Italy for other works. In 1909, production of cyanamide at Piano d'Orta was transferred to the Carburo facility at Papigno, near Terni (Fig. 5.14). Soon after, manufacture was taken up at San Marcel, in the Aosta Valley (Val d'Aosta) in north-western Italy, in Ascoli Piceno (Marche region; Società Industriali Carburo), and at the Sebenico carbide works, in Dalmatia. Other early cyanamide factories were at Fiume (Rijeka) in Istria, and Dugi Rat, near Almissa (Omiš), in Dalmatia. The latter facility was opened in 1910. The intended markets were the Balkans, Asia Minor, and Egypt.

The Franks, father and son, and Caro placed great emphasis on promoting their process before technical and scientific audiences, and not only in London. They, along with others engaged in nitrogen capture, appeared at specialist congresses that during the first decade of the twentieth century represented more than at any time before the internationalization of science. These events were no less valuable in describing new technologies to a wide range of audiences. In April 1906, Adolph Frank lectured on cyanamide in Rome at the Sixth International Congress of Applied Chemistry, and in May 1907 in Vienna, where he addressed the Eighth International Agricultural Conference (Internationaler Landwirtschaftlicher Kongress) [56] (Fig. 5.15).

5.9 North-Western Cyanamide Company

A strong British connection with cyanamide followed the founding in 1904 by Alfred Ernst Barton of the Sun Gas Company, to manufacture acetylene. To ensure a supply of carbide, Barton took over the Swedish Alby Carbid Fabrik, and he soon after became interested in the manufacture of cyanamide. With support from Samuel Eyde and the Wallenbergs, Barton set up in London the North-Western Cyanamide Co. in July 1906, with a capital of 120,000 pounds sterling. In 1907, a manufacturing and market-sharing arrangement was concluded with the assignee of Frank-Caro patents, La Società Generale per la Cianamide, whereby North-Western

Fig. 5.14 Advertisement for calcium cyanamide, Società Italiana per la Fabbricazione di Prodotti Azotati ed [e di] altre sostance per l'Agricoltura (SIPA), from an unidentified Italian journal, 1919. At that time cyanamide was manufactured in Italy by SIPA at Papigno, near Terni, and San Marcel. (Author's collection)

Cyanamide controlled markets in Britain, Norway, Sweden, Denmark, Belgium, and all British "colonies, protectorates, and dependencies, except Egypt and Canada." The Norwegian factory was at Odda (Hordaland), some distance southeast of Bergen. It was located at the south end of the Sørfjorden, a branch of

Fig. 5.15 Adolph Frank's portrait on a commemorative medallion. (Evonik Industries AG, Corporate Archives, Hanau/Marl)

Hardanger Fjord, and received hydro-electricity generated at Tyssedal, a few miles to the north, on the east side of the fjord. (In 1908 North-Western Cyanamide acquired from Elektrokemisk water power rights amounting to 50,000 hp at Tyssedal.) At Odda, 196 cyanamide ovens were installed, each one constructed for a 300-kg charge. Thirty tons of cyanamide containing 18% nitrogen were produced every twenty-four hours. Initial annual production of nitrolime was 12,500 tons. Carbide furnaces were also installed. The principal markets were Sweden and Belgium, with Denmark, Germany and Holland accounting for 30% of consumption. On 9 June 1908, North-Western Cyanamide started operations at Odda; on the same day, Albert Frank described the new factory before the Faraday Society in London [57, 58].

On November 2, Adolph spoke at Burlington House, London, before the Society of Chemical Industry on "Chemical Industry in Relation to Agriculture," which included a biographical account of his potash work [59]. In 1909, Caro was in London to attend the Seventh International Congress of Applied Chemistry and to promote the cyanamide process [60].

In April 1910, Barton acquired British rights to the Ostwald process for the catalytic oxidation of ammonia, no doubt with the intention of converting cyanamide-derived ammonia into nitric acid. However, this proved less than successful in practice [61].

The Sun Gas Company, by 1910 known as Alby United Carbide Factories, Ltd, promoted the activities of North-Western Cyanamide. The Alby United board included Albert Vickers (1838–1939), of the engineering firm Vickers Ltd of Sheffield, Sir Vincent Gaillard, and Barton. Alby United constructed new carbide works near Alby in Sweden. This was followed in 1912 with the setting up of the subsidiary Nitrogen Fertilisers, Ltd, to which ownership of the Odda and Alby factories was assigned, and in 1913 of Nitrogen Products & Carbide Company Ltd. Another notable individual involved in this "English Consolidation" was Sir Richard Awdrey of the Nobel-Dynamite Trust Company. The technical director was Albert Johan Peterssen (1870–1914), who had studied at Zurich's Eidgenössischen Polytechnikum (Eidgenössische Technische Hochschule, or ETH). He was responsible for all development work.

Waterfalls in Norway and Iceland were acquired to enable further expansion. Unlike the arc processes, the cyanamide process required pure nitrogen. The Odda cyanamide plant was claimed to be the first bulk customer for nitrogen obtained by fractionation of air according to the process of former locomotive engineer and refrigeration inventor Carl von Linde (1842–1934: ennobled 1897). Linde's cryogenic process relied on development of a suitable cooling technology and separation by distillation, as achieved by 1908. The nitrogen was produced at Odda in the largest Linde pump manufactured until that time, liquefying 100 tons of air every day, providing about 77 tons of nitrogen.

At the Odda works the latest technical developments were introduced. Meters were used to control the reaction. Great care was required to ensure that the oven did not overheat, preventing reversal of the reaction. A steady reading indicated that all the carbide had taken up nitrogen; the resulting block of solid cyanamide was easily removed. Later a continuous process was developed, though it suffered from technical problems. Large-scale manufacture relied on the synergy between carbide manufacture and air liquefaction. The Odda facility was expanded and incorporated the rival Claude cooling process for producing nitrogen, which was found to be purer (0.1% of oxygen compared with 0.4% by the Linde process), and to be simpler to work.

Samuel Eye had intended that the Norwegian Elektrokemisk firm embark on manufacture of calcium carbide, and in turn of calcium cyanamide, following purchase in 1913 of the Bjølveffosen power company at Ålvik, Hardanger. However, he dropped these plans and focused on metallurgical developments, with experimental work carried out at the former BASF laboratory in Kristiansand.

In Switzerland, the Société Suisse des Produits Azotés opened a Frank-Caro works in the Rhône Valley. Cyanamide was also taken up in Japan during 1908–1909. Large-scale manufacture in Germany by Frank and Caro was delayed until 1910, when the Bayerische Stickstoff-Werke AG at Trostberg, Upper Bavaria, was opened (Sect. 5.13). By then Albert and Adolph Frank, and Nikodem Caro, had developed a process for the conversion of cyanamide to ammonia, and the oxidation of ammonia to nitric acid.

5.10 Cyanamide in the United States

An important but untapped market for cyanamide was the United States, where Adolph Frank had useful family contacts. During 1905–1906 Frank corresponded with relatives concerning the introduction of cyanamide to farmers in the American West. Soon after, La Società Generale per la Cianamide provided a license to Cornell-trained civil engineer and entrepreneur Frank Sherman Washburn (1839–1922).

Frank Washburn already had a background in nitrogen products, as well as in the railway and dam industries. In 1889, he joined W.R. Grace & Co., the leading US importer of Chilean nitrate. Later while working in the southern states he became involved in hydro-electric power, as president of the Alabama Power Company. In 1907, he travelled to Norway to meet with Eyde in order to discuss a licensing arrangement for the electric arc process. Eyde, apparently, pointed out the high cost of power and recommended that Washburn instead investigate the cyanamide process, which required just one quarter the amount of energy. A meeting was arranged in Germany with Albert Frank. The outcome was that Washburn and partners founded the American Cyanamid Company on 22 July 1907 to work the Frank-Caro process. Washburn's intention was to construct a factory at Muscle Shoals, northern Alabama, which had the advantages of water power, good rail transport, and access to coal, and lime, as well as local markets. However, this required a private-public partnership, for which Congress would not approve. He then opted for the Canadian side of Niagara Falls [62]. Frank-Caro ovens of Italian design were acquired, and a plant was erected for the conversion of coal to coke, which was used to manufacture carbide. From late 1909 (officially 18 January 1910) the Frank-Caro process was worked at Niagara Falls. The initial capacity was 5000 tons per annum, soon increased to 25,000 tons (Figs. 5.16, 5.17, 5.18 and 5.19).

The writer Williams Haynes (1886–1970), best known for his series *American Chemical Industry: A History*, in an incomplete (and sometimes whimsical) manuscript intended for publication to mark the hundredth anniversary of American Cyanamid, described the early construction work thus [63]:

> Theirs was not a simple, easy job, for they were building a plant the like of which had never been seen on this continent to operate a process with which none of them had any experience. They were not even sure how to pronounce the name of their company and its sole product – *cy - an - amid*, with the accent on the second syllable. They had to work with foreign blueprints measured in meters, not inches and feet, whose instructions and specifications seemed curiously worded....They built the first cyanamide plant in America in eleven months, thirty days ahead of schedule. The first shipment – 52,240 pounds – rolled away from Niagara Falls, Ontario, bound for Baltimore, Maryland, on December 4, 1909.... Even within the American Cyanamid Company nobody really knew, firsthand, anything about the uses of cyanamide, and so while [Kenneth F.] Cooper [a Stanford trained engineer] and his gang were building the plant, a couple of boatloads of crude cyanamide, unhydrated, right out of the furnaces, were imported from the Odda works in Norway. It was landed at Baltimore.... Baltimore and Charleston were the then headquarters of the fertilizer trade... That crude cyanamide from Norway had to be refined and even then was an exceptionally dusty fine powder, black as ink.

Fig. 5.16 Banks of calcium cyanamide furnaces (ovens), American Cyanamid Company, Niagara Falls. They are similar to the oven invented in 1906 by Adolph Frank. (C. L. Mantell, *Industrial Electrochemistry*, 2nd ed. New York: McGraw-Hill, 1940, p. 520. American Cyanamid Company photograph)

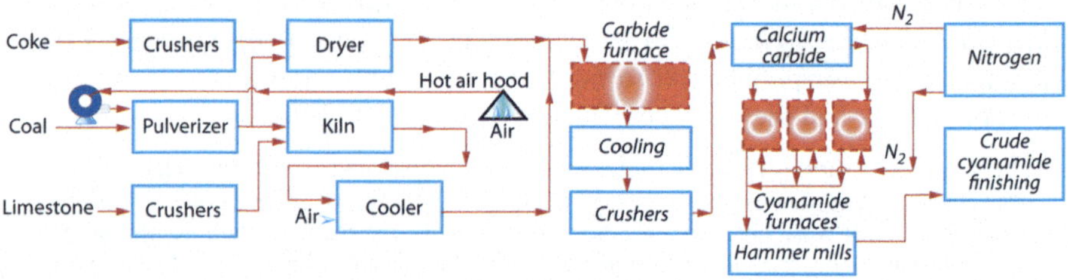

Fig. 5.17 Scheme for the manufacture of calcium carbide and its conversion into calcium cyanamide, both by electrothermal processes. (Based on C. L. Mantell, *Industrial Electrochemistry*, 2nd ed. New York: McGraw-Hill, 1940, p. 521)

The pure nitrogen required in the cyanamide process was obtained by a process similar to that of Brins. As Haynes tells us, the gas was obtained at Niagara by passing air over copper briquettes held in heated retorts. This removed the oxygen of the atmosphere and left the unreactive, pure nitrogen. The oxidized copper was then reduced, using readily available natural gas, which took up the oxygen of the oxide leaving the copper in a metallic state available for reuse:

Fig. 5.18 The Niagara Falls, around 1905. (From John G. Leigh, "The Water-Power Stations of Niagara Falls." In: Archibald Williams, ed., *Engineering Wonders of the World*. London: Thomas Nelson and Sons 1909, vol. II, pp. 295–311, on p. 295. Ontario Power Company photograph)

Fig. 5.19 Interior, electricity generating station of the Ontario Power Company on the Niagara River, opened in 1905. The six generators shown here, each connected to two turbines, produced in total the equivalent of 60,000 horsepower. From 1909 the station provided power to American Cyanamid's Niagara Falls factory. (From John G. Leigh, "The Water-Power Stations of Niagara Falls." In: Archibald Williams, ed., *Engineering Wonders of the World*. London: Thomas Nelson and Sons 1909, vol. II, pp. 295–311, on p. 308. Ontario Power Company photograph)

> [T]he process [was] repeated over and over again. Since the oxidation of copper took up only oxygen the rare gases in the air were left with the nitrogen, and in 1913 [US] General Electric suggested that, if the argon could be stripped out of this gas mixture, they would buy it. Irving Langmuir [1881–1957] had just developed [at General Electric] his orginal

idea that instead of striving to get a more perfect vacuum in electric light bulbs, their efficiency might be increased if they were filled with an inert gas. Cyanamid accepted what was the first industrial order for any of the rare gases and between December 1914 and May 1915 delivered 8,000 tons of argon.

Electrodes were made by calcining anthracite in an electric furnace. In 1914, a new design of carbide oven was introduced, and in the following year Claude plant was installed for the liquefaction of air. Despite new innovations, the cyanamide process involved considerable manual labour.

As elsewhere, farmers were reluctant to use Washburn's dirty-looking product, which caused discomfort in use, particularly to the hands, in addition to uncertainties in application. Several difficulties were experienced in formulating a suitable cyanamide fertilizer product until chemist Walter Savage Landis (1881–1944), an expert in electrothermics at Lehigh University, was taken on. In around 1912, Landis steamed "out the nitrogen in cyanamide under pressure to produce ammonia." The ammonia was reacted with phosphoric acid to give ammonium phosphate, a compound fertilizer. Plans for a calcium cyanamide factory in the Cotton Belt, backed by British investors, were cancelled after Congress vetoed a Bill authorizing the construction of a dam for generating electricity on the Coosa River.

In the late spring of 1914, "Frank Washburn and James Duke met by chance on The Strand in London. These hydro-electrical enthusiasts had known each other and they went off to lunch at the old Northumberland hotel on Trafalgar Square, famous for its whitebait and fillet of sole" [64]. Washburn's description over lunch of work on the new fertilizer certainly appealed to Duke, who, as we have seen, had at that time an interest in the Pauling process [65]. Moreover, Duke owned a phosphate mine in Florida and apparently was also interested in producing a nitrogen-phosphorus fertilizer. As a result of the meeting, Duke's Amalgamated Phosphate Company, of Brewster, Florida, became the Ammo-Phos Corporation and eventually led to Duke becoming a major investor in the American Cyanamid Company. The site chosen for the Ammo-Phos factory was at Warners, New Jersey. Meantime by 1915 the annual cyanamide capacity of the Niagara Falls plant had been raised to around 32,000 tons. The cost of producing the Ammo-Phos product was high due to transportation charges for cyanamide from Niagara Falls and phosphate from Florida.

While the fertilizer did not generate great profits, its development gave American Cyanamid invaluable experience in converting cyanamide, via ammonia, into nitric acid, a process that would bring about much needed income during World War I. Significantly, immediately after the outbreak of war in Europe in August 1914, Landis sailed for Germany to collect two autoclaves—part of an order placed in the spring of that year—for converting cyanamide into ammonia. This was probably the result of his close contacts with the Franks. According to Haynes, Landis also wished to visit the Frank-Caro Bayerische Stickstoff-Werke factory at Trostberg in Upper Bavaria to observe the ammonia oxidation process used there. He was fortunate to get the autoclaves out of Europe just before Germany imposed an embargo on the export of chemical manufacturing equipment. This was achieved with the help of "good friends in high places"—no doubt the Franks, who arranged for shipment via a "Scandinavian port," presumably in Norway, perhaps with bills of lading giving the destination as the Odda works. The balance of the order remained in Germany [66]. The autoclaves were installed at Warners, and with the aid of German blueprints enabled the design of home-built vessels. In the spring of 1916, work began on a pilot plant for oxidation of cyanamide-derived ammonia to nitric acid at Warners. Ammo-Phos production ceased in 1917 for the duration of the war.

5.11 Cyanamide in Japan

Around 1900, the Japanese chemical industry was negligible—mainly the manufacture of sulphuric acid by Sumitomo for superphosphate, produced from 1887 at Tokyo Artificial Fertilizer (later Dai Nippon Artificial Fertilizer, Dai Nippon Jinzō Hiryō, then Nissan Chemical Industries). Nitrogen fertilizer relied on imported bean cake mainly from China and, increasingly, ammonium sulphate from Britain. This business flourished due to the scarcity of arable land, and the small size of farms. The nitrogen fertilizer supply situation began to change during the early 1900s, after young, technically qualified entrepreneurs developed hydro-electricity and exploited this resource to introduce Western electrochemical technologies, notably carbide and cyanamide manufacture.

5.11.1 Noguchi Shitagau and Fujiyama Tsuneichi

The first calcium cyanamide facility in Japan was opened in 1909 by Noguchi Shitagau (1873–1944), who as a result of his electrochemical enterprises would create one of the leading new zaibatsu, the industrial conglomerate known as Nitchitsu. Noguchi studied engineering at Tokyo Imperial University, graduated in 1896, and in 1903 joined the Japanese branch of Siemens & Halske (around the time that the power division was named Siemens-Schuckert) where he would have become familiar with modern electrical engineering through the firm's turbines, generators, and electric motors.

Noguchi's technical expert was Fujiyama Tsuneichi, who in 1898 graduated in engineering from the same Tokyo institute. Following Fujiyama's experiments carried out during 1901, Noguchi designed and set up a carbide factory in Koriyama, Fukoshima Prefecture, that began operating in early 1902. Subsequently, using Siemens electrical equipment, Noguchi opened at Kagoshima, Kagoshima Prefecture, on the west coast of Kyūshū, a hydro-electric power station (Sogi Electric, Inc., 1906), to serve local gold mines and, with Fujiyama and Ichekawa Seiji, using surplus power, a carbide factory nearby at Minamata, Kumamoto Prefecture (Japan Carbide Company, 1907).

No doubt Noguchi's connection with Siemens, an investor in the Rome-based Frank-Caro patent holder, enabled him in April 1908 to acquire a license for working the cyanamide process in Japan. Sogi Electric and Japan Carbide were merged in August 1908 to create Japan Nitrogeneous Fertilizer, Inc. (Nippon Chisso Hiryō KK; Nitchitsu), near Osaka. Capital was raised through a network of family members and contacts, some connected with Mitsubishi Bank. A small plant was installed during 1908–1909 in Osaka. The Italian Frank-Caro patent holder and the Japanese Siemens-Schuckertwerke Electric Company (Siemens-Schuckertwerke Denki KK), in consultation with Noguchi, worked closely in protecting the Frank-Caro patents in Japan, including Albert Frank's 1909 patent for preparing fines of calcium carbide in high yield.

In 1909–1910, Noguchi commenced cyanamide production at the Minamata carbide factory. Several technical difficulties were encountered, and yields were lower than expected. These problems were overcome with redesigned ovens and other innovations. Fujiyama developed a continuous cyanamide process, introduced in September 1911. Adolph Frank was so impressed with the details that he sent his son, Albert, to Japan before the end of the year to observe the process in action. Despite financial and other difficulties—one setback was the destruction in July 1912 of a plant for conversion of cyanamide to ammonium sulphate—and Fujiyama's leaving earlier in the year, Noguchi by 1914 had opened another cyanamide facility, at Kagami, northern Kyūshū [67]. The cyanamide was generally converted into ammonium sulphate that suited the wet soil conditions (Fig. 5.20). Meantime during 1912 the Sogi Electric generating plant and part of

Fig. 5.20 Rice cultivation, Hida Folk Village, Takayama, Japan. Ammonium sulphate fertilizer was well suited to the wet soil cultivation of rice and was preferred to Chilean nitrate and cyanamide. As a result cyanamide was converted into ammonium sulphate

the Minamata factory were acquired by the government railways in connection with proposed railway electrification. This actually provided useful government funding at a time of uncertainty, particularly following extensive damage caused by floods [68]. The Minamata facilities were purchased back by Noguchi in 1915.

In 1912, following his departure from Noguchi, Fujiyama set up a carbide business, Hokkai Carbide, with the help of Oji Paper—associated with Mitsui Mining Co.—at Tomakomai, a village on Hokkaido. In March 1913, Fujiyama patented his continuous cyanamide process, which, despite his having introduced new features, led to decade-long litigation with Noguchi. In April, the manufacture of calcium cyanamide began, with enhanced nitrogen content. In April 1915, backed by several venture capitalists, Fujiyama set up the reorganized Hokkai Carbide as Denka (Electro-Chemical Co.), an affiliate of Mitsui. A second plant was opened in 1916 at Ōmuta, in Fukuoka Prefecture on Kyūshū [69]. Four other cyanamide manufacturers entered the business, one in 1917, three in 1918.

Cyanamide represented the most important early phase in the development of the modern Japanese chemical industry, which continued to be stimulated by European technological innovations. It would lead in the early 1920s to other, more far reaching, synthetic nitrogen processes. We return to Noguchi's enterprise in Chapter 15.

5.12 Ferdinand Polzenius

In 1901 Austro-Hungarian chemist Ferdinand Eduard Polzenius (1862–1918), while working at the Polack Accumulator Works in Frankfurt am Main, overcame the main cyanamide oven difficulty—the high temperature requirement—by introducing calcium chloride as a "catalyst" or flux. This enabled reduction of the operating temperature from 1100 °C to 700–800 °C. After the Polack company went into liquidation around this time, Polzenius and chemist Constantin Krauss (1864–1928) developed a cyanamide manufacturing process and obtained financial backing, mainly from Metallurgische Gesellschaft (Metallgesellschaft) of Frankfurt am Main. In 1904 the process was taken up, also at Westeregeln, but it proved uneconomic, perhaps due to problems with the ovens. Following improvements to the process, Deutsche Karbid Aktien Gesellschaft was founded on 31 May 1906 in Frankfurt am Main by Metallgesellschaft, whose Dr Rudolf de Neufville was the first chairman. A new factory was erected, at Knapsack, south-west of Cologne in the brown coal (lignite) district. It began operating with three ovens in October 1907 using purchased carbide. A brown coal power plant and a calcium carbide production facility were added in 1908. Several difficulties were encountered, including with the supply of labour. The conditions were hardly salubrious. The work was arduous: "The furnaces were tapped with 6 m long rods as the carbide was discharged from the furnaces at a temperature of more than 2000 °C. It was exhausting work that only a few could do for any length of time, in spite of the high wages offered... the company looked more like a camp of desperadoes than a factory" [70].

In 1909, Aktiengesellschaft für Stickstoffdünger was co-founded by Metallgesellschaft and Consolidierte Alkali-Werke to take over the Knapsack factory and rights to the Polzenius process. In the following year Krauss was appointed manager. In 1911, Knapsack began the manufacture of ammonia and ammonium sulphate from cyanamide, which was made in a continuous process, using a horizontal cast iron cylinder, some 150 feet in length and 6 feet in diameter, through which a rail track was laid. Twenty-eight trucks loaded with carbide were pushed into the cylinder, which was then closed and made gas tight, and the reaction started.

Polzenius's hygroscopic calcium chloride "catalyst" was later replaced with, for example, calcium fluoride, as introduced by Oskar Fredrik Svante Carlson (1844–1916), founder of Stockholms Superfosfat Fabriks Aktiebolag (other alkali and alkaline earth salts were also specified). Oskar was assisted by his son, Birger Carlson (1873–1928), who was educated at the Technische Hochschule Charlottenburg. They worked a small-scale cyanamide process at Mansbo, near Avesta, Sweden, around 1912, and constructed a large works shortly after at Johannesbergfors, near Alby. The Carlson process was continuous, employing a tower fitted with shelves rather than a tunnel. Carbide was charged at the top and pushed down mechanically from one shelf to the next. The process was at first problematic when, as a result of the mechanical action, the product caked up.

5.13 The Frank-Caro Process in Germany

Large-scale Frank-Caro operations in Germany had been delayed until January 1908, when Frank and Caro came to an arrangement with Polzenius interests in Milan over the latter's cyanamide patent [71]. On 6 November 1908, Cyanid-Gesellschaft, with the backing of Deutsche Bank, created a subsidiary, Bayerische Stickstoff-Werke AG, in Munich, for the manufacture of calcium cyanamide in Upper Bavaria. A large factory was built at Trostberg, on the River Alz. Hydro-electric power was supplied from what became known as power station I, located on the new canal excavated next to the Alz by the Bavarian government in order to withstand periodic floods (Fig. 5.21). Manufacture of calcium cyanamide began in 1910. Soon after, cyanamide-derived ammonia was converted into nitric acid. Possibly the reason why the BASF electric arc factory on the Alz was rejected had to do with the preference of the Bavarian authorities for cyanamide, or the terms agreed with its promoters.

Annual output of cyanamide from the works was 6000 tons. In 1913, the company began operating a tunnel (horizontal) furnace with countercurrent flow of nitrogen and carbide for calcium cyanamide manufacture. However such furnaces were costly, and yield and quality were not up to expectations, which did not justify their widespread installation.

Carbide was manufactured nearby, upstream at Schalchen, close to the site of power station II, at Tachetring. Adolph Frank and Nikodem Caro were the principals of Bayerische Stickstoff-Werke. Albert Frank worked in administration until

Fig. 5.21 Turbine hall, Bayerische Stickstoff-Werke Trostberg hydro-electric power station, located on the canal excavated next to the River Alz. Dating from 1910, it was known as power plant I at the time when this photograph was taken in the mid-1920s. (Evonik Industries AG, Corporate Archives, Hanau/Marl)

1916, when following the death of his father he joined the board. The rulers of Bavaria, Prince Luitpold until 1912, then Prince Ludwig during 1912–1913 (both on behalf of King Otto), and Ludwig III, 1913–1918, followed developments at the company, in which the Bavarian state held a substantial interest.

Albert and Adolph Frank and Nikodem Caro collaborated with Carl von Linde in developing the Linde-Frank-Caro process for preparing pure hydrogen from water gas. With the engineering firm Bamag in 1909 they founded a company to market the technology [72]. Fritz Rothe continued with his investigations into the electrothermal fixation of atmospheric nitrogen and in 1911 filed an electric arc process patent with Elektrochemische Werke.

In 1912, fourteen calcium cyanamide factories were operating worldwide (Table 5.1). By 1914, annual production of calcium cyanamide was 220,000 tons, one-quarter of which was manufactured in Germany, mainly at Trostberg and Knapsack. Production would increase considerably during World War I.

With few exceptions, all electrothermal processes continued to be located close to sources of cheap electricity, mainly hydro-electricity. A cyanamide factory at Brühl in the Rhineland relied on lignite for power, as did a factory near Bromberg, Province of Posen, which produced 2500 tons per year, much needed in an area of poor soil quality. The power supply situation would change only in the 1920s, when

Table 5.1 Manufacturers of calcium cyanamide and reported annual capacities, in 1912. Capacities in 1909 shown in parentheses

Company	Place	Capacities annual in tons, approx.
Nitrogen Fertilizers (1912). (North-Western Cyanamide Co.)	Odda, Norway	15,000 (12,000)
Nitrogen Fertilizers (1912). (North-Western Cyanamide Co.)	Alby, Sweden	15,000
Società Italiana Prodotti Azotati (SIPA)	Piano d'Orta, Italy (1905–1909)	4000 (manufacture moved to Papigno, 1909)
Società Italiana per il Carburo di Calcio (Carburo)	Papigno, near Terni, Italy	15,000
Società Piemontese per il Carburo di Calcio	San Marcel, Aosta Valley, Italy	3000
Le Société Française des Produits Azotés	Martigny, Switzerland (French owned)	7500 (3750)
Le Société Française des Produits Azotés	Notre-Dame-de-Briançon, Hautes-Alpes, France	7500
Bayerische Stickstoff-Werke	Trostberg, Bavaria	15,000
Ost-Deutscher Stickstoffcalc und Chemische Werke	Bromberg, Province of Posen, Prussia	2500 (2500)
AG für Stickstoffdunger	Knapsack, near Cologne, Westphalia	18,000 (5000)
Società per l'Utilizzazione delle Forze Idrauliche della Dalmazia	Cernica, near Sebenico (Šibenik), Dalmatia, Austro-Hungary	4000
Società per l'Utilizzazione delle Forze Idrauliche della Dalmazia	Dugi Rat, near Almissa, Dalmatia, Austro-Hungary	[8,000]
Japan Nitrogenous Fertilizer (Japanese Nitrogen Products Co.; Nitchitsu)	Minamata, Kyūshū, Japan	4000
American Cyanamid Company	Niagara Falls, Ontario	25,000 (5000, in 1910)

Source: Ernest Kilburn Scott, "The Manufacture of Nitrates from the Atmosphere," *Nature*, vol. 89, no. 2227 (4 July 1912):463–465. American Cyanamid drew up plans for a 4000 ton-per-year cyanamide facility at Nashville, Tennessee.

large-scale electrical transmission networks, capable of handling thousands of kilowatts, were introduced.

The significance of the various developments in electrothermic chemistry as related to nitrogen fixation was not lost on the Nobel Prize committee at the Swedish Academy of Sciences [73]. The committee considered the fixation of atmospheric nitrogen from around 1909, when Peter Klason of the Stockholm Institute of Technology guided members in their deliberations. Klason initially favoured Adolph Frank, alone or possibly with Caro, but later expressed more interest in Birkeland. Difficult relations between Sweden and Norway at the time, the fact that Birkeland had formed a commercial partnership with Eyde, and more, that Eyde wanted to be a co-nominee, may have militated against Birkeland being awarded the prize. The committee, unable to reach a consensus, dropped nitrogen fixation from its agenda, at least for the time being. In 1912, both Birkeland and Otto Schönherr were nominees. For his several achievements in physics and chemistry between 1907 and 1917, Birkeland was nominated for the Nobel Prize seven times in all [74].

References

1. Crookes W (1898) The world's wheat supply. Reports of the British Association for the Advancement of Science, London, p 4.
2. Crookes W (1900) The wheat problem. GP Putnam's Sons, New York. John Murray, London.
3. Stock JT, Orna MV (eds) (1989) Electrochemistry: past and present. ACS Symposium Series 390. American Chemical Society, Washington.
4. Mader U (2001) Emil und Walther Rathenau in der elektrochemischen Industrie, 1888–1907: eine historische Studie. Gesellschaft, Geschichte, Gegenwart, Band XX. Trafo Verlag, Berlin.
5. Bäumler E (English version) (1968) A century of chemistry [Hoechst]. Econ Verlag, Düsseldorf, pp 49–52.
6. People. Markets. Chemistry. The Wacker Chemie formula for success, 1914–2014 (2014). Piper, Munich, pp 10–47.
7. Trescott M (1989) The rise of the American electrochemicals industry, 1880–1910. Contributions in economics and economic history. Greenwood Press, Westport, CT.
8. Mierzecki R (1998) The Polish chemical industry: isolated developments in a divided country. In: Homburg E, Travis AS, Schröter HG (eds), The chemical industry in Europe, 1850–1914: industrial growth, pollution, and professionalization. Kluwer, Dordrecht, pp 59–71, on 70.
9. Egeland A, Leer F (1986) Professor Kr. Birkeland: his life and work. IEEE T Plasma Sci PS-14 (6)(December):666–677.
10. Grimnes OK (2001) Sam Eyde: Den grenseløse gründer. Aschehoug, Oslo (in Norwegian).
11. Sogner K (2014) Creative power. Elkem 110 Years, 1904–2014. Elkem, Oslo.
12. http://www.hydro.com/en/About-Hydro/Our-history/1900-1917/1905-Three-remarkable-men (accessed 12 August 2014).
13. Sagatos OJ (2005) Progress of a different nature: Hydro 1905–2005. Pax Forlag, Oslo, pp 43–44.
14. Witt ON (1905) Das neue technisch-chemische Institut der Königlich-Technischen Hochschule zu Berlin und die Feier seiner Eröffnung am 25. Nov. 1905. Weidmann, Berlin.
15. Sagatos OJ (2005) Progress of a different nature: Hydro 1905–2005. Pax Forlag, Oslo, p 45.
16. Ramsay W (1906) Special articles: the fixation of nitrogen. The Times, Engineering Supplement, no. 52 (21 February).
17. Guye PA (1906) The electro-chemical problem of the fixation of nitrogen. J Soc Chem Ind 25 (12)(30 June):567–578.
18. Birkeland K (1906) On the oxidation of atmospheric nitrogen in electric arcs. Discussion. T Faraday Soc 2:116–119, on 116–117.
19. Harker JA (1922) Post-war progress in the fixation of nitrogen. J Soc Chem Ind 41(18) (30 September):387R–390R, on 387R.
20. Lunge G (1916) Handbuch der Schwefelsäurefabrikation und ihrer Nebenzweige, Band I. Friedr. Vieweg & Sohn, Braunschweig, pp 257–261.
21. Eyde S (1912) Oxidation of atmospheric nitrogen and development of resulting industries in Norway. Ind Eng Chem 4(10)(1 October):771–774.
22. Crookes W (1914) Presidential address. J Soc Chem Ind 33(14)(31 July):720–722, on 720.

23. Lunge G (1916) Handbuch der Schwefelsäurefabrikation und ihrer Nebenzweige, Band I. Friedr. Vieweg & Sohn, Braunschweig, pp 261–262.
24. Taugbøl T, Andersen EM, Grønn U, Moen BF (2014) Nomination dossier: Rjukan-Notodden industrial heritage site. Nomination to the UNESCO World Heritage List, Norway, pp 31–42.
25. Kilburn Scott E (1915) Production of nitrates from air, with special reference to a new electric furnace. J Soc Chem Ind 34(3)(15 February):113–126, on 121.
26. Behrman EJ (2005) Karl Elbs, 1858–1933. Bull Hist Chem 30(1):19–22.
27. Scharff M (1927) Dr. Otto Schönherr zum Gedächtnis. 1861–1926. Z angew Chem 40 (14)(7 April):393–394.
28. Mantell CL (1940) Industrial electrochemistry, 2nd ed. McGraw-Hill, New York, p 576.
29. Lunge G (1916) Handbuch der Schwefelsäurefabrikation und ihrer Nebenzweige, Band I. Friedr. Vieweg & Sohn, Braunschweig, pp 263–269.
30. Reinhardt C (1993) Ueber Wissenschaft und Wirtschaft: Fritz Habers Zusammenarbeit mit der BASF 1908–1911. In: Albrecht H (ed), Naturwissenschaft und Technik in der Geschichte. 25 Jahre Lehrstuhl für Geschichte der Naturwissenschaft und Technik am Historischen Institut der Universität Stuttgart. Universität Stuttgart, Stuttgart, pp 287–315.
31. Johnson JA (2004) The power of synthesis, 1900–1925. In: Abelshauser W, von Hippel W, Johnson JA, Stokes RG, German industry and global enterprise. BASF: the history of a company. Cambridge University Press, New York, pp 144–145.
32. Lunge G (1916) Handbuch der Schwefelsäurefabrikation und ihrer Nebenzweige, Band I. Friedr. Vieweg & Sohn, Braunschweig, pp 253–256.
33. Kilburn Scott E (1912) The manufacture of nitrates from the atmosphere. Nature 89(2228) (11 July):490–492, on 491.
34. Wisniak J (2007) Phillipe-August Guye. Revista CENIC Ciencias Químicas 38(3):415–422.
35. Mościcki to Einstein, 27 September 1932, Albert Einstein Archives, The Hebrew University of Jerusalem, AE 71–782.
36. Lichocka H (2014) Swiss experiences of Ignacy Mościcki. Tech Transactions. Fundamental Science (Czasopismo Techniczne: Nauki Podstawowe [CzT]) 1-NP/2014, pp 130–149. http://www.ejournals.eu/pliki/art/3420/Nauki-Podstawowe-Zeszyt-1-NP(Z)2104-12.pdf (accessed 20 January 2015).
37. Pitschikoff N (1908) Chimie technique en Russie. Revue d'Electrochimie et d'Electrometallurgie 2(7)(July):168–169.
38. Schweitzer H (1907). In: Meldola R, Green AG, Cain JC (eds), Jubilee of the discovery of mauve and of the foundation of the coal-tar colour industry by Sir W. H. Perkin, F.R.S., D.Sc., LL.D., Ph.D., Dr.Ing. Perkin Memorial Committee, London, p 62.
39. Kastens ML, McBurney WG (1952) Calcium cyanamide. Modern chemical processes: a series of articles describing chemical manufacturing plants, vol. II. Reinhold Publishing, New York, pp 97–110, on 100.
40. Nikodem Caro, chemist and entrepreneur, http://history.evonik.com/sites/geschichte/en/predecessor-companies/skw-trostberg/caro (accessed 19 August 2014).
41. Eissler M (1902) The cyanide process for the extraction of gold and its practical application on the Witwatersrand goldfields & elsewhere. Crosby Lockwood & Son, London.
42. Adolph Frank to Siemens, 14 January 1901; Siemens to Adolph Frank, 16 January 1901. The Adolph and Albert Frank Collection of papers and correspondence, Leo Baeck Institute, New York, AR 7176/MF772. Series II. Cyanid-Gesellschaft, 1892–1915.
43. Cyanid-Gesellschaft, German patents nos 152,260 (1901), and 154,505 (1902).
44. Frank A[dolph] (1906) Über die direkte Verwertung des Stickstoffes der Atmosphäre zur Gewinnung von Düngemitteln und anderen chemischen Produkten. Z angew Chem 19:835–840.
45. Correspondence and general arrangement drawings held with the Adolph and Albert Frank Collection of papers and correspondence, Leo Baeck Institute, New York, AR 7176/MF772.
46. Travis AS (2014) The emerging role of titrimetry in late nineteenth-century industrial problem solving: the example of trace analysis for perchlorate in Chile saltpetre. Ambix 61(1):67–94.

47. Caro N (1906) Über einheimische Stickstoffquellen. Z angew Chem 19:1569–1581.
48. Pranke EJ (1913) Cyanamide: manufacture, chemistry and uses. Williams & Norgate, London.
49. Cafanga L (1973) The industrial revolution in Italy,1830–1914. In: Cipolla CM (ed), The Fontana economic history of Europe: the emergence of industrial societies. Fontana/Collins, London, pp 279–328, on 316–317.
50. di San Filippo PA (1998) The Italian chemical industry from 1861 to 1918. In: Homburg E, Travis AS, Schröter HG (eds), The chemical industry in Europe, 1850–1914: industrial growth, pollution, and professionalization. Kluwer, Dordrecht, pp 45–57, on 48–49.
51. Benegiamo M (2013) Bussi e la grande chimica in Abruzzo. Un'ambizione fallita. Textus Edizioni, L'Aquila, pp 114, 189, 193–198, 270–271.
52. Draft statutes, La Società Generale per la Cianamide. The Adolph and Albert Frank Collection of papers and correspondence, Leo Baeck Institute, New York, AR 7176/MF772, Cyanid Gesellschaft, 1892–1915, series II, box 2, folder 26.
53. Fixation of atmospheric nitrogen (1906) Electrochemical and Metallurgical Engineering 4(August):327–328.
54. di San Filippo PA (1998) The Italian chemical industry from 1861 to 1918. In: Homburg E, Travis AS, Schröter HG (eds), The chemical industry in Europe, 1850–1914: industrial growth, pollution, and professionalization. Kluwer, Dordrecht, pp 45–57, on 49.
55. Kilburn Scott E (1912) The manufacture of nitrates from the atmosphere. Nature 89(2227) (4 July):463–465, and 89(2228)(11 July 12):491–492.
56. Frank A (1907) Ersatz des Chilisalpeters in der Zukunft – Kalkstickstoff. VIII Internationaler Landwirstschaftlicher Kongress, Wien 1907. Printed by Leonhard Simion, Berlin.
57. Frank AR (1908) On the utilisation of the atmospheric nitrogen in production of calcium cyanamide, and its use in agriculture and chemistry. T Faraday Soc 4:99–114.
58. A new Norwegian calcium carbide and cyanamide works (1909). Electrochemical and Metallurgical Engineering 7(May):212–216.
59. Frank A (1908) Chemical industry in relation to agriculture. J Soc Chem Ind 27(22)(30 November):1093–1100.
60. Caro N (1909) Industrien des Kalkstickstoffes und verwandte Verfahren. Z angew Chem 22 (24)(11 June):1178–1182.
61. Kilburn Scott E (1915) Production of nitrates from air, with special reference to a new electric furnace. J Soc Chem Ind 34(3)(15 February):113–126, on 126.
62. Haynes W (1939) Chemical pioneers. The founders of the American chemical industry. D. Van Nostrand, New York, pp 245–247.
63. Haynes W [1957] On the chemical frontier: the Cyanamid story. Copy held at Sidney M. Edelstein Library for the History and Philosophy of Science, Technology and Medicine, National Library of Israel, pp 22, 24, 25–26, 48–49.
64. Haynes W [1957] On the chemical frontier: the Cyanamid story. Copy held at Sidney M. Edelstein Library for the History and Philosophy of Science, Technology and Medicine, National Library of Israel, p 33.
65. Massell DP (2000) Amassing power: J. B. Duke and the Saguenay River, 1897–1927. McGill-Queens University Press, Montreal, pp 114–119.
66. Haynes W [1957] On the chemical frontier: the Cyanamid story. Copy held at Sidney M. Edelstein Library for the History and Philosophy of Science, Technology and Medicine, National Library of Israel, p 40.
67. Molony B (1990) Technology and investment: the prewar Japanese chemical industry. Council on East Asian Studies, Harvard University. Distributed by Harvard University Press, Cambridge, MA, pp 71–84.
68. Molony B (1990) Technology and investment: the prewar Japanese chemical industry. Council on East Asian Studies, Harvard University. Distributed by Harvard University Press, Cambridge, MA, pp 87–89.
69. Denka Milestones, www.denka.co.jp/eng/corporate/about/history/ (accessed 8 September 2015).
70. Bäumler E (English version) (1968) A century of chemistry [Hoechst]. Econ Verlag, Düsseldorf, p 61.

71. Waeser B (1932) Die Luftstickstoff-Industrie, mit Berücksichtigung der chilenischen Industrie und des Kokereistickstoffs. Springer, Berlin, p 141.
72. Hans-Liudger D (2004) Linde: history of a technology corporation, 1879–2004. Palgrave Macmillan, Basingstoke, chapter 2.
73. Crawford E (1987) The beginnings of the Nobel institution: The science prizes, 1901–1915. Cambridge University Press, Cambridge, pp 181–183.
74. Egeland A, Burke WJ (2005) Kristian Birkeland: the first space scientist. Astrophysics and Space Library. Springer, Dordrecht, pp 139–141.

Chapter 6
The Direct Synthesis of Ammonia

6.1 BASF

The market for fertilizers in industrialized countries expanded greatly after 1900. Artificial nitrogen products, if they could be manufactured at the right cost, offered the potential for substantial profits from the agricultural sector. The interest in nitrogen capture was especially strong at BASF, the leading dye manufacturer at the turn of the century. This interest had to do with overcoming Germany's shortage of natural resources and its considerable dependence on imports from other countries. In the case of fixed nitrogen, there were concerns over not just British participation in the Chile saltpetre monopoly, and how long the supply of natural nitrate would last, but also external shocks such as economic and political events, including the outbreak of war. Moreover, nitrogen offered an opportunity for diversification at German firms away from dyes. To provide a clear picture of how BASF came to lead in nitrogen fixation developments in the early 1900s it is necessary to review the firm's rise to a leading place in the realm of science-based industry.

On 30 March 1856, within a few days of William Henry Perkin in London discovering the first synthetic dye, aniline purple, later known as mauve, the Treaty of Paris brought an end to the first Crimean War (1853–1856). The main beneficiary of this final break up of post-Napoleonic Europe was a non-combatant, Prussia, that soon was carving out a new empire by unifiying the German states and principalities. This was particularly the case during 1864–1870, which coincided with the rapid emergence of the German synthetic dye industry. The German firms included BASF, established in 1865, at Mannheim. It soon relocated to nearby Ludwigshafen, on the west bank of the River Rhine, in the Bavarian Palatinate. Its early success owed much to the technical leader, Heinrich Caro. During 1869–1870, he developed a process for the manufacture of synthetic alizarin—the important red dye previously isolated from the root of the madder plant—drawing on academic-industrial collaboration in Germany, and an exchange of technical information with William Perkin's firm. After 1872, German firms led in the

A.S. Travis, *Nitrogen Capture*, https://doi.org/10.1007/978-3-319-68963-0_6

production of synthetic alizarin, causing Perkin to sell his business in 1873. Inadequate patent law and failure in science-based research and development brought about the decline of much of the British synthetic dye industry by around 1880.

Synthetic alizarin also brought about the collapse of the massive trade in the natural red dye, and placed the German dye industry on a secure footing. Novel brilliant synthetic dyes without analogues in nature entered the marketplace. Caro and his co-workers at BASF, as well as chemists at Hoechst, Bayer, and AGFA, undertook extensive research into synthetic colorants. German chemists also contributed to the drawing up of a patent system that was without equal when it came to protecting chemical inventions. The dye industry became an important contributor to the economy of the German Empire. Heinrich Caro was involved in all these endeavours. Exploiting new theories and methods of synthesis in chemistry, with the aid of academic and industrial chemists, he created a revolution in the production of novel materials, based on scientific discovery and invention [1]. In 1874, Adolf Baeyer (1835–1917; ennobled 1885), at the then new German University of Strasbourg (Kaiser-Wilhelm-Universität), and Caro together published the modern structural formula of alizarin. Caro inaugurated an industrial programme for synthetic indigo, working closely with Baeyer, following the latter's move to Munich. The natural product was imported into Europe from India. Britain held the monopoly on its trade. In 1883 Baeyer sketched, in a private letter to Caro, the almost correct chemical structure of the colouring matter. This opened the way to the synthetic product, which was produced in 1897 by both BASF and Hoechst. The natural indigo industry collapsed. Caro was responsible for the design and construction of the Central Research Laboratory at the BASF Ludwigshafen factory. Completed at the end of the 1880s its activities encompassed not only research and development projects, but also the critically important support service required in patent litigation. BASF became adept at undertaking risky and expensive ventures based on close collaboration with academic inventors. Most importantly, especially for later developments, the major German dye firms created well equipped in-house facilities for research and development, and actively pursued the acquisition of new inventions and know-how from external sources, mainly academic scientists.

By 1900, bolstered by strong reserves and with the confidence of the banking industry behind them, these firms had diversified, including, at Hoechst, into serum therapy, at Bayer, into pharmaceuticals and, at AGFA, into photo products. The most difficult challenge, in which BASF and Hoechst in particular were involved, was the capture of atmospheric nitrogen. All possible routes were explored. In addition to electric arc, cyanide, and cyanamide processes, there were occasional attempts to combine hydrogen with nitrogen to produce ammonia. The latter was investigated at the University of Leipzig by the leading physical chemist Wilhelm Ostwald (1853–1932) [2].

On 12 March 1900, Ostwald advised the BASF management that he had devised a method for preparing ammonia directly from its elements by applying a high temperature to a closed apparatus containing nitrogen and hydrogen and an iron wire catalyst [3]. He also pointed out that the yield increased with the application of pressure to the reactant gases. Ostwald had filed a patent, specifying an iron or copper catalyst, confident that his process worked and could be exploited industrially. However, when the method was tried at BASF it was found that ammonia was formed not directly from the nitrogen and hydrogen gas mixture, but from the reaction of hydrogen with an iron-nitrogen compound in the catalyst. The scientist in charge of the evaluation and who exposed the failure of Ostwald's method was Carl Bosch. Little wonder then, that for the next few years so much faith was placed in the electric arc.

6.2 Carl Bosch

Carl Bosch (1874–1940) grew up in Cologne, where his father owned a metalworking business that served the needs of local coal gas suppliers and distributors. His uncle, Robert Bosch, founded the Bosch electrical engineering firm in 1886, in Stuttgart. In the Cologne workshop, Carl made his own equipment, with the assistance of a foreman. In 1893, he underwent an apprenticeship in a metallurgical factory, and during 1894–1896 studied metallurgy and mechanical engineering at the Technische Hochschule Charlottenburg. In 1896, he embarked on research in chemistry at the University of Leipzig. Among the topics that attracted his attention was spectral analysis, which later usefully aided his understanding of the structures of metals and probably led to his long-term interest in astronomy. In 1898, he received his doctorate in organic chemistry under Johannes Wislicenus (1835–1902) [4, 5].

Bosch joined BASF in April 1899, at first working in the Indigo Department with Rudolf Knietsch, who came to Ludwigshafen in 1884 from the dye firm Bindschedler & Busch of Basel. Around 1900, Bosch was assigned to the problem of nitrogen fixation; it was here that he quickly showed why Ostwald's ammonia method was flawed. From 1902, Bosch investigated metal cyanides and nitrides as possible sources of stable nitrogen products. The cyanide method involved the high temperature reaction of barium oxide (from barytes) with carbon (from coal) and nitrogen to give barium cyanide. With steam, the cyanide decomposed to ammonia and barium hydroxide. It was not altogether different from Mond's method, and even showed some similarities with the early carbide work of Frank and Caro. A pilot plant for barium cyanide was erected in 1907, but it was not viable as a technical route, and was abandoned in 1908. The failure of Ostwald's method and the huge energy requirements of the cyanide method certainly emphasized the great challenges posed to academic and industrial chemists and inventors in their attempts to fix atmospheric nitrogen. The complexities and technical difficulties pushed the boundaries of scientific and technical knowledge and expertise, and within firms of their management, to the limits for those who sought to master nitrogen capture.

The challenges of nitrogen fixation, and their implications, were expressed in no uncertain terms during the summer of 1903 by Ostwald, writing in the *Schwäbische Merkur*. "The significance of bound nitrogen. . .is especially high for both war and peace. . .Without saltpeter the best military is almost helpless. . .Were a war to break out today between two great powers, one of which was able to prevent the export of saltpeter from Chile's few harbours, that ability alone would allow it to render its opponent almost incapable of fighting" [6]. Therein lay the nitrate problem. Its solution was a strategic necessity, quite apart from the critical importance to agriculture.

This was also recognized in the United States, where in 1909 chemist Charles Edward Munroe (1849–1938) drew the problem to the attention of the Naval Institute, and in Britain, where electrical engineer Ernest Kilburn Scott (1868–1941) advised the British Association for the Advancement of Science in 1911 and the Royal Society of Arts in 1912 of the strategic need to develop technical methods of nitrogen fixation. In 1912, Kilburn Scott pointed out that "Nitric acid is, of course, the main constituent of gun-cotton, dynamite and smokeless powders, &c., and at the present time we are mainly dependent on overseas supplies of raw material from which to make the acid. In case of war we should undoubtedly be in a very serious position, for whereas most Continental countries have plants for the fixation of nitrogen from the air, this country does not make a single ounce" [7]. Such a situation, he reminded his audience, had strong parallels with the Napoleonic wars, when the French faced shortages of saltpetre. Kilburn Scott believed that the electrochemical and electrothermal industries would in time become more significant than electric traction, which had made great strides in the early 1900s, with electrification of street tramways, underground railways and tunnels, long distance main lines, and power generation and distribution.

In 1915, Kilburn Scott repeated his earlier calls for action, this time reminding members of the Society of Chemical Industry that he had "urged the necessity of starting the production in this country of nitrates from air by electricity" [8]. Kilburn Scott was promoting his three-phase electric arc process for capturing nitrogen as nitric oxide, for which, by 1915, a laboratory scale apparatus had been assembled at Northampton Polytechnic Institute, Clerkenwell, London, where he was a lecturer [9]. A three-phase arc process for capturing nitrogen was worked in Zurich by the Nitrumwerke AG (Verträge Nitrumwerke AG (Bodio) Salpetersäurelieferung und Vermietung Kesselwagen) between 1916 and around 1920. The main inventor was Werner Siebert, who later collaborated with Nikodem Caro and Albert Frank in development of improved processes for the manufacture of concentrated nitric acid. In 1925, Nitrumwerke AG adopted the Claude synthetic ammonia process, for which SA Nitrammonia was founded by financial interests in Bavaria and Switzerland.

6.3 Fritz Haber

Haber was later accused by the cynics of having provided Germany with ammonium nitrate for explosives and making the war inevitable. He was later accused of providing poison gas to the army. This takes no account of what men, including chemists, [do] in the name of patriotism for the Country to which they belong, especially when both combatants are sure they have God on their side!

Ralph C. Chirnside (colleague of Robert Le Rossignol), 1980

Though several studies and numerous texts, as well as plays, novels, films, biographical articles and more have appeared on the life and times of Fritz Haber (1868–1934), the central figure here, it is worth recording the salient facts of his early life [10–22]. His parents were the cousins Siegfried and Paula Haber, of Breslau. They were members of a typical emancipated Jewish family. Fritz, born on 9 December 1868, was their only child. His mother, Paula, died a few weeks later from complications following the birth. Siegfried remarried in 1874 to Hedwig Hamburger, with whom he had three daughters. The three half sisters became close to Fritz, who was several years their senior. Until the age of eleven, Fritz attended the Johanneum School, where his classmates came from Catholic, Protestant and Jewish homes. In 1878, he entered the Elisabeth High School, where there was a strong emphasis on the classics, though mathematics and science were also taught. Siegfried Haber owned a business that traded in chemicals, including dyes and pharmaceutical products, and intended that Fritz would eventually take over its day-to-day running. In 1886, at age 18, Fritz entered Friedrich-Wilhelms-Universität, Berlin (the present-day Humboldt University of Berlin), where he attended lectures of the ageing organic chemist A. Wilhelm von Hofmann and physicist-philosopher Hermann von Helmholtz (1821–1894). There was also a course in German philosophy. This was followed by studies at Heidelberg, with lectures in chemistry by Bunsen and in differential and integral calculus by mathematician Leo Königsberger (1887). Haber returned to Berlin around 1889 and began studies at the Technische Hochschule Charlottenburg, prior to serving for one year with the Sixth Field Artillery Regiment in Breslau. There he met Clara Immerwahr (1870–1915), whose father, Philipp, had studied chemistry at Breslau and obtained a doctorate under Bunsen. Immerwahr senior then managed a beet sugar processing factory at Trachenberg, around 40 km north of Breslau, and took charge of a family estate at Polkendorf. Sugar beet production and sugar refining were by then major European commercial activities, with important branches in Hungary and Russia, as well as Germany, as we have seen with Adolph Frank. It was a mainly Jewish preserve in nineteenth-century Poland. Fritz and Clara kept in touch over the following years. In the autumn of 1890, Fritz Haber embarked on research at the Technische Hochschule Charlottenburg under organic chemist Carl Liebermann, co-inventor with Carl Graebe (1841–1927) of the original method for synthetic alizarin. Liebermann also held a post at the Friedrich-Wilhelms-Universität. Among the lectures that Fritz attended were those given by the dye chemist Otto N. Witt.

Fritz Haber graduated *cum laude* in 1891 at Friedrich-Wilhelms-Universität with a doctoral thesis on an indigo derivative. Through exposure to Hofmann and Liebermann, who had both made massive contributions to science-based industrial organic chemistry, Haber would have received a thorough grounding in the role played by aromatic chemistry in Germany's emergence as a leading industrial power.

However, aromatic organic chemistry was a little too much like cookery for the ambitious Haber: his thesis was "miserable. One and a half years of new substances prepared like a baker's bread rolls" [23]. Following his doctoral studies, there was a working grand tour of chemical industry: an alcohol distillery in Budapest (from spring 1891), where liquors from molasses, produced from beet sugar, were fermented; an Austrian-owned Solvay factory, probably near Krakov, served by a local salt mine; and from early 1892 a paper and cellulose works [24]. Though of shorter duration, these were not unlike the activities in which Adolph Frank had participated prior to the early 1890s.

Of more interest was a semester during 1892 at the chemical institute of Zurich's ETH, under the German-born Georg Lunge (1839–1923). Lunge, a distant relative of the Habers, was an industrial chemist and leading expert on gas analysis, which was critical to several areas of industrial application, including the products of coal gas works. By the 1890s, the manufacture, storage, transportation and analysis of gases were important for the chemical industry, including the dye-making firms. In 1888, Knietsch at BASF developed chlorine liquefaction, enabling the corrosive gas to be compressed, stored and transported in steel cylinders. In 1901, of 25,000 tons of chlorine produced in Germany, 17,000 tons came from the electrolysis of sodium chloride, the chlor-alkali process [25].

After the short stay in Zurich, Haber joined his father's business, but not for long. The reason was the timing of the purchase of a large amount of chloride of lime (calcium chloride) intended to prevent the spread of a cholera epidemic in Hamburg which in 1892—as a result of poor hygiene and contaminated water—took the lives of almost 9000 people. The lime arrived too late to be of any use—the selling price fell as the outbreak subsided—and the large amount of unsold stock brought Fritz Haber's commercial career to a quick end.

Through his friendship with physicist Siegfried Czapski (1861–1907), at the University of Jena, Haber was enabled to join the university in 1892 as an assistant under the youthful organic chemist Ludwig Knorr (1859–1920). Fritz's interests were now moving towards the emerging field of physical chemistry, stimulated in part by attending lectures of the theoretical physicist Rudolf Straube (1858–1935), designer of the first anastigmatic lens for Carl Zeiss. It had been Haber's intention to work with physical chemist Wilhelm Ostwald, but this did not work out. Meantime, in 1893, in his mid-20s, Haber underwent conversion to Christianity, an act intended to improve his career prospects in the German Reich.

In 1894, Knorr gave Haber a letter of recommendation to leading organic chemist Carl Engler at the Technische Hochschule Karlsruhe, Germany's oldest technical institute [26]. Karlsruhe, 55 km south of Ludwigshafen, was the capital of the Grand Duchy of Baden. The attraction was probably Engler's involvement from the mid-1880s in petroleum technology and the general interest in applied chemistry at Karlsruhe. Engler was active in the National Liberal Party, for which from 1887 to 1890 he was a member of the Reichstag (from 1890 to 1904 he was a member of the First Chamber of the Baden Estate).

Haber accepted the post of research assistant at Karlsruhe and was assigned to Hans Bunte (1848–1925). His research increasingly involved thermodynamics and reactions in the gaseous phase, particularly fuel technology. In 1896 Haber received his *Habilitation* working under Bunte on "Experimentaluntersuchungen über Zersetzung und Verbrennung von Kohlenwasserstoffen" (Experimental Studies of the Decomposition and Combustion of Hydrocarbons), and became a Privatdozent. He taught textile dyeing and printing, and in the autumn of 1896 undertook a trip to Silesia, Saxony and Austria to investigate dyeing practices in industry. In the same year, Haber began to take an interest in the application of electricity to chemistry, including experiments on reduction of nitrobenzene, the hydroquinone/quinone equilibrium, and on electrodes, aided at Karlsruhe by the recently arrived Austrian electrochemist Hans Luggin (1863–1899), who had previously worked with physical chemist Svante Arrhenius (1859–1927).

As a result of this new interest, Haber in 1898 published *Grundriß der technischen Elektrochemie auf theoretischer Grundlage* ("Outline of Technical Electrochemistry on a Theoretical Basis"). At the annual meeting of the German Electrochemical Society (the Bunsen Society) in Leipzig he gave a stunning presentation on electrochemistry as applied to organic chemistry, including reduction of nitrobenzene to aniline. Ostwald was impressed. With the support of Bunte and Engler, Haber was appointed associate professor (ausserordentlicher Professor, Extraordinarius) on 6 December 1898. In 1901, Haber had hoped to gain the post of professor and head of the new Karlsruhe Institute of Physical Chemistry and Electrochemistry, but was deeply disappointed when it went to Max Julius Louis Le Blanc (1865–1943), formerly associated with Ostwald and the Hoechst dyeworks.

Also in 1901, Fritz met up with Clara Immerwahr at a conference. Clara had converted to Christianity in 1897. She received her doctorate in chemistry, the first one to be awarded to a woman in the German Empire, in 1900 under Richard Abegg (1869–1910) at Breslau. Abegg, an old and close friend of Fritz Haber, received his doctorate under Hofmann in Berlin in 1891, on derivatives of chrysene, after which he immediately moved to physical chemistry, assisting, successively, Ostwald, Nernst and Arrhenius. In 1891, Abegg had recommended Fritz to Ostwald. A decade later, in 1901, Abegg, famous for his work on valency theory, was appointed editor of *Zeitschrift für Elektrochemie*, a journal in which Fritz Haber's name was already appearing among the more prolific contributors. Abegg's early death in 1910 arose from a ballooning accident.

Fritz and Clara were married in August 1901 [27, 28]. A son, Hermann, was born on 1 June 1902, by which time Fritz Haber was an acknowledged expert in electrochemistry. In mid-August he departed on a four-month trip to the United States, first to represent the German Electrochemical Society at a conference of the American Electrochemical Society held at Niagara. This provided an opportunity to inspect the various manufacturing concerns that relied on hydro-electric power from the Niagara River, including the Bradley and Lovejoy electric arc facility, where he met with Bradley. Afterwards Haber travelled around the United States gathering information on chemical industry and teaching at universities.

6.4 Nitrogen Fixation: Haber's Studies

In 1904, Fritz Haber received a request from the Margulies brothers, Otto (1867–1926) and Robert (1873–1906), owner and managing director, respectively, of the Österreichische Chemische Werke, in Vienna, to investigate conditions for the combination of nitrogen with oxygen, and also with hydrogen to form ammonia. The latter involved measuring the equilibrium between ammonia and nitrogen and hydrogen on heating, as studied in the 1880s by William Ramsay and Sydney Young (1857–1937) [29]. This was taken up again at University College Cardiff in 1904 by Edgar Philip Perman (1866–1947), who, following earlier researchers—including Marcellin Berthelot (1827–1907)—used a high potential electric discharge to bring about the decomposition of ammonia [30]. In 1904, Haber and his assistant Gabriel van Oordt measured the ammonia equilibrium at temperatures of around 1000 °C. Ammonia was found to be present at from 0.102 to 0.024%. Since it decomposed on heating, starting at around 500 °C, any synthesis from its elements had to be carried out below that temperature. At normal pressure in the presence of an iron catalyst the yield was far too low to suit industrial application [31, 32]. The conclusion was that the equilibrium would favour formation of ammonia only if a suitably active catalyst could be found. This proved to be impossible and the work was abandoned in 1905.

Details of experiments undertaken for the Margulies brothers were included in Haber's *Thermodynamik technischer Gasreaktionen*, published in 1905. Clara Haber assisted Fritz's chemical work by translating articles appearing in English-language journals. An English translation of the book by Haber's American assistant Arthur Becket Lamb (1880–1952), *Thermodynamics of Technical Gas Reactions*, appeared in 1908. Haber emphasized the difficulties associated with fixing nitrogen [33]:

> We may here appropriately discuss the preparation of nitric oxide from the air on a technical scale. This process, which we may call the "burning of air," is of the very greatest technical importance, because from nitric oxide, air and water we can make nitric acid, and from this the nitrates which are of such fundamental importance in agriculture and the manufacture of explosives. It is estimated that the saltpetre beds of Chile will be exhausted by 1940. No other natural deposits of importance are known. Attempts to fix the atmospheric nitrogen by heating calcium carbide and to then obtain a fertilizer [calcium cyanamide] which could take the place of saltpetre have been successful so far as the process itself is concerned, but it is doubtful how far this fertilizer can be substituted for saltpetre, and its transformation, first into ammonia salt and further into nitrates, does not stand perhaps on so favourable an economic basis as does the burning of air. It is possible to get nitric acid by the oxidation of ammonia with oxygen, and ammonia may be obtained in great quantities from the distillation (Mond) of coal. But it seems that this change is always connected with an appreciable loss of ammonia transformed into nitrogen, so that here too the economic basis is less favourable than in the burning of air, where all that is needed to produce nitric acid is air, water and power.

The electric arc methods, the outcomes of an ambitious collective research effort, at that time offered the greatest hope for capturing nitrogen. Haber's new interest in nitrogen capture had led him to assist the Pauling brothers with their electric arc.

In 1906, Max Le Blanc left Karlsruhe to take up an appointment at the University of Leipzig. In his place, Fritz Haber, then aged 38, became professor and head of the Institute for Physical Chemistry and Electrochemistry. Involvement in electrochemistry and his book on the thermodynamics of gas reactions, with a special emphasis on technical uses, had made his reputation. Within a short time, his laboratory attracted an international group of researchers from Europe, the United States, and Japan that undertook investigations into gas reactions and their practical applications.

It was also in 1906 that physical chemist Walther Nernst (1864–1941), at Berlin (he had left Göttingen one year earlier) formulated his heat theorem (later known as the third law of thermodynamics), enabling calculation of yield from thermochemical data. Nernst disagreed with Haber, who had reported a 0.005% yield of ammonia at 1000 °C with an iron catalyst at atmospheric pressure. According to Nernst, this was higher than that predicted by the heat theorem. Nernst synthesized ammonia from its elements under pressure in a ceramic apparatus. Nernst's claim spurred Haber to resort to synthesis under pressure, starting at around 30 atmospheres. Haber confirmed that his earlier results were correct [34, 35].

Applying pressure to gas reactions was quite novel. The French chemist Henri Le Chatelier (1850–1936) at the end of the 1880s formulated the principle that chemical changes take place in directions that tend to counteract the physical changes used to bring them about. The reaction between nitrogen and hydrogen in the synthesis of ammonia exemplifies this principle.

$$\underset{\text{4 volumes}}{N_2 + 3\ H_2} \rightleftharpoons \underset{\text{2 volumes}}{2\ NH_3} \qquad \Delta H = -92\ \text{kJ mol}^{-1}$$

As pressure is applied the equilibrium moves in the direction which reduces its effect, that is, to the ammonia side of the equation.

The temperature of the reaction has also to be taken into account. Since the reaction is exothermic, lowering of the temperature encourages formation of product ammonia. But this slows down the reaction. To activate the synthesis and encourage the rapid formation of ammonia a high temperature is required, though not sufficient to cause the ammonia to decompose. High pressure and elevated temperature act in opposition in ammonia synthesis. The barrier to reaction could be overcome by introducing a suitably active catalyst. Further, to prevent decomposition of the ammonia its rapid withdrawal and cooling were required.

In 1900, Le Chatelier understood that a compromise between the rate of reaction and the yield of ammonia could be obtained by adjusting the pressure and temperature and introducing a catalyst. He investigated the synthesis of ammonia under such conditions and filed a patent for his method in 1901. However, his apparatus blew up, due to the presence of oxygen in the gas mixture, and research was brought to an abrupt halt [36]. While Perman confirmed the validity of Le Chatelier's work, it was Fritz Haber who took up ammonia synthesis where Le Chatelier and Ostwald had left off.

Carl Engler, who had joined the board of BASF in 1903, closely followed Haber's work at Karlsruhe. At this time Schönherr was developing his electric arc and Carl

Bosch was investigating nitrogen fixation by metal cyanide and nitride routes. Studies on nitrides, focusing on barium nitride, were carried out in the Inorganic Department of BASF jointly with Alwin Mittasch (1869–1953), who joined the firm in 1904. Mittasch received his doctorate in 1901 under the supervision of Max Bodenstein (1871–1942) at Wilhelm Ostwald's institute in Leipzig [37]. Nitrides, including of titanium and silicon, gave poor results, as did cyanides.

Engler recommended Fritz Haber to BASF as an expert in nitrogen fixation on 16 February 1908, emphasizing nitric oxide studies using electrical discharges [38]. BASF Central Research Laboratory director August Bernthsen (1855–1931) visited Karlsruhe, as a result of which two research contracts were drawn up with Haber on March 6. One contract was between BASF and Haber and his colleague Adolf König for synthesis of nitric oxide. The other was between BASF and Haber, allocating 6000 marks a year until 1911 towards investigations into the combination of nitrogen with hydrogen. Methods and patents on inventions were to be assigned to BASF [39, 40]. Non-disclosure restrictions on publication protected the firm's corporate interests. It was the standard way in which private industrial organizations supported academic research and promoted information and technology transfer.

König subsequently joined BASF, where he undertook technical work on the electric arc method and collaborated with Haber, who in 1909 filed his first patent for the fixation of nitrogen, based on a cool or silent electric arc.

Until the summer of 1909 the electric arc remained the most viable route to nitrogen fixation. That this was generally agreed was made clear in London at the Seventh International Congress of Applied Chemistry, formally opened on May 27 by the Prince of Wales (from 1910 King George V) at the Royal Albert Hall. The sessions were held at University College London. Sir William Ramsay chaired the sessions concerned with electrothermal methods for the capture of nitrogen, which unlike many others attracted good audiences. Among those present were Kristian Birkeland, Samuel Eyde, Nikodem Caro, BASF research leader August Bernthsen, Otto N. Witt, former consultant to Birkeland and Eyde, and Russian high-pressure chemist Vladimir Nikolayevich Ipatieff (1867–1952) [41]. It was the largest ever gathering of inventors and representatives of firms interested in electrothermal processes for capturing nitrogen, including the cyanamide route. Birkeland reviewed progress in the application of his furnace. Bernthsen described the Schönherr furnace in which BASF held high hopes and let it be known that he did not think much of calcium cyanamide as a fertilizer.

In Karlsruhe, however, Haber was about to achieve success in another direction, the high-pressure synthesis of ammonia from its elements.

6.5 Reaction Variables

Haber's second 1908 agreement with BASF included further investigation of the direct combination of nitrogen with hydrogen. It also provided funds for supporting until mid-1909 his young English research assistant, Robert Le Rossignol

(1884–1976), recently arrived from William Ramsay's laboratory at University College London. Le Rossignol, the son of a physician in St. Helier, on the Isle of Jersey, achieved excellent results in the laboratories at University College, a far cry from the comforts and conveniences of a modern German research institute. In London he had taught himself how to operate machine tools, and he achieved great success in making metal components required in working with gases under pressure, and in experimental work in general, which was not Ramsay's strong point. In 1906, according to one account, Ramsay recommended research under Richard Abegg at Breslau, but Le Rossignol opted for Haber's laboratory because he had heard that there were already "too many Englishmen" at Breslau. According to another account, given by Rosssignol later in life, the attraction was Haber's reputation as a good man to work with [42]. Also arriving at Karlsruhe from Ramsay's laboratory was Joseph E. Coates. Together they worked on a pressure flame for the fixation of atmospheric nitrogen (Fig. 6.1).

Fig. 6.1 Fritz Haber (*front row, seated at centre*), colleagues, and guests from overseas, at Karlsruhe, 1909. Included in this group of 40 are: technician Friedrich Kirchenbauer, seated, cross-legged, on the ground; *in the first, seated, row*, Robert Le Rossignol, *second from left*, and Adolf König, *second from right; in the second row*, Friedrich Bergius, *at extreme left*, Joseph E. Coates, *third from left*, and Tamaru Setsuro from Tokyo, *fourth from left*. Also from Japan is Kotera Fusajiro, *in the fourth row, third from left*. The lady is a Ms Unkowskaja, perhaps a departmental secretary. Seated to the right of Haber are his assistants Gerhardt Just (with beard), and Alfred Leiser. (Bild-Nr VII/3 Haber, Archiv der Max-Planck-Gesellschaft, Berlin-Dahlem)

Le Rossignol found that there was a no more inspiring workplace than Fritz Haber's laboratory. He was soon assigned to a reinvestigation of the synthesis of ammonia. Haber was greatly impressed by the newcomer's efforts, especially after Le Rossignol confirmed the results of van Oordt's work on ammonia. Haber calculated that an estimated 8% equilibrium yield of ammonia could be achieved at 600 °C and 200 atmospheres, with a suitable catalyst.

The application of pressure to chemical reactions was not without dangers. There had been explosions and fatalities in French and English dye factories when using pressures of 50 atmospheres, or more. High pressures were used in steam-powered plants, for electricity generation, and in locomotive and ship boilers. They were also applied in the laboratory for gas liquefaction. At the end of the 1890s, Sir James Dewar (1842–1923) at London's Royal Institution achieved a pressure of around 200 atmospheres. In 1900, Morris William Travers (1872–1961) at University College liquefied hydrogen at around 180 atmospheres. Travers in the 1890s had worked on the synthesis of acetylene from calcium carbide, and the fractional distillation of air from which argon and krypton were obtained. He had early on used a compressor donated by Ludwig Mond and controlled the flow of gas with a special valve supplied by the Brins Oxgygen Company [43].

The successful use of very high pressures in bringing about chemical reactions, as was now being considered by Haber, was achieved in the laboratory by Vladimir Ipatieff at the Mikhaylovskya Artillery Academy in Saint. Petersburg, in 1903. Ipatieff, who had received experience of materials and explosives in the military academy, applied high pressures in part to test the activity of catalysts under varying conditions. His "chief difficulty," he recalled in this early work [44],

> consisted in the construction of an apparatus which could be heated to 600° C while under several hundred atmospheres pressure. Also, the apparatus had to be constructed in such a way that the pressure developed within could be measured continuously throughout the experiment. . ..After many trials, I finally succeeded in constructing a high pressure apparatus which enabled me to perform thousands of experiments. The temperature in this apparatus sometimes reached 600° and the pressure at this temperature was about 400 atmospheres. . ..Such an apparatus (bomb) was first constructed according to my drawings in the private workshop of Malwistrem, a mechanical engineer.

Unlike Fritz Haber, Ipatieff was mainly interested in organic reactions, such as the catalytic decomposition (dehydrogenation) of alcohols. However they both faced similar problems in their chosen pioneering areas of study. High pressure studies were also conducted by physicist Percy William Bridgman (1882–1961) from 1905 at Harvard. But the credit for a method that could be applied in industrial chemical synthesis belongs to Fritz Haber and Robert Le Rossignol.

In 1908, Haber decided to resort to a high pressure at an elevated temperature in the presence of an active catalyst in an effort to improve the combination of hydrogen with nitrogen far beyond trace levels. Le Rossignol's assignment was to devise a steel apparatus, of his own design, incorporating a novel compressor newly acquired by Haber that was capable of achieving a pressure of 200 atmospheres. Le Rossignol supervised the laboratory technician Friedrich Kirchenbauer (1876–[1936]), and in the machine shop they fashioned an ultra-strong reaction vessel capable of withstanding both high pressures and high temperatures.

Le Rossignol was well suited to every aspect of the assignment. He also succeeded in introducing several novel innovations. His first major contribution was a special valve, the "conical valve," made in the Karlsruhe workshop with the help of Kirchenbauer, though it was not originally intended for the ammonia synthesis. Later called the Le Rossignol valve, it withstood the high pressure and enabled control of the flow of hot, compressed gases. In March 1908, he registered a patent-like document that protected such inventions, a gebrauchsmuster, or utility model. He was rewarded with the equivalent of 500 pounds sterling and at that time promoted to Haber's private assistant at an annual salary the equivalent of 50 pounds. Their joint publication in the *Zeitschrift für Elektrochemie* revealed the direction of Le Rossignol's collaboration with Haber [45].

The formation of ammonia took place with a manganese catalyst at 200 atmospheres. but in excess of 700 °C. Conversion at a lower temperature, around 500 °C, was desirable. Iron, chromium, and nickel catalysts were also tried. But there was little enhancement until osmium, and uranium, as uranium carbide, were used. In March 1909, finely divided osmium was found to be an effective catalyst [46, 47]. This rare and valuable metal was received with the help of banker and industrialist Leopold Koppel (1843–1943), for whom Haber acted as a scientific consultant. Koppel, whose brother Arthur co-founded the Orenstein & Koppel engineering firm, had taken over the troubled enterprises built up by Austrian chemist Carl Auer von Welsbach (1858–1929). These concerns dealt with gas mantles (the main invention, patented in 1885), lamps, and rare earth alloys. The gas mantle, made of cotton fabric impregnated with a rare earth metal, in a gas flame gave a greatly increased illumination from a given volume of gas. Haber's services were mainly as consultant to Auer-Gesellschaft, associated with Deutsche Gasglühlicht-Aktiengesellschaft, founded in 1892 by Koppel and Welsbach. Important inventions at the Auer-Gesellschaft included an electric lamp that incorporated metal ribbons based on osmium, the Auer-Oslight, and the wolfram (tungsten) light. This led to a novel electric lamp, known as the Osram, a contraction of osmium and wolfram, the two metals used in the filament. The name was registered in 1906. Soon after, the British General Electric Company joined with German partners, mainly Auer-Gesellschaft, to establish the Osram-Robertson Co. for the manufacture of Osram lamps in England (Fig. 6.2).

In 1908, Tamaru Setsuro (1879–1944), a graduate of the Imperial University of Tokyo, joined Haber's group. He undertook experiments on the synthesis of ammonia with osmium as catalyst in order to gather precise data on specific heats, used to calculate the dissociation equilibrium for the reaction. After pursuing this work at Karlsruhe, he continued with it at the Auer-Gesellschaft and from 1912 at Haber's Kaiser Wilhelm Institute, where he was one of the first three researchers [48].

Le Rossignol designed the ammonia apparatus such that the compressed mixture of reactant gases was successively dried, directed through the steel converter, where the reaction took place, and cooled, enabling liquefied ammonia to be separated from unreacted gases in a trap. With osmium from Koppel's Auer firm the yield was around 5%. After observing a successful run, Haber advised Engler: "Es tropft, Herr Geheimrat!"[49]. On 23 March 1909, Haber advised the BASF

Fig. 6.2 Osram Lamp Division, General Electric Company (from April 1962 Osram (GEC) Ltd), on the GEC North Wembley estate, north-west of London, in the early 1980s. (Osram is derived from *Os*mium-Wolf*ram*.) From 1923, Robert Le Rossignol worked at the adjacent research laboratories

management of the results and recommended that BASF buy up the stock of osmium at Auer-Gesellschaft. Three days later, Bosch, Brunck and Bernthsen turned up to look over the laboratory apparatus. Haber was understandably anxious. After the BASF delegation had fielded a number of technical questions, and were satisfied with the answers, Bosch expressed the opinion that suitable steel equipment could be constructed to enable work on a larger scale.

This was enough to encourage further investigations aimed at technical application and improvement of the apparatus. The Haber-Le Rossignol set-up was designed for continuous operation. Innovations included the pre-heating of reactant gases by the outgoing reaction mixture while still in the converter, and the recirculation of unreacted gases. Le Rossignol designed and built a double acting circulation pump for the ammonia apparatus that forced unreacted gases back to the converter.

Hydrogen was obtained by electrolysis. The benchtop demonstration apparatus was less than one metre high. In one version, platinum asbestos (platinized asbestos) catalysed the reaction between hydrogen and any oxygen that might have entered the system; the gases were then thoroughly dried of resulting water before entry into the converter (Figs. 6.3, 6.4 and 6.5).

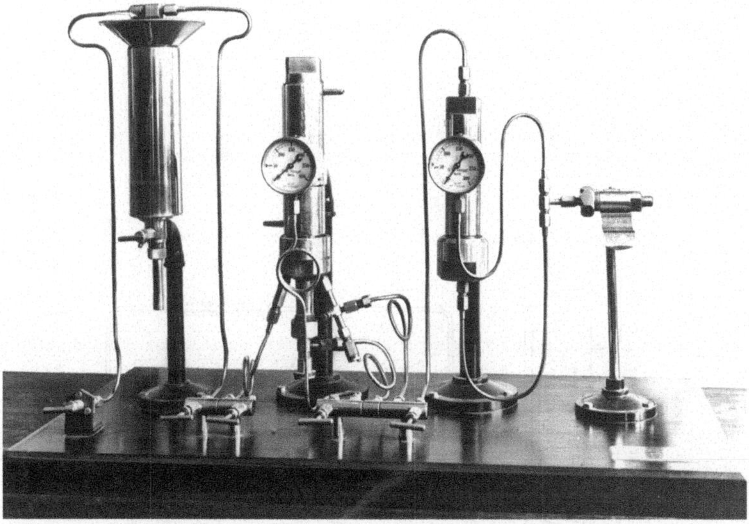

Fig. 6.3 Haber-Le Rossignol benchtop apparatus for the synthesis of ammonia, as originally designed and constructed by Robert Le Rossignol, with assistance from technician Friedrich Kirchenbauer, shown without the double-acting circulation pump. The mixture of hydrogen and nitrogen under pressure was delivered into the horizontal tube at right, where a heated platinum catalyst removed traces of oxygen, and then into the drying chamber (the first vertical column). The gas then entered the electrically heated converter—the second, central, column—where in the presence of catalyst the ammonia was formed. The mixed gases were cooled in the third column. There, unreacted nitrogen and hydrogen were separated from ammonia and directed via the pipe at the extreme left back to the converter (piping connection not shown). (Robert Le Rossignol/Ralph C. Chirnside/Edelstein Center)

There was a further advantage: the process, if viable on a manufacturing scale, and unlike the electric arc and cyanamide processes, did not require a cheap source of electricity [50–52].

In sum, optimum conditions were established at around 175 atmospheres, with compressed gases (a 3:1 mixture of hydrogen and nitrogen) introduced into the electrically heated converter (at around 550 °C) that contained the active catalyst. It was a major breakthrough, since the process was entirely continuous with the rapid withdrawal of product ammonia and recirculation of unreacted hydrogen and nitrogen.

The catalyst enabled the breaking of the stable triple bond between the two atoms in the nitrogen molecule. The percentage of ammonia in the combined gases leaving the converter at a single pass was around 5%.

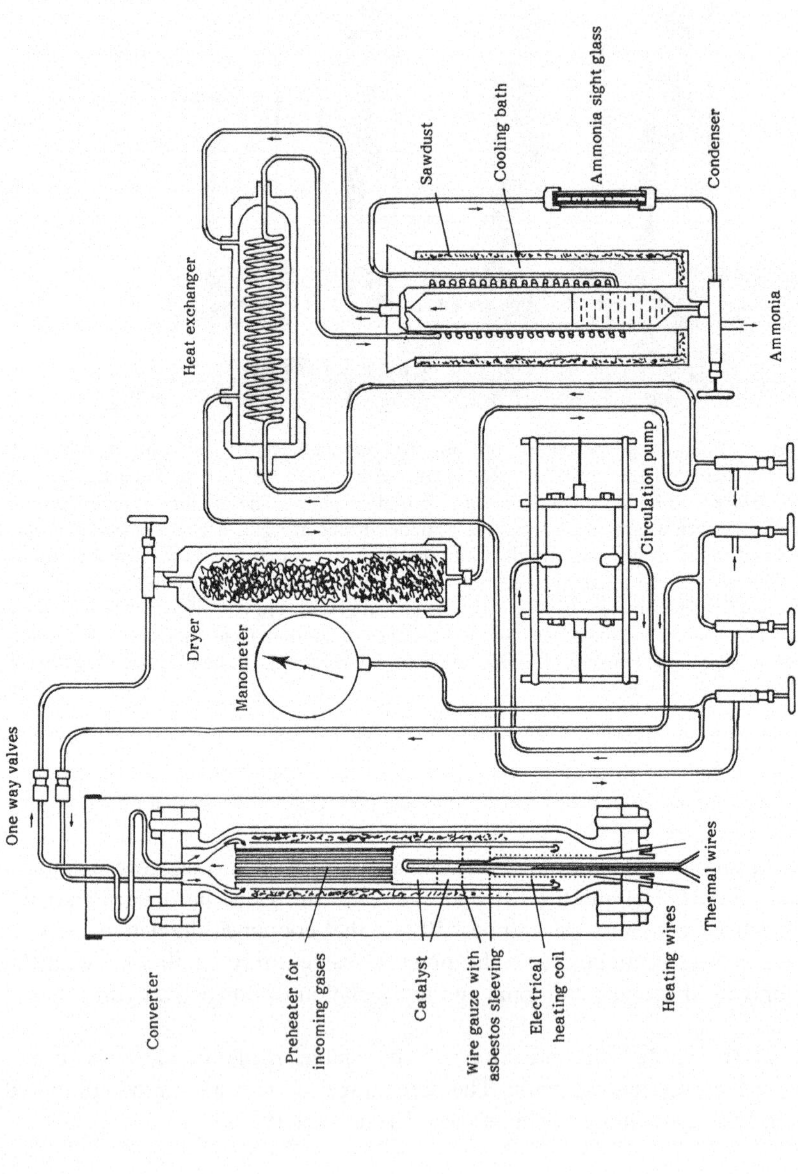

Fig. 6.4 Diagram of laboratory apparatus for the continuous synthesis of ammonia from nitrogen and hydrogen, with complete loop (after Fritz Haber and Robert Le Rossignol)

Fig. 6.5 Apparatus used for laboratory demonstration of the ammonia synthesis at Haber's Berlin institute, around 1914. Next to the apparatus is Gerhardt Just, Haber's first scientific assistant at Karlsruhe, who, along with Richard Leiser and Setsuro Tamaru, moved from Karlsruhe to Berlin in 1912. (Tamaru Setsuro photograph/Tamaru Family Collection)

6.6 Bosch and Haber

Though Haber had euphorically advised Engler and colleagues at Karlsruhe of the success of his innovation, and impressed the visitors from BASF, August Bernthsen decided that BASF was not particularly interested in the ingenious and certainly unconventional, if not overly complex, setup. BASF, it seemed, remained committed to the electric arc. Engler, however, on behalf of Haber, wrote to BASF suggesting that the experimental work was worthy of following up. Carl Bosch, as BASF nitrogen-fixation specialist, was instructed to return to Karlsruhe to observe an improved apparatus in operation. According to the reminiscences of one participant, on 1 July 1909, Bosch, Alwin Mittasch, and BASF chief mechanic Julius Kranz turned up at Haber's laboratory. The first run of the apparatus was hardly auspicious. A bolt from one of Le Rossignol's carefully made seals sprang under the great pressure. The circulation pump was shut down and repairs put in hand. Bosch was unable to wait and returned to Ludwigshafen. Mittasch and Kranz remained until the seal was put right, and the next day they saw cooled, liquefied ammonia drip slowly into a glass collector. On July 3, Haber advised BASF [53]:

> Yesterday we began operating the large ammonia apparatus with gas circulation in the presence of Dr. Mittasch and were able to keep its uninterrupted production for about five

> hours. During this whole time it had functioned correctly and it produced continuously liquid ammonia. Because of the lateness of the hour, as we were all tired, we had stopped the production because nothing new could be learned from the continuing experiment. All parts of the apparatus were tight and functioned well, so it was easy to conclude that the experiment could be repeated. . .. The steady yield was 2 cm^3/minute and it was possible to raise it to 2.5 cm^3/minute. This yield remains considerably below the capacity for which the apparatus has been constructed because we have used the catalyst space very insufficiently.

Heinrich von Brunck and his colleagues were impressed.

After much discussion in the boardroom, BASF agreed to invest in further studies based on Haber's method. For the early experimental work at Ludwigshafen, around 100 grams of osmium was purchased—probably at the time most of the world supply of the hard, bluish-white metal. Mittasch was assigned the challenge of developing an inexpensive and equally active catalyst, one that might make the method commercially viable [54]. In the works laboratory of the BASF Acid Department, his colleague Georg Stern (1883–1959) devised an air-cooled, internally heated laboratory apparatus—a vertical tube with external air cooling—for screening catalysts. Thirty identical experimental units were manufactured; each test sample of two grams was introduced in a small cartridge. One unit is preserved at the Deutsches Museum, Munich (Fig. 6.6).

Bosch recognized that the complicated emission spectrum of iron suggested that the metal might be a good candidate as catalyst. This aided the investigations. Swedish magnetite, a porous oxide of iron, examined on 6 November 1909 by

Fig. 6.6 At right: BASF catalyst test apparatus, with vertical pressure tube, as used by Alwin Mittasch. At left: replica Haber-Le Rossignol benchtop apparatus for the synthesis of ammonia. (Photo by author at the Deutsches Museum, Munich, 1982)

Mittasch's colleague Hans Wolf (1881–1937) was found to give a good yield, of around 5%. Iron-containing compounds were soon activated with the presence of promoters. On 9 January 1910, Bosch, Mittasch, Stern, and Wolf filed a patent for mixed catalysts. Mittasch supervised over 6500 screening experiments, involving 2500 mixtures of metals. The study of ammonia catalysts continued until 1922, by which time some 20,000 experiments had been undertaken (Fig. 6.7) [55].

No less problematic were the metallurgical and engineering challenges. Various designs for the converter were drawn up (Fig. 6.8). Carl Bosch assembled a pilot plant, aided by engineer Franz Lappe (1878–1950), head of a workshop dedicated to the ammonia process. They had to construct a converter suited to continuous operation at the required high pressure and high temperature. Early experiments with two tubular converters based on Haber's work revealed some of the dangers of working at high pressures. Though not intended, they effectively tested reaction vessels to destruction. The special 2.5-metre long hot rolled seemless steel tubes acquired from Mannesmann in 1910 for this work burst under the forcing conditions, at first believed to be the result of embrittlement caused by nitrogen. Replacement converters were then encased in concrete, minimizing damage to the surroundings from explosions.

By studying the wear and tear Carl Bosch (Fig. 6.9) used his training in metallographic research to establish the cause. At the high pressure and elevated temperature, it was found that hydrogen diffused into the iron of the steel wall, reacting with carbon in its crystal structure to form methane, causing the steel to blister as a result of decarbonization; explosions were inevitable. The converter was re-designed to prevent, or at least minimize, this difficulty. The development work continued with experimental and pilot-plant scale converters, again encased in concrete. Finally in February 1911, Bosch, with the help of Lappe, succesfully overcame the problem by designing a double-wall converter [56]. In this, the inner wall, of soft iron or low-carbon steel, which did not undergo decarbonization, allowed hydrogen to diffuse through it, with rapid loss of pressure, under which conditions the outer wall of ordinary steel did not become brittle. This was aided by narrow grooves machined on the outside of the inner wall as the tube was turned. Small holes in the outer wall enabled the release of some hydrogen to the surroundings (Fig. 6.10). Early on, when the converters were externally heated, there was also the problem of large temperature gradients that magnified stresses on the pressure shell. This was overcome by burning some of the reaction gas mixture with air inside the converter, producing the so-called "inversed" flame [57]. One important advantage of increasing the dimensions of converters was that external heating was no longer required. A later innovation was to flush the space between the catalyst basket and shell with nitrogen in order to prevent attack by hydrogen, and in general to reduce stress on the shell.

Compressors capable of providing the very high synthesis pressure—not previously used in chemical industry—had to be designed. They had to be free of leaks and reliable in operation over long periods. These were of the reciprocating type, using pistons driven by crankshafts to deliver reactant gases at 200 atmospheres and powered by gas engines, due to the insufficient power then available from electric motors. Tests carried out on 5 March 1911 with a 4-metre high converter were

17,951. Johnson, J. Y., [*Badische Anilin & Soda Fabrik*]. Aug. 5. **1909**

Catalytic materials, processes, and apparatus.—Ammonia is produced from nitrogen and hydrogen by treatment with a catalytic material under pressure at a raised temperature, the ammonia is removed at a lower temperature for instance by cooling or by an absorbent or both, and the remaining gas with the addition of more nitrogen and hydrogen is treated to produce ammonia as before. The catalytic substance is contained in a tube A, from which the treated gas passes through a heat interchanger which may consist of a double coil c, c^2, or which may be of other form, to a condenser d^2, the liquid ammonia being collected in a vessel E, from which it is withdrawn by a tap e^3. The uncondensed gas passes by a pipe e to a pump F and thence by the path f, c^2, B to the tube A, nitrogen and hydrogen being added by a tap g^2 on a pipe g. The tubes A, B and the heat interchanger c, c^2 are inclosed in insulating packing in an outer tube H. The tube B has a small opening a^3 to equalize the pressures within and without, and is surrounded by an electrical heating-coil J.

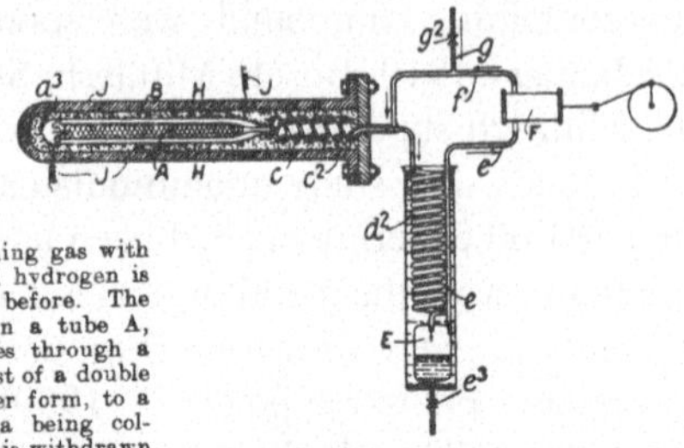

When a catalytic material such as osmium is employed adapted to produce a relatively high concentration of ammonia, the heat interchanger may be omitted. If the ammonia is not required in the liquid form, it may be allowed to vaporize and to cool the gas leaving the catalytic material. The pressure used may be 150 atmospheres, and the temperature of the tube A 500–1000° C.

1910

19,249. Johnson, J. Y., [*Badische Anilin & Soda Fabrik*]. Aug. 16. [*Cognate Applications*, 19,250/10 *and* 19,251/10.]

***Catalytic materials.* — Better yields are obtained in the synthetic production of ammonia by the addition of substances termed "promoters" to the iron, nickel, or cobalt used as the catalytic agent. These substances are generally compounds of magnesium, beryllium, aluminium, or of the alkali, alkaline-earth, or rare-earth metals, or of zirconium, vanadium, tantalum, chromium, manganese, molybdenum, or tungsten. The addition of the oxide or a compound of magnesium or aluminium is particularly favourable. It is found, however, that certain bodies have the property of diminishing or destroying the catalytic property of the iron &c. Such bodies are, for example, arsenic, sulphur, selenium, tellurium, phosphorus, boron, antimony, bismuth, lead, tin, or zinc, or their compounds, and the presence of carbon monoxide and dioxide, or of oil in the gases should be avoided. Iron or the like, which is unsuitable for use as a catalyst owing to the presence of contact poison and for other reasons, can be rendered effective by heating to a high temperature in admixture with a promoter, so that a mixture of oxides is obtained, which is reduced by means of hydrogen, preferably at as low a temperature as possible. It is preferred** that the first heating should be sufficient to melt the mixture. This treatment may be repeated until the desired result is obtained. The reaction gases may be freed from contact poison by passage over a material capable of absorbing it, and this material may be a further quantity of the catalytic agent placed in advance. The mixture of iron and a promoter may be supported on a carrier such as asbestos or chamotte, and one constituent of the carrier may constitute the promoter, but it is necessary in this case that the iron and promoter should be actually mixed together. In use, a mixture of nitrogen and hydrogen, preferably in combining proportions, is passed over the heated catalytic agent, and the reaction may be effected under great pressure, as described in Specification 14,023/10.

14,023. Haber, F. Sept. 13, 1909, [*Convention date*].

Catalytic materials and processes.—**The catalytic process for the production of ammonia from nitrogen and hydrogen is effected under a pressure of 100 atmospheres, or higher pressures. Suitable catalytic materials are finely-divided uranium or osmium, or even iron. Specification 17,951/09 is referred to.**

100,216. Wade, H., (*Pier, M.*) April 18, 1914.

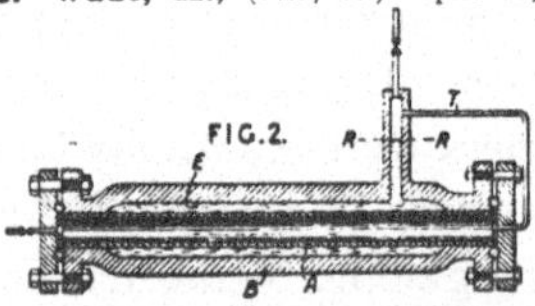

Catalytic apparatus; effecting reactions between gases.—**For effecting the reaction of hydrogen and another gas under pressure and at high temperature, for instance in the catalytic production of ammonia, the reaction vessel, which may be of a porous material such as unglazed porelain or magnesia, or of metal, for instance iron having a low carbon-content, is enclosed in an outer pressure-resisting vessel, the space between being occupied by a fused material, for instance Woods metal, which will not allow the diffusion of** hydrogen through it. The inner vessel A may be suspended in the outer vessel B, perforations X allowing communication between the interior of the vessel A and the space around it for the metal W. When the lid has been screwed on, the gas mixture is admitted and the apparatus is heated, but the part above the level P is not allowed to rise above 200° C. In a modification, Fig. 2, the inner vessel is welded at its ends to the outer vessel, the space E between being filled with molten material to the level R, and a pipe Γ communicating between the space above the level R and the interior of the vessel A. The outer vessel may be of steel, and if the temperature is to exceed 550°, it is preferred to treat it by heating to a temperature suitable for hardening, quenching it in oil, heating to above the working temperature, and cooling slowly. The most suitable steels for subjecting to this treatment are nickel steel and nickel chrome steel. Specifications 15973/10 and 9661/14 are referred to.

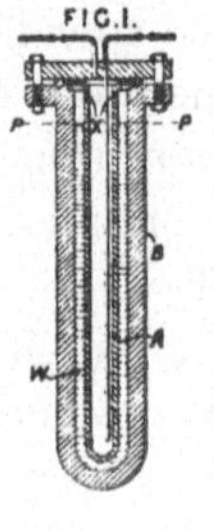

Fig. 6.7 British patent abstracts for synthetic ammonia, filed by Fritz Haber, BASF, and (below) Mathias Pier (who joined BASF in 1920)

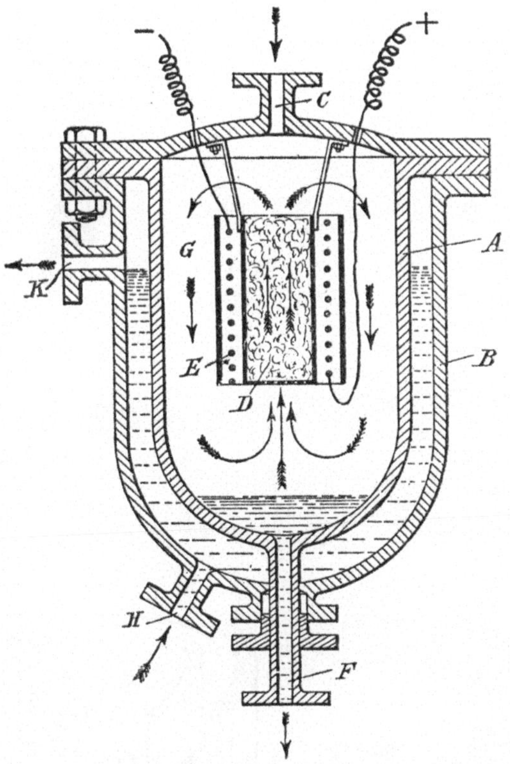

Fig. 6.8 Apparatus for the synthesis of ammonia, as appearing in BASF British patent 13,097 of 30 May 1910. "… ammonia is condensed by abstracting heat from a surface in or forming part of the walls of a vessel in which the combination takes place…the reaction vessel A, capable of withstanding pressure, is surrounded by a cooling jacket B. The reacting gases enter the apparatus at C, pass up through the catalyst D, which is electrically heated and the condensed ammonia collects at the bottom and is withdrawn through pipe F." Such patents revealed sufficient to ensure protection to the patentee, but not enough to enable technical working of the method (Figure from patent abstract)

considered sufficiently successful for the commissioning of a full-size manufacturing plant. What might have been a setback was an attempt by various parties, including Hoechst—like other rival German dye firms also interested in nitrogen fixation—to disallow the ammonia patents. Fortunately for BASF, the claim made by Hoechst, whose expert witnesses included Ostwald, was rejected on 4 March 1912. Nernst's expert opinion supported BASF.

On 11 September 1912, Bernthsen, now no longer enamoured of the electric arc fixation of nitrogen, and with great pride and confidence in Bosch's achievement, revealed an outline of the new BASF synthetic ammonia process with special reference to catalyst promoters and poisons, at the Eighth International Congress of Applied Chemistry, held in New York [58]. The presentation included a demonstration apparatus, one of two taken to America. It was on the same occasion that Samuel Eyde described the latest developments in the Birkeland-Eyde process. In

Fig. 6.9 Carl Bosch. (*Erzeugnisse unserer Arbeit*. Frankfurt am Main: I.G. Farbenindustrie, 1938, p. 17)

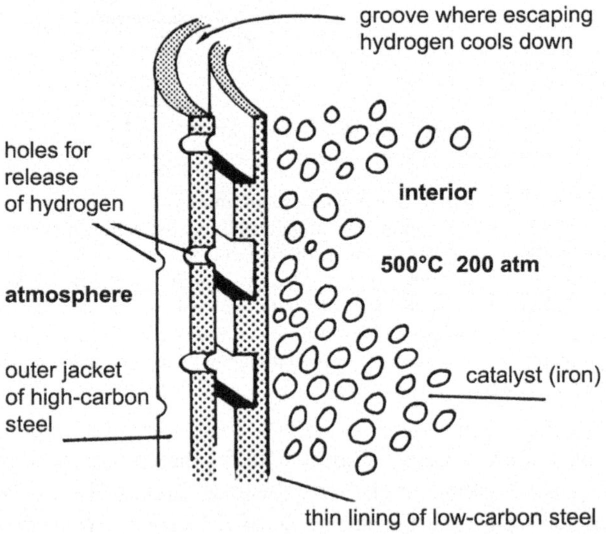

Fig. 6.10 Simplified diagram of double wall converter, as invented by Carl Bosch for the synthesis of ammonia from nitrogen and hydrogen. The Bosch holes were used on early converters, before improved steels became available after 1918

the audience was the Japanese engineer Ōya Jun, a government employee, who sent back reports on progress in nitrogen capture to colleagues in Tokyo.

Continuous production of ammonia was based on a kit of units, with each unit serving a different function—such as compression, gas transfer, gas reaction, heat exchange, and the recirculation of unreacted gases—and an extensive network of interconnected pipes through which gases were pumped. For the BASF pilot plant, nitrogen was at first obtained by Carl von Linde's process for liquefaction and distillation of air, though it was restricted in output and expensive. Hydrogen, as in earlier studies, came from electrolysis of brine. For full-scale production, the supply of large amounts of extremely pure nitrogen and hydrogen gases required improvement of existing coal-based processes, as well as methods of cleaning gases. Nitrogen was obtained by blowing air over coke to form the mixture of carbon monoxide and nitrogen known as producer gas (not to be confused with hydrogen-rich Mond producer gas). Hydrogen was obtained from water gas, the mixture of carbon monoxide and hydrogen resulting from the action of steam on incandescent coke, by the Bamag process (Fig. 6.11). Unwanted constituents were separated from hydrogen by liquefaction [59]. At first, this made use of the Linde-Frank-Caro process, which turned out to be problematic.

Carbon monoxide, a catalyst poison, was removed from both producer gas and water gas by conversion to carbon dioxide. Removal of the final traces of monoxide was difficult. Cuprous oxide under pressure worked but corroded the iron surface of the reaction apparatus. The problem was solved by using a copper solution containing ammonia, which did not corrode iron, as worked out at BASF by Carl Krauch

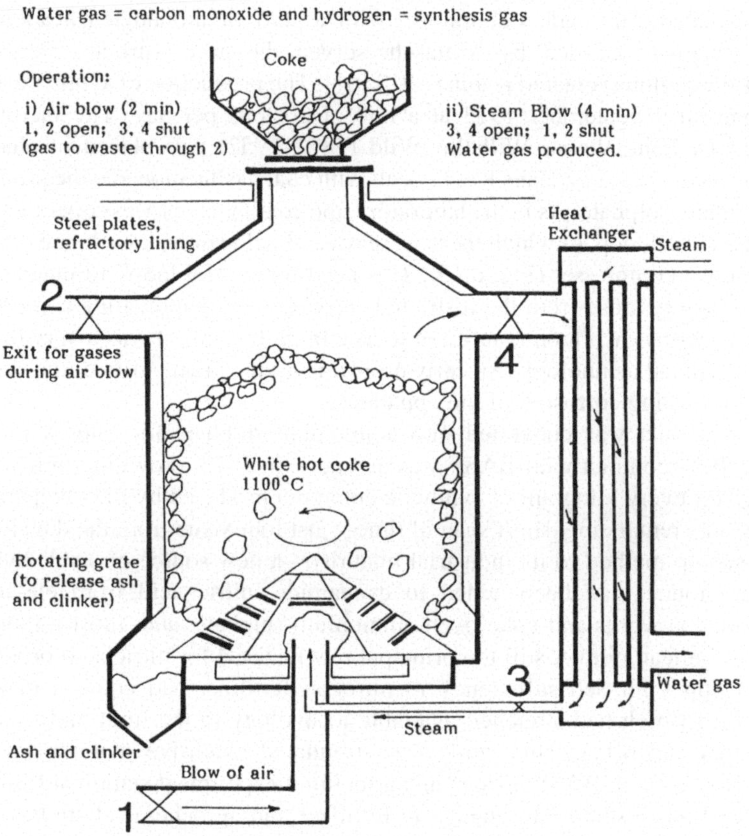

Fig. 6.11 Manufacture of water gas (carbon monoxide and hydrogen) from coke, steam and air, by the Bamag (Berlin-Anhaltische Maschinenbau AG) process

(1887–1968), a former student of Nernst. The gas mixture was then adjusted to 75% hydrogen and 25% nitrogen, as required by the equation for the formation of ammonia. (The Linde liquefaction process was later used for making up the final hydrogen-nitrogen gas mixture obtained from the coke-based processes.)

Pilot plant experiments continued throughout 1911. In November, the decision was taken to build a new factory, three km north of the Ludwigshafen facility, at Oppau. Construction, under the supervision of Johannes Fahrenhorst, started in the spring of 1912 (Figs. 6.12 and 6.13) [60]. Operation of the new equipment—including compressors; eight-metre high-pressure converters, each weighing eight-and-a-half tons and supplied by Krupp; and circulation pumps—called for the design of new types of instruments, many designed and manufactured within the firm. A system of semi-automatic control was devised. Gas flows were controlled by magnetically operated quick-acting, self-closing stop valves, with spring releases and slide-valves for cutting off gas flow in cases of leakage or breakage. Pressures, temperatures, and densities had to be monitored, calling for further developments in instrumentation (Fig. 6.14). Storage tanks and piping were no less important.

A dedicated Ammonia Laboratory, with a staff of around ten academically trained chemists, headed by Mittasch, served the new Nitrogen Department (Stickstoffabteilung) created in June 1912 [61]. The production of synthetic ammonia began on 9 September 1913 at a rate of 20 tons per day. The facility was managed by Fahrenhorst. Wilhelm Wild (1872–1952) was placed in charge of purification of the gases; Hans Keller dealt with coal gasification and the production of ammonium sulphate. As in the laboratory, the continuous process operated on the basis of a closed loop, of which there were several, all identical, each one driven by an individual compresser (Fig. 6.15). This need for several loops arranged side by side in a battery arose from the restricted capacities of reciprocating compressors. In modern terms each loop is referred to as a train. Overall, it was a very different kind of chemical technology. An early problem was the great purity of the synthetic ammonia, causing corrosion of the apparatus.

The ammonia was converted into ammonium sulphate for sale as fertilizer. Haber's 1908 contract with BASF was amended in 1913, providing him with 1.5 pfennigs for every kilogram of synthetic ammonia produced by the company.

Now, in a remarkably short span of time—just four years from the development of a benchtop method to its industrial maturity—a new source of stable synthetic nitrogen product had been added to cyanamide, nitric acid from electric arc processes, gas works and coke oven ammonium sulphate, and, most importantly, imported Chilean nitrate, still the principal raw material for nitric acid production. German hopes for self-sufficiency in nitrogen fertilizer had come a giant step closer. Carl Bosch had stretched available technology to the limit and succeeded (Figs. 6.16 and 6.17). This work also stimulated extensive materials research (Fig. 6.18). In 1914, BASF opened an agricultural experiment station at Limburger Hof, near Ludwigshafen, to engage in fertilizer and agrochemical studies. In the same year Bayer's Carl Duisberg (1861–1935), referred for the first time to the Haber-Bosch process. It was a process whose development was more daring than anything previously achieved in chemical industry.

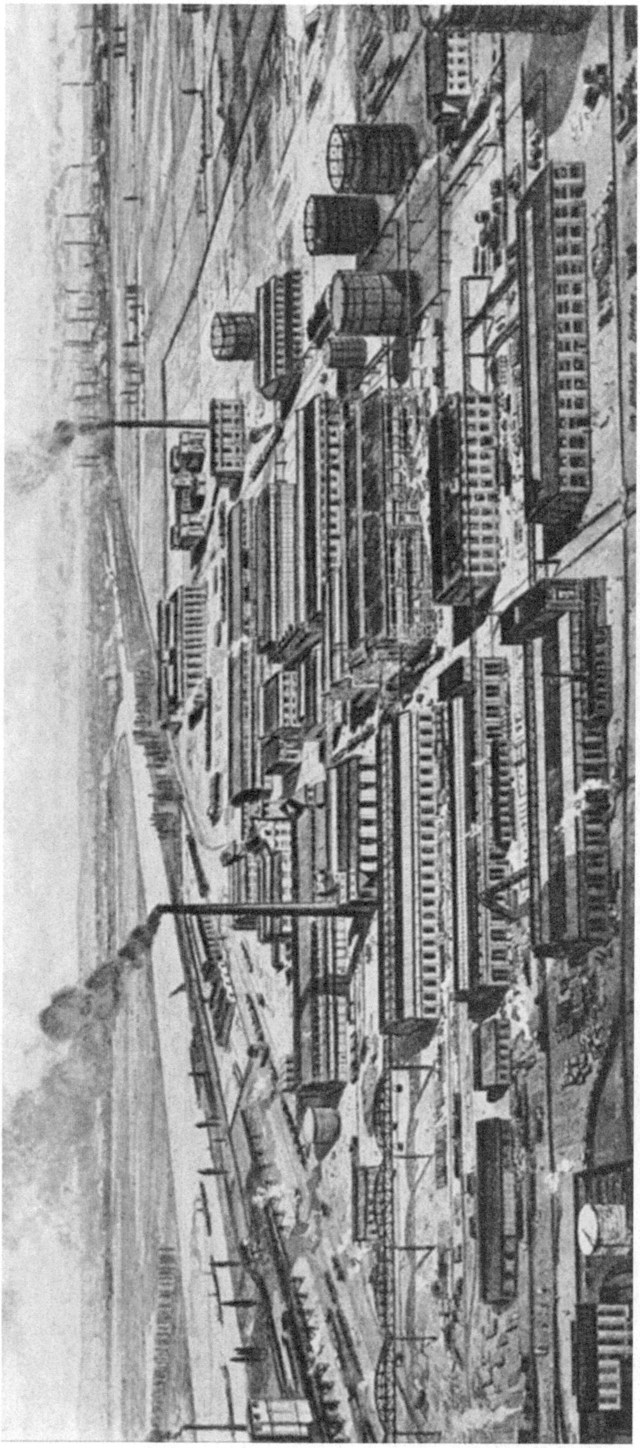

Fig. 6.12 General view of the Oppau works, around 1914. From a painting by the industrial artist Otto Bollhagen (1861–1924) (BASF Unternehmensarchiv)

Fig. 6.13 Maintaining a compressor used in gas generation, Oppau, 1917. (BASF Unternehmensarchiv)

In 1911, Haber—by then a leading physical chemist, noted for his ingenuity and organizational skills, and supported by Koppel—was appointed head of the new Kaiser-Wilhelm-Institut für physikalische Chemie und Electrochemie, in Berlin, which opened in 1912 [62]. From 1914, the institute hosted Albert Einstein (1879–1955) while waiting for his institute of physics to be established (which took until 1917) and developing his field equations (1915) (Fig. 6.19). The Habers had acted as mediators during the separation of Einstein from his first wife in July 1913 [63].

Travelling on business in Europe in the autumn of 1913, around the time of the inauguration of the Oppau ammonia facility, was American chemist Louis Cleveland Jones (1870–1945). After studying chemistry at Yale, he had joined the Solvay Process Company in 1896, where he would remain until 1922. On 2 October 1913, while in Paris, he drew up a report on developments in nitrogen capture processes. Among others, Jones had met with Otto N. Witt, who confided that he was the only person outside of BASF to have witnessed the synthetic ammonia process at Oppau

Fig. 6.14 Instrument laboratory, Ludwigshafen, 1921. (BASF Unternehmensarchiv)

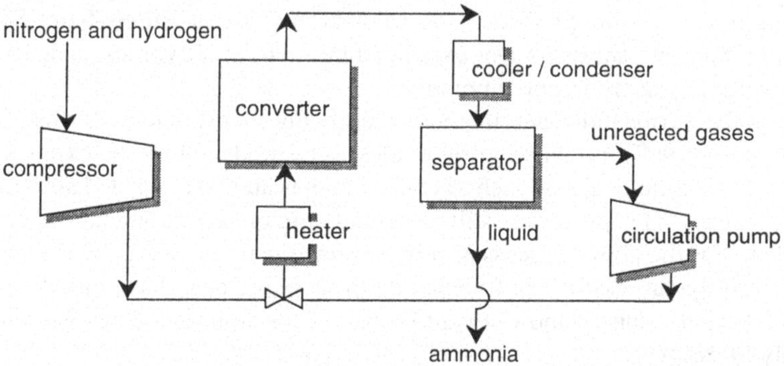

Fig. 6.15 Block diagram for the synthesis of ammonia

in operation. Without special permission, entry to Oppau was strictly forbidden. Jones had written to BASF regarding conditions for acquiring a license. He did not get very far [64]: "Their determination to pursue independent ways, however, is indicated by their refusal to see me, on the ground that they would probably not favorably consider an American license." Through his European contacts, Jones was advised that the Semet-Solvay concern in Brussels had closed its anhydrous ammonia plants and was buying ammonia from Oppau. A similar situation prevailed at the German Solvay works. At first, BASF anhydrous ammonia sold more readily than ammonium sulphate (until BASF found a market through its

Fig. 6.16 Early BASF Haber-Bosch converter, showing the massive bolts and studs required to secure the cover plate. This converter is now preserved at the Wroughton outstation of the Science Museum London

purchase of a firm engaged in preparing and distributing fertilizers). Significantly, the French were not purchasers of BASF or Semet-Solvay anhydrous ammonia, preferring to rely on the product made from cyanamide at Villers-Saint-Sépulcre and Notre-Dame-de-Briançon. An experimental plant at Vilvordes, near Brussels, also converted cyanamide into ammonia.

Among the various nitrogen methods then being investigated was that of Friedrich Häusser (1907), in which nitric oxides were produced by the explosion of a compressed mixture of gases, such as coke oven gas and "oxygenated air." Like the Serpek process, "This process is still far from being on a commercial basis." Jones concluded: "Of the above processes only two have really arrived: the Haber seems to be arrogantly successful, the Cyanamide Process seems to have quietly come to stay. The Serpek, Hauser and Ostwald Processes for ammonia conversion are still evidently in trouble."

6.7 Nitric Acid

The production of synthetic ammonia was a technological triumph. As introduced to the agricultural sector, synthetic ammonia was available as ammonium sulphate, a product already produced from gas and coke works ammonia and immediately accepted by farmers, particularly in Germany. For many purposes, including the production of nitro compounds for dyes and explosives, a means for converting ammonia into nitric acid was required. Throughout the nineteenth century, nitric acid was manufactured by reacting Chilean nitrate and sulphuric acid in cast iron

Fig. 6.17 Installing a converter, at Oppau, 1920. (BASF Unternehmensarchiv)

pots. The resulting nitrous fumes were passed through a series of absorption towers. For certain nitrations, nitric acid of medium to high ("strong") concentration was essential.

Means for the oxidation of ammonia to nitric acid relied on newly available catalytic processes. They were used on a limited scale for the conversion of cyanamide-derived ammonia and coke works ammonia. The first large plant for the catalysed air oxidation of ammonia to nitric acid was constructed in 1908 by the Gewerkschaft des Steinkohlenbergwerks Lothringen. The ammonia came from a coke-oven plant at Gerthe, Westphalia (Gerthe gave its name to the plant), near Bochum in the Ruhr. This used Wilhelm Ostwald's method, developed jointly with Eberhard Brauer, his future son-in-law, during 1900–1901, using a platinum catalyst, solid or spongy, as foil or wire, for which a patent application was filed in 1902. Ostwald's first apparatus was a vertical nickel tube into which was inserted a roll of platinum. A mixture of ammonia and air was introduced at the bottom and

Fig. 6.18 Materials research laboratory, Ludwigshafen. (BASF Unternehmensarchiv)

preheated by the outgoing gas mixture. For large-scale production, a consortium of local coal producers was brought together. Brauer and Köln-Rottweiler Sprengstoff-Gesellschaft co-operated with Kohlenzeche Lothringen, also in Gerthe, to work on improvements to the process. Notable was the work during 1906 of German engineer Friedrich Uhde (1880–1966) at Kohlenzeche Lothringen; he worked from 1905 to 1914 as an employee of Dr. C. Otto & Co., a firm well known for coke oven developments. In 1921, Uhde set up as an engineering contractor.

Though difficulties were encountered due to the impermeability of the platinum, the output of nitric acid from the ammonia process at Gerthe in 1911 reached 1495 tons [65]. According to Louis Jones of Solvay, in his letter from Paris, Gerthe in 1913 was producing 150 tons of 5% nitric acid each month.

As a result of BASF successfully arguing that prior publication by Kuhlmann and others invalidated Ostwald's main patent, Ostwald in 1907 withdrew his application in Germany (though his patent was accepted in other countries). Researchers in Germany then investigated the action not only of platinum but also of other catalysts. Georg Lunge's comprehensive overview of the nitrogen industry prior to World War I, including processes for the oxidation of ammonia to nitric acid and concentrating the acid, is an important source for following developments [66]. Thus, Nikodem Caro worked on the catalytic conversion of

Fig. 6.19 Fritz Haber, at left, and Albert Einstein, at Haber's Berlin institute in the summer of 1914. The photograph was taken by Haber's Japanese colleague Tamaru Setsuro. (Tamaru Family Collection/The Albert Einstein Archives, The Hebrew University of Jerusalem, call no. 64-007)

cyanamide-derived ammonia to nitric acid, using thorium compounds as catalysts at atmospheric pressure. In 1907, Caro and Albert Frank jointly with the engineering firm Bamag developed ammonia oxidation in the presence of air using platinum as catalyst. Special tanks were designed for bulk transportation of the acid. In 1909, Karl Kaiser of the Technische Hochschule Charlottenburg suggested preheating air prior to mixing it with ammonia. In 1911 he proposed the use of platinum gauze. A pilot plant was erected at Spandau, Berlin, around 1912. It attracted little interest, apart from a single use in Russia at the Kharkov industrial centre, around 400 miles south of Moscow. Kaiser's gauze became a later standard, and he advised Bamag on plant design [67]. In early 1914, probably to avoid reliance on platinum, a process employing Caro's thorium oxide catalyst was about to be introduced at Bayerische Stickstoff-Werke. For Caro, H. Hackl and H. Bunzel investigated catalysts based on barium nitrate [68].

During 1914, Caro and Albert Frank filed patents for the oxidation of ammonia with an electrically heated platinum gauze catalyst. With Bamag, they worked on an improved apparatus for the oxidation, a water-cooled aluminium box fitted with a conelike hood of aluminium incorporating a mica sight glass. Baffles distributed the incoming gases that were directed over an electrically heated single layer of platinum gauze. The resulting nitrogen oxides were absorbed in a stream of water, to give nitric acid. The catalyst arrangement was a considerable improvement over Ostwald's corrugated platinum strip, which had a lifetime of just four to six weeks [69, 70].

Severe corrosion problems in absorption areas created difficulties that would be resolved only after the introduction of special steels. Under normal economic conditions this process could not replace the large-scale manufacture of nitric acid from Chilean nitrate. However, this was not the case in a situation of war emergency. By 1916, Bamag had supplied more than 70 units for oxidation of ammonia incorporating platinum gauzes, and as a result of war-related demand was called upon to manufacture several more.

Synthetic ammonia and the oxidation of ammonia to nitric acid would come to the fore as the result of a world-shattering event: the assassination of Archduke Franz Ferdinand of Austria-Este, eldest son of Archduke Karl Ludwig and heir presumptive to Franz Josef and the Austro-Hungarian Empire, and his wife, at Sarajevo on 28 June 1914. On July 28, there was the first declaration of what was to become the "All-Europe Conflict" or Great War. Until then there had been no general mobilization of industry for war. Oppau was producing (according to mainly post-war figures) around 30 tons of synthetic ammonia every day by the Haber-Bosch process, for conversion to the fertilizer ammonium sulphate. Hans Wolf at BASF had examined metals of the platinum group as catalysts for the oxidation of ammonia. Studies on nitrogen fixation and the oxidation of ammonia to nitric acid were conducted at Bayer, and at Hoechst.

References

1. Reinhardt C, Travis AS (2000) Heinrich Caro and the creation of modern chemical industry. Kluwer, Dordrecht.
2. Felck G (1993) Wilhelm Ostwald, 1853–1932. In: James LK (ed), Nobel laureates in chemistry 1901–1992. History of modern chemical sciences. American Chemical Society, Washington DC, pp 61–68.
3. Stoltzenberg D (1994) Fritz Haber: Chemiker, Nobelpreisträger, Deutscher, Jude: Eine Biographie. VCH, Weinheim, pp 141–143.
4. Holdermann K (1949) Carl Bosch und die Naturwissenschaft. Naturwiss 36(6):161–165.
5. Krauch C (1940) Carl Bosch zum Gedächtnis. Angew Chem 53:285–288.
6. Stoltzenberg D (2004) Fritz Haber: Chemist, Nobel laureate, German, Jew. Chemical Heritage Press, Philadelphia, quoting Ostwald, on p 130.
7. Kilburn Scott E (1912) The manufacture of nitrates from the atmosphere. Nature 89(2228) (11 July):490–492, on 492.
8. Kilburn Scott E (1915) Production of nitrates from air, with special reference to a new electric furnace. J Soc Chem Ind 34(3)(15 February):113–126, on 113.
9. Kilburn Scott E (obituary) (1941) Engineer, London 172(4462)(18 July):41.
10. Goran MH (1967) The story of Fritz Haber. University of Oklahama Press, Norman.
11. Wille HH (1969) Der Januskopf: Leben und Wirken des Physikochemikers und Nobelpreisträgers Fritz Haber. Verlag Neues Leben, Berlin.

12. Harris H (1992) To serve mankind in peace and the fatherland in war: the case of Fritz Haber. Ger Hist 10(1):24–38.
13. Stoltzenberg D (1994) Fritz Haber: Chemiker, Nobelpreisträger, Deutscher, Jude: Eine Biographie. VCH, Weinheim.
14. Szöllösi-Janze M (1998) Fritz Haber 1868–1934: Eine Biographie. VCH-Beck, Munich.
15. Dunikowska M, Turko L (2011) Fritz Haber, the damned scientist. Angew Chem Int Edit 50:10050–10062.
16. Hoffmann R, Laszlow P (2001) Coping with Fritz Haber's somber literary shadow. Angew Chem Int Edit 40:4599–4604.
17. Smil V (2001) Enriching the earth: Fritz Haber, Carl Bosch, and the transformation of world food production. MIT Press, Cambridge, MA, pp 41–48.
18. Stoltzenberg D (2004) Fritz Haber: Chemist, Nobel laureate, German, Jew. Chemical Heritage Press, Philadelphia.
19. Charles D (2005) Master mind: the rise and fall of Fritz Haber, the Nobel laureate who launched the age of chemical warfare. Harper Collins, New York. In the UK: Between genius and genocide: the tragedy of Fritz Haber, father of chemical warfare. Cape, London.
20. Friedrich B (2005/2006) Fritz Haber (1868–1934). Angew Chem Int Edit 44:3957–3961, and 45:4053–4055.
21. Thomas H (2008) The alchemy of air: a Jewish genius, a doomed tycoon, and a scientific discovery that fed the world but fueled the rise of Hitler. Harmony Books/Crown Publishing Group/Random House, New York.
22. Ragussis D (2008) Haber: the father of gas warfare. Film directed by Ragussis (United States).
23. Stoltzenberg D (2004) Fritz Haber: Chemist, Nobel laureate, German, Jew. Chemical Heritage Press, Philadelphia, quoting Haber, on p 22.
24. Stoltzenberg D (1994). Fritz Haber: Chemiker, Nobelpreisträger, Deutscher, Jude: Eine Biographie. VCH, Weinheim, 28–29.
25. Müller G (1909) Die Chemische Industrie. BG Teubner, Leipzig, p 199.
26. Schuster C (1976) Wissenschaft und Technik. BASF, Ludwigshafen, pp 108–127.
27. von Leitner G (1993) Der Fall Clara Immerwahr: Leben für eine humane Wissenschaft. CH Beck, Munich.
28. Sicheritz H (2014) Die Häfte des Lebens. Femina doctissima. Clara Immerwahr. Television film, directed by Sicheritz (Austria).
29. Ramsay W, Young S (1884) The decomposition of ammonia by heat. J Chem Soc 45:88–93.
30. Perman EP (1905) The direct synthesis of ammonia. Proc R Soc London 76:167–174.
31. Haber F, van Oordt G (1905) Uber Bildung von Ammoniak aus den Elementen. Z anorg Chem 43:111–115; 44:341–378; 47:42–44.
32. Dronsfield AT, Morris PJT (2007) Who really discovered the Haber pocess? http://www.rsc.org/Education/Eic/issues/2007May/WhoReallyDiscoveredHaberProcess.asp (accessed 19 August 2014).
33. Travis AS (1984) The high pressure chemists. Brent Schools and Industry Project, Wembley, quoting Haber, on p 15.
34. Gal J (2015) Remembering Fritz Haber in the year 2015. L'actualite chimique, nos. 397–398 (June–July), pp 114–121.
35. Barkan DK (1999) Walther Nernst and the transition to modern physical science. Cambridge University Press, Cambridge, pp 129–131.
36. But see Uline MJ, Corti DS (2008) The ammonia synthesis reaction: an exception to the Le Chatelier principle and effects of nonideality. J Chem Ed 83(1)(January):138–144.
37. Holdermann K (1957) Alwin Mittasch, 1869–1953, in Memoriam. Chem Ber 90:xli–liv.
38. Schuster C (1976) Wissenschaft und Technik. BASF, Ludwigshafen, pp 124–125.
39. Smil V (2001) Enriching the earth: Fritz Haber, Carl Bosch, and the transformation of world food production. MIT Press, Cambridge, MA, pp 75–82.
40. Stoltzenberg D (2004) Fritz Haber: chemist, Nobel laureate, German, Jew. Chemical Heritage Press, Philadelphia, p 89.

41. Ipatieff VN (1946) In: Ipatieff J, et al. (eds), The life of a chemist: memoirs of Vladimir N. Ipatieff (trans. Haensel V, Lusher RH). Stanford University Press, Stanford, pp 154–155, 327–328.
42. Sheppard D (2017) Robert Le Rossignol, 1884–1976: engineer of the 'Haber' process. Notes Rec R Soc London. DOI: https://doi.org/10.1098/rsnr.2016.0019 (accessed 18 March 2017).
43. Travers MW (1901) The experimental study of gases. Macmillan, London, pp 196–206.
44. Ipatieff VN (1936) Catalytic reactions at high pressures and temperatures. Macmillan, New York, p 31.
45. Haber F, Le Rossignol R (1908) Bestimmung des Ammoniakgleichgewichtes unter Druck. Z Elektrochem 14:181–196; 513–514.
46. Topham SA (1985) The history of the catalytic synthesis of ammonia. In: Anderson JR, Boudart M (eds), Catalysis: science and technology, vol. VII. Springer, Berlin, pp 1–50.
47. Tamaru K (1991) The history of the development of the ammonia synthesis. In: Jennings JR (ed), Ammonia synthesis: fundamentals and practice. Plenum, New York, pp 1–18.
48. Tamaru Oyama H (2015) Setsuro Tamaru and Fritz Haber: links between Japan and Germany in science and technology. Chem Rec 15(2):535–549.
49. Schuster C (1976) Wissenschaft und Technik. BASF, Ludwigshafen, p 126.
50. Haber F, Le Rossignol R (1913) Ammonia; technical preparation of....from its elements. J Soc Chem Ind 32(3)(15 February):134–138.
51. Le Rossignol R (1928) Zur Geschichte der Herstellung des synthetischen Ammoniaks. Naturwiss 186:1070–1071.
52. Haber F (1920) The synthesis of ammonia from its elements, Nobel lecture, June 2, 1920. Nobel lectures including presentation speeches and laureates biographies, chemistry, 1901–1921. Nobel Foundation/Elsevier Publishing, Amsterdam, pp 326–340.
53. Translation of Fritz Haber to BASF, 3 July 1909, in Zardi U, Zardi F (2009) 100 not out. History of the birth of the modern synthetic ammonia industry. Paper presented at the Nitrogen & Syngas International Conference and Exhibition, Rome, Italy (22–25 February 2009).
54. Mittasch A (1951) Geschichte der Ammoniak-synthese. Verlag Chemie, Weinheim.
55. Mittasch A (1950) Early studies of multi component catalysts. In: Komarewsky VI, Rideal RK (eds), Advances in catalysis, vol. II. Academic Press, New York, pp 81–104.
56. Holdermann K (1954) Im Banne der Chemie: Carl Bosch, Leben und Werk. Econ-Verlag, Düsseldorf, p 95.
57. Appl M (1982) The Haber-Bosch process and the development of chemical engineering. In: Furter WF (ed), A century of chemical engineering. Plenum, New York, pp 29–53, on 39–43.
58. Bernthsen A (1912) Synthetic ammonia. Ind Eng Chem 4(10)(1 October):760–767.
59. Teed PL (1919) The chemistry and manufacture of hydrogen. Longmans, Green & Co., New York, pp 113–125.
60. Travis AS (1998) High pressure industrial chemistry: the first steps, 1909–1913, and the impact. In: Travis AS, Schröter HG, Homburg E, Morris PJT (eds), Determinants in the evolution of the European chemical industry, 1900–1939: new technologies, political frameworks, markets and companies. Kluwer, Dordrecht, pp 3–21, on 11–12.
61. Reinhardt C (1998) Basic research in industry: two case studies at I.G. Farbenindustrie AG in the 1920's and 1930's. In: Travis AS, Schröter HG, Homburg E, Morris PJT (eds), Determinants in the evolution of the European chemical industry, 1900–1939: new technologies, political frameworks, markets and companies. Kluwer, Dordrecht, pp 67–88, on 81.
62. James J, Steinhauser T, Hoffman D, Friedrich B (2011) One hundred years at the intersection of chemistry and physics: The Fritz Haber Institute of the Max Planck Society, 1911–2011. Walter de Gruyter, Berlin/Boston.
63. Stern F (1999) Einstein's German world. Princeton University Press, Princeton, pp 59–164.
64. Nitrogen products-miscellaneous nitrogen fixation-ammonia manufacture. The Solvay Process Company. Nitrogen products department. Rep. no. 750. Works at European. October 2nd, 1923. Nitrogen fixation. Ammonia manufacture. Haber, Cyanamide, Serpek, Hauser, Ostwald

processes. Dr L. C. Jones. Report drawn up on 2 October 1913, Paris, France. Solvay Process Company archives, microfilm roll 17. Sidney M. Edelstein Library, Israel National Library.
65. Travis AS (1984) The high pressure chemists. Brent Schools and Industry Project, Wembley, p 50.
66. Lunge G (1916) Handbuch der Schwefelsäurefabrikation und ihrer Nebenzweige, 2 vols. Friedr. Vieweg & Sohn, Braunschweig.
67. McDonald D, Hunt LB (1982) A history of platinum and its allied metals. Johnson Matthey, London, pp 387–389.
68. Lunge G (1916) Handbuch der Schwefelsäurefabrikation und ihrer Nebenzweige, Band I. Friedr. Vieweg & Sohn, Braunschweig, pp 284–285.
69. Hunt LB (1958) The ammonia oxidation process for nitric acid manufacture. Platin Met Rev 2(4):129–134, on 133.
70. Inskeep GC, Henry TH (1954) Nitric acid in Great Britain: oxidation of ammonia. In: Modern chemical processes: a series of articles describing chemical manufacturing plants, vol. III. Reinhold, New York, pp 218–227, on 218.

Chapter 7
A Time of Guns and Grain

7.1 War and Fixed Nitrogen

> The story still is told of a Minister, a member of the War Cabinet, who, finding the conversation at a certain dinner turning to the sinister menace of the submarine campaign, then at its height, and its effects especially on the Chile communications, turned to his neighbour with the enquiry: 'Tell me, what is this nitrate they are all making such a fuss about?'
>
> Stanley I. Levy, "The Status of Chemists and Chemistry," *Chemistry and Industry*, no. 11, 14 March 1924, pp. 285–286.

Apocryphal or not, this extract from the correspondence columns of the then new British journal *Chemistry and Industry* in 1924 exposes the apparent general ignorance in Britain, and also for a time in Germany, of a crucial and often desperate episode in the conduct of World War I. "Nitrate," a commodity essential to the production of all modern explosives used in warfare—mainly the aromatic nitro compounds such as TNT (trinitrotoluene) and picric acid—was common currency to all belligerents (and was often called "nitrates"). Nevertheless outside of scientific and industrial circles the critical roles of what was in fact Chilean nitrate (Chile saltpetre), and other nitrogen-containing chemicals of commerce such as calcium cyanamide and ammonia, as sources of vast and violent destructive power, was generally given little, if any, prominence at the start of the war in early August 1914. The conflict was, however, to become a war of industries—of industries far too modern for, and beyond the comprehension of, ruling and political classes on both sides—most particularly the chemical industry, in which the nitrogen sector would play a principal role. It is astonishing that even today histories of World War I pay little or no heed to this fact.

Within weeks of the outbreak of war the German Army, pursuing the campaign drawn up by Count Alfred von Schlieffen, occupied Luxembourg, crossed neutral Belgium, and entered northern France, laying waste to towns and cities. The Schlieffen plan was aimed at the rapid surrender of Paris and capitulation of France, while Russia was held at bay by a small force. Acts of aggression were reported.

A.S. Travis, *Nitrogen Capture*, https://doi.org/10.1007/978-3-319-68963-0_7

Belgian civilians were shot, including in groups, in response to trumped up charges of sabotage, and the ancient library at Louvain was reduced to ashes. In France, Lille and Douai were lost, the cathedral at Rheims damaged, and Arras almost destroyed. The French army was pushed back to within 40 miles of Paris. Anticipating a rapid German victory, several of the nation's leading scientists and scholars volunteered for service at the front, some well over conscription age. They included Walther Nernst and the sociologist Alfred Weber. Scientific prowess played no part at this stage.

The mood in Britain was not much different. Brunner, Mond chemists George P. Pollitt and Francis Arthur Freeth signed up for service shortly after war was declared. Herbert Humphrey's patriotic fervour encouraged him to invent a one-man torpedo, which, he advised Winston Churchill at the Admiralty, would cause "certain destruction to the torpedo and man as well as the Dreadnought, yet I would willingly be the first to put the idea into practice" [1]. Had he done so the story of the British synthetic ammonia industry might have been very different.

Reality soon set in. The Schlieffen plan failed during the first two weeks of September 1914. The German advance was brought to a halt at the First Battle of the Marne (September 6–8) by an Allied counter-attack on a front extending from Verdun to the Oise. The Germans retreated a short way to the north across the River Aisne. Both sides then raced north to the coast. The Kaiser's army gained the ridge that overlooked Flanders, at the First Battle of Ypres (October–November).

All belligerents were unprepared for a prolonged war; hardly any effort had been made to mobilize industry in a support role. Meantime, channels of global trade had been abruptly severed, bringing the threat of shortages on both sides. Only a few, and mainly chemists, had considered the critical role of the chemical industry in the supply of modern explosives. In particular, the availability of nitrogen products was seriously underestimated by the over-optimistic German military. That changed with the stalemate that began on the banks of the River Marne and the ensuing systems of long, winding trenches excavated by both sides from early 1915. This Western Front, a result of the German occupation of French and Flemish soil, rapidly stabilized, and remained as such for well over three years. It stretched 440 miles, from Dunkirk, across Belgium and northern France, to the Swiss border.

This situation stimulated new approaches to large-scale warfare based on nitrogen products, later capturing the public's attention in ways quite unlike that in which the very same or similar products, as fertilizers, serving to sustain so much of the world's growing population, would do so. It represented a major step in the near-total reliance of warfare on science and technology, and brought on the rapid emergence of the first major modern academic-industrial-military complexes.

The German army and navy had now to grapple with the ramifications. To maintain military strength, new strategies had to be called into play, particularly for the supply of the materials of war—not least of which were aromatic nitro compounds for explosives, and also for the intermediates from which were synthesized dyes, now mainly field grey and navy blue. The availability of nitro compounds in turn relied on adequate supplies of nitric acid and the expansion of a science-based chemical industry. Nitrogen products came to the fore. According to the French chemist Camille Matignon, the Germans came into possession of over 32,000 tons of Chilean nitrate that had been abandoned by the retreating Belgian army at Antwerp, Ostend, Ghent and Bruges. In Germany, stocks held on 1 August 1914, included 45,000 tons of nitrate at various ports, and 100,000 tons of ammonium sulphate from gas and coke works, mainly the latter. Various consumers, chemical companies mainly, accounted for another 55,000 tons of Chilean nitrate [2]. These supplies were hardly adequate for meeting the demands of large-scale prolonged warfare. Synthetic nitrogen products were required for all forms of firepower, including machine gun and rifle bullets, mines, mortars, shells and torpedoes. Within weeks, there were already threats to the sea lanes that stretched across the Atlantic. The consequences of the vulnerability of nitrogen to any disruption in shipping were not appreciated by the military leaders in Berlin.

On 13 August 1914, the Prussian War Minister Erich von Falkenhayn placed Walther Rathenau of AEG in charge of the new War Raw Materials Office (Kriegsrohstoffabteilung, KRA). (In September, von Falkenhayn was appointed Chief of General Staff of the German Army, replacing Helmuth von Moltke.) Rathenau's leading assistant was Wichard von Mollendörff (1881–1937), an AEG engineer who had formulated the draft plan for the KRA. Both the German military and the chemical industry grossly underestimated the nitrogen requirements. Thus late in August, Carl Bosch advised Haber, who was working with Rathenau and the supplies section of the Prussian War Ministry, that he did not anticipate early expansion of the Oppau ammonia plant, apart from pre-war plans to achieve annual production of 130,000 tons of ammonium sulphate fertilizer by the end of 1915.

Just over one month later, on 30 September 1914, the War Chemicals Corporation, the Kriegschemikalien AG (Kriegschemikalien AG zur Bewirtschaftung chemischer Rohstoffe), under the presidency of Gustav Aufschläger (1853–1954), was established for procurement of chemical raw materials and organization of their usage. It was one of several large government-supervised corporations created to ensure adequate supplies of war-related materials. Bosch was present at the inaugural meeting. He was encouraged to increase the BASF output of ammonia, and by oxidation convert it into nitric acid necessary for the manufacture of explosives. BASF had made good progress with the oxidation, though Bosch was not prepared at first to share details with other firms.

A review of the nitrogen situation for Rathenau's office had already been drawn up by organic chemist Emil Fischer (1852–1919) (Fig. 7.1). In a presentation he gave in Essen on October 1, Fischer recommended the large-scale production of synthetic ammonia and its conversion into nitric acid. With Haber he warned of the impending crisis in nitric acid supply, a matter that could not be lightly dismissed [3, 4]. Rathenau, with Haber's support, sent Carl Bosch to Berlin to meet with the War Ministry. Bosch, struck by the lack of scientific and technical knowledge among senior military personnel, described the various processes involved in the production of high explosives, in particular the need for nitric acid. According to BASF chemist Karl Holdermann, after Bosch explained that "When supplies of Chilean nitrate are finished, we are finished," the representatives of the War

Fig. 7.1 Emil Fischer. (Edelstein Center)

Ministry declared: "But we have the large potash deposits at Stassfurt!" in response to which Bosch had to explain that potash, unlike South American nitrate, was of no use in the manufacture of nitro compounds [5]. Germany, he emphasized, relied almost entirely on nitrate from distant Chile. The only export route was across the Atlantic. The journey took three months [6].

The military needs of finished explosives, per month, were calculated to be 12,000 tons or more, a twentyfold increase on peacetime production. This required 20,000 tons of nitrate [7].

The nitrate situation would now depend on the tide of war at sea. In September, the Germans won advantages in commerce-raiding and submarine operations. On 1 November 1914, a British Royal Navy squadron in the South Pacific Ocean off the Chilean coastal city of Coronel suffered a major defeat in a duel with the German East Asia Squadron, commanded by Vice Admiral Maximilian Reichsgraf von Spee. Admiral Spee's squadron was now free to attack merchant ships loaded with consignments of nitrate destined for the Allies and to enable strategic supplies to reach Germany. The situation was soon reversed. On December 8, at the Battle of the Falkland Islands, Spee went down with his flagship, the *Scharnhorst*, and all but one of the German squadron (Fig. 7.2). From then until the end of the war Chilean nitrate was almost completely cut off from Germany by the British blockade. The

Fig. 7.2 The battles of Coronel and the Falkland Islands, 1914

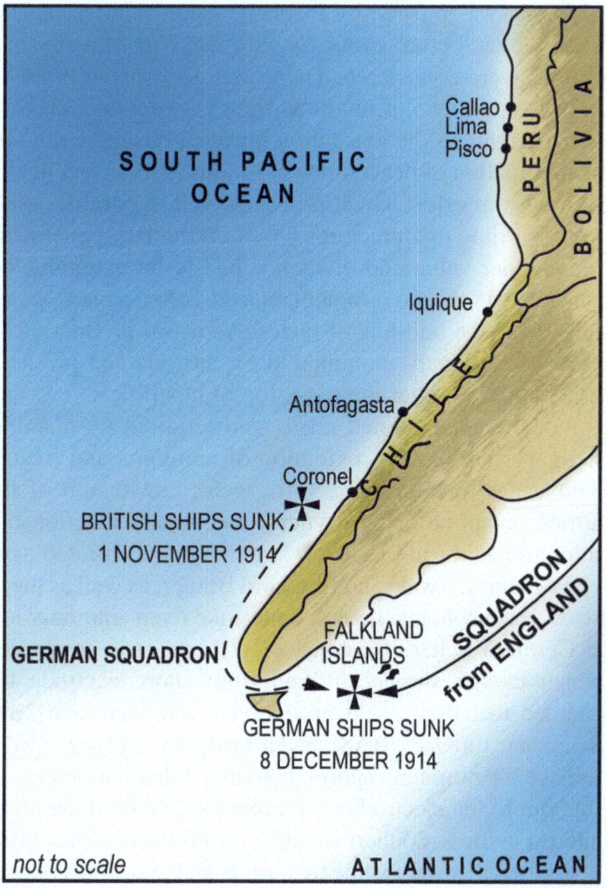

export of nitrate, which had fallen considerably from August (Germany was a major customer), now recovered, to the benefit of the Allies. Though the Allies continued to be hampered at sea, including by an intensive programme of submarine attacks between late February and mid-May 1915, and especially during 1916–1917, they were not denied access to nitrate.

It was now vital for Germany to replace the mineral nitrate used to make nitric acid with products from the new nitrogen processes already worked at Oppau, Trostberg and Knapsack, and to maintain supplies from the Norwegian electric arc factories. To accelerate the expansion of synthetic ammonia production, Haber, on behalf of the Reich, had already on October 24 drawn up a draft contract with BASF at Ludwigshafen. The final version was ready for signing in mid-November, but was then renegotiated to include further expansion, no doubt as a result of the loss of Chilean nitrate. The contract was made ready for signing in mid-January 1915. The total subsidy offered to BASF was raised from an initial four to six million marks.

Bosch was pushed by Fischer to start adding capacity to the Oppau ammonia facility even before contracts with the War Ministry had been finalized [8]. Other options were considered. There was widespread behind the scenes rivalry, as well as over-optimistic, if not downright exaggerated, claims by manufacturers anxious to participate in the expanding nitrogen business, not only during the war, but also in anticipation of peacetime conditions [9]. A great deal was at risk, and not just the German war effort. Great profits as well as royalties on innovations were to be had from wartime manufactures based on the latest chemical inventions. Then as now, shareholder value and dividends had to be maintained, which would certainly be enhanced if the government placed orders, made guarantees for purchases, and subsidized the building of factories for war production. This also held out possibilities of previously unthinkable expansions and profits that would continue to be generated long after the cessation of hostilities.

For the concentrated nitric acid required by manufacturers of explosives four steps were necessary: oxidation of ammonia and treatment of the resulting oxide with water to give dilute nitric acid; conversion of the dilute acid into sodium nitrate; evaporation to dryness; and finally reaction of nitrate with concentrated sulphuric acid (oil of vitriol), whereby concentrated acid was obtained.

Wilhelm Ostwald and Eberhard Brauer, as well as the coal industry for which they acted as consultants, favoured the coke oven ammonia-to-nitric acid process adopted at Gerthe. Fischer also considered the electric arc process of the Pauling brothers, which though simple suffered from short electrode life (however, the Paulings claimed that they could produce concentrated acid), and the Birkeland-Eyde and Schönherr furnaces. BASF had already, in 1911, dropped its interest in the Schönherr furnace for nitrogen capture, convinced that the process based on Haber's work was superior to the electrothermal processes. Only in the mid-1930s did BASF renew an interest in the Schönherr furnace, for production of acetylene from methane [10, 11].

Before the war, Chilean nitrate and gas and coke works ammonium sulphate represented over 90% of nitrogen consumption in Germany. Based on an account appearing in the *Frankfurter Zeitung* on 23 November 1919, consumption in 1913 was made up as follows: Chile saltpetre, 750,000 tons (nitrogen content 116,000 tons); ammonium sulphate from gas and coke works, 460,000 tons (nitrogen, 92,000 tons); Norwegian nitrate from the electric arc processes, 35,000 tons (nitrogen 4500 tons); calcium cyanamide, 30,000 tons (nitrogen 6000 tons); and Haber-Bosch derived ammonium sulphate, 20,000 tons (for September to December; nitrogen 4000 tons). The total, calculated as nitrogen, was 222,500 tons [12].[1]

In November 1913, Fritz Haber, when addressing the British Society of Chemical Industry in Liverpool, had stated that 36,000 tons of cyanamide were manufactured annually in Germany, for use as fertilizer or conversion to ammonia [13]. His opinion was that BASF's synthetic ammonia process would have little impact on other producers of nitrogen products, since the demand for nitrogen was

[1]In 1913 Germany may have imported 48,000 tons of the Norwegian nitrate.

so great. In late 1913, the monthly output of Oppau, in contrast to the press report, was 600 tons of ammonia, converted into 3000 tons of ammonium sulphate.

These figures emphasize the overriding importance of Chilean nitrate. Until 1914, Germany was the leading importer, mainly for the agricultural sector. The supply of nitrogen was under the remit of the Ministry of Agriculture, and investigation of its uses the concern of soil chemists. This was the situation in September 1914, by which time the Haber-Bosch process had been in operation for just one year.

7.2 Nitrogen Rivalries

In the early months of the war, even before the supply of South American nitrate was threatened, leading men of science and industry negotiated in German boardrooms and government offices over contracts and subsidies for the increased production of nitrogen products, including cyanamide, nitric acid, and synthetic ammonia. This soon turned into a battle over competing technologies. The adversaries included personalities as well as companies.

At first, the main emphasis was on calcium cyanamide. Nikodem Caro negotiated with Emil Fischer on expanding electrical power facilities for the Bayerische Trostberg works in order to meet the needs of the agricultural sector. Caro intended to increase production at Trostberg at least fourfold. Caro lobbied for construction of a new power station (Margarethenberg) and carbide factory both located to the north of Trostberg. Eberhard Ramm, of the Prussian Ministry of Agriculture, discussed with Caro even greater expansion of cyanamide production. There were good reasons. The cyanamide industry though not large was well established and the manufacturing process not particularly complex. The conversion of cyanamide to ammonia with steam had been achieved, and ammonia oxidation had been integrated into the Trostberg facility. Though cyanamide energy requirements were far greater than for the BASF synthetic ammonia process, the cyanamide process required relatively simple equipment.

BASF claimed that its synthetic ammonia was superior to that obtained from calcium cyanamide. It had already antagonized cyanamide manufacturers with public statements to the effect that the grey-black cake of calcium cyanamide was just intractable muck. Other interests came into play. Thus Karl Helfferich, who was appointed secretary of the Reich Treasury Office in December 1914, was a member of the board of Deutsche Bank which had invested in cyanamide. Walther Rathenau, who recognized nitrogen supply as a vital strategic project, also favoured cyanamide and pushed for construction of new facilities, particularly for supplying the agricultural sector. However, there was the challenge of promoting cyanamide use among farmers, most of whom were unfamiliar with the product. Unlike ammonium sulphate, they found cyanamide difficult to apply during the spring dressing. The outcome was a competition for the design of a suitable spreading device, which became available by the end of 1915 [14].

A nitrogen commission, whose purpose was to draw up contracts with manufacturers, was established on 28 November 1914 (Commission for the Preparation of Contracts for the Purpose of Increasing Nitrogen Production; Kommission zur Vorbereitung der Verträge zwecks Steigerung der Stickstoffproduktion). The Ministry of Agriculture was closely involved in the work of the commission, whose policy was invariably formulated, understandably, by those with the greatest experience and also the greatest vested interests. Rivalry for lucrative, long-term nitrogen contracts pitched Carl Bosch and BASF against Caro and Bayersiche Stickstoff-Werke. Fritz Haber, in his work as a government official—as head of the Chemical Section—quite naturally favoured BASF, whose chemists and engineers were working on the oxididation of synthetic ammonia to nitric acid, using catalysts other than the rare platinum, and on concentrating the acid without conversion first to sodium nitrate.

Rathenau, as head of the KRA of the Prussian War Ministry, oversaw the work of Haber and other academic chemists, including Fischer and Richard Willstätter (1872–1942). However he had little time for them, or economists for that matter. Following a falling out with Haber, Fischer, and Willstätter, at the beginning of October 1914, Rathenau took over total control of planning for war raw materials, in particular nitrogen products. His emphasis on cyanamide-derived ammonia led the Reich Treasury in early 1915 to provide funds and guarantees for an unprecedented programme of cyanamide expansion. Nikodem Caro, once armed with sales guarantees, commenced enlargement of Trostberg in 1915. To satisfy anticipated demand, new cyanamide factories in which the state was closely involved were deemed essential.

On 5 March 1915, the Reich Treasury, with Rathenau's backing, commissioned Bayerische Stickstoff-Werke to erect and manage state-owned cyanamide production facilities at Piesteritz (Reichsstickstoffwerke Piesteritz; the Imperial Nitrate Works), on the River Elbe, near Wittenberg; and at Chorzów (German: Königshütte), near Katowice (German: Kattowitz), north-west of Krakov, in Prussian Upper Silesia. Work began almost immediately at Piesteritz. Both factories started to produce cyanamide late in 1915. Piesteritz, the world's largest operating cyanamide factory, with an annual capacity of 150,000 tons, began large-scale production in early 1916 [15]. For the isolation of nitrogen, the Linde company supplied four plants of unprecedented size, with a total capacity of 14,000 cubic metres of gas per hour. Power came from a large brown coal-fired electricity generation facility, commissioned from Braunkohlen Golpa-Jessnitz AG. Brown coal was also the source of electrical power for Chorzów. It is evident that even allowing for vested interests among government officials responsible for commissioning these works, they were hardly based on bad decisions. The Bayerische, based in Munich, opened a new office in Berlin to handle the increased business.

In March 1915, shortly after arranging the increases in cyanamide production with the Bayerische, Rathenau resigned. He returned to the management of AEG [16]. The wartime organization that he created became a model that was followed in the militarising and autocratic regimes of all political and economic persuasions, during and after the war.

By this stage, Haber had secured his association with the military and business elite. With Fischer, he enabled additional and very favourable financial arrangements to be made with BASF. Fischer believed from early on that the Haber-Bosch process was superior to the other nitrogen fixation processes. The product was cleaner—certainly more so than that from the coal gas and coking processes—and labour requirements were less. Once large-scale production by the various nitrogen-capture and nitric acid processes was successfully under way he began to promote the Haber-Bosch process, at least for military explosives. By January 1916, the output of Oppau converters had been increased fourfold compared with 1914 to 80 tons of ammonia every day [17].

Electric arcs also received some support. Fischer, as noted, had considered the Pauling process in September 1914, when he was advised that it was capable of producing a concentrated nitric acid. The Pauling brothers' firm subsequently erected a new factory near Bitterfeld at Muldenstein, drawing power from a coal-fired generating station that served Germany's first long-distance electrified railway (from Bitterfeld to Dessau, opened in 1911; it reverted back to steam traction in August 1914 as a result of the war). The Pauling nitric acid plant ceased production early in 1917, mainly as the result of shortages of brown coal. In any case, its output was not great [18]. Problems of coal and electricity supply also reduced capacities at the cyanamide works.

Max Sering (1857–1939), of Berlin's Friedrich-Wilhelms-Universität, who headed the War Ministry's Scientific Commission (Wissenschaftliche Kommission), established in 1915, wrote in that year, perhaps somewhat optimistically and more with an eye on the propaganda impact: "The complete cutting off of the supply of Chili saltpetre during the war has been made good by our now taking nitrogen directly out of the air in large factories built during and before the war. With extraordinary rapidity the question has been solved of how the enormous quantities of the needed ammunition were to be produced, a question which in England still meets with difficulties, in spite of the help from America" [19].

Much of this expansion in nitrogen was in fact then still in the planning stages or at best about to be implemented. Certainly German-made ammonia available from both the cyanamide and Haber-Bosch processes was converted into nitric acid and ammonium nitrate, both required in the production of explosives, far more than into fertilizer. Overall control and production had, almost, become a state monopoly, directed by Julius Bueb (1865–1944), Reich Commissioner for the Nitrogen Economy (Reichcommissar für die Stickstoffwirtschaft).

7.3 Enemy Aliens: Le Rossignol and Tamaru

When the war broke out in August 1914, Robert Le Rossignol and Tamaru Setsuro were working in Germany and as a result were classified as enemy aliens. Their fates were to some extent dictated by Fritz Haber. Le Rossignol, while at Karlsruhe, married a local girl. Funding for his work with Haber expired, it appears, around the time that the benchtop ammonia process was perfected. In August 1909 he moved to Berlin, where, on Haber's recommendation, he joined the laboratories of the Auer-Gesellschaft, manufacturer of the Osram lamp. For a time he continued to deal with matters related to the synthetic ammonia method, probably in connection with patent issues abroad (and, later, scientific publications, once agreement was reached with BASF in 1910). As an enemy alien living in Germany, Le Rossignol was interned on 6 November 1914 at the makeshift Ruhleben camp in Berlin. Haber intervened and arranged Le Rossignol's release on 22 March 1915, apparently in exchange for a similar arrangement in Britain concerning a German national. Le Rossignol then returned to the laboratories of the Auer-Gesellschaft. He left Germany for England on 6 December 1918. In 1919 he joined the Hammersmith (London) laboratories of the British Osram Company Ltd, a firm sequestered as enemy property and from 1916 wholly owned by the British General Electric Company. From 1923, General Electric research was carried out at purpose-built laboratories located in North Wembley. There, Le Rossignol developed large high-powered transmitting valves for radio communication. Haber had in 1908 arranged to assign 40% of any future royalties on a successful new nitrogen fixation process to Le Rossignol, which acted as a great incentive to the young assistant. Though events in Germany prevented Le Rossignol from receiving full royalty payments, he received a substantial amount until around 1924 [20]. Royalties were boosted by the wartime production of ammonia.

Tamaru Setsuro, one of Haber's most important assistants in Berlin, escaped to England, no doubt with Haber's connivance, and was at first a guest of Sir William Ramsay, whose son, the chemist William George Ramsay (1880–1927), had also worked with Haber. In Liverpool on 24 November 1915, members of the Society of Chemical Industry were treated to a long, detailed lecture by Tamaru on calorimetric measurements pertaining to ammonia that he had undertaken during his seven-year stay in Germany (1908–1914), mainly with Haber's group. Shortly after, he joined Theodore W. Richards (1868–1928), recipient of the 1914 Nobel Prize in Chemistry, at Harvard. While in the United States, he worked in the laboratory of Takamine Jōkichi (1854–1922), co-founder of superphosphate manufacturer Tokyo Artificial Fertilizer who made his fortune in America from his enzyme work. Takamine, no doubt advised by Tamaru, investigated the high-pressure synthetic ammonia process of the General Chemical Company for possible use in Japan. The General Chemical process was based on the designs of Frederick W. de Jahn, and in many respects it closely imitated the Haber-Bosch process. Patents were filed by de Jahn in 1915. An experimental plant was erected during 1915–1916 at Laurel Hill, on Long Island, and an inspection was probably arranged for the benefit of Tamaru. Despite several shortcomings, this was the only synthetic ammonia process worked outside Germany until after the war. Through Takamine's network at least four Japanese firms expressed interest in the process. Research carried out by the US government during and after the war, and developments at General Chemical, would enable the process to inaugurate, on a small scale, the US synthetic ammonia industry in 1921.

Tamaru and Takamine also discussed the founding in Japan of a modern scientific research institute on German lines—in particular Fritz Haber's Kaiser Wilhelm Institute, where Tamaru had worked. There, high-pressure ammonia synthesis could be studied. Shortly after his return to Japan in 1917 Tamaru joined the Society for the Investigation of Ammonium Sulphate (Ryūan Chōsakai), where, along with colleagues, he continued to promote the Haber-Bosch process and reject the Serpek process that had received support in America [21].

7.4 The Explosives: Nitro Compounds and Nitrates

Thus far, this chapter has been mainly concerned with the production of nitrogen compounds that were required for the German war economy—namely, ammonia and, in turn, nitric acid. The need was to overcome the complete loss of access to South American nitrate. Among the Allied nations, there was less concern, at least until intensified attacks by U-boats on merchant ships on the high seas brought home the need to establish alternative sources. Whatever the form of fixed nitrogen—Chilean nitrate, ammonia, nitric acid, or cyanamide—these compounds mainly served for one common purpose, the production of nitro compounds as explosives. Moreover, new and expanded processes for the production of nitro compounds, and also ammonium nitrate, brought forth challenges comparable in scale and scope to those encountered in Germany at Oppau and the cyanamide works. No less than in Germany, industrial production on an unprecedented scale in the service of warfare was brought about through the emerging connections between the military and science-based industrial concerns, particularly from 1915 on. Here the main emphasis is on the British response.

Britain and its allies from late 1914 mainly ruling the waves of the South Pacific and Atlantic oceans—despite attacks on merchant shipping—relied on Chilean nitrate and gas and coke works ammonia, and on Canada and the United States for finished explosives. There was also cyanamide from Odda and Alby, and electric arc products from Norway, when the supply lines were not threatened.

In September 1914, the prominent lawyer Lord Moulton (he had been given the title Baron Moulton in 1912) was appointed chair of the Committee on Supply of High Explosives and shortly after became director-general of the Department of Explosives Supply within the War Office. After almost a lifetime spent in litigation, including matters related to the synthetic dye industry, he had a more than adequate background in aromatic organic chemistry. He soon instigated a programme of TNT (2,4,6-trinitrotoluene) manufacture in both private factories and at a new state-owned factory located at Oldbury, near Birmingham, next to the facility of Chance & Hunt, Ltd, manufacturer of acids (acquired by Brunner, Mond in 1917). Construction of the factory commenced in December 1914 and production of TNT began in May 1915. This was the first of the large national factories engaged in the manufacture of explosive materials [22]. They included Queensferry, near Chester, and Gretna, on the Sottish border.

On 9 June 1915, the Ministry of Munitions was created in response to the British "shell crisis." To the new ministry was transferred Lord Moulton's Department of Explosives Supply. In July, the Committee of Scientific and Industrial Research was created (from December 1916 known as the Department of Scientific and Industrial Research). Its main pupose was to bring about cooperation between industrial firms in conducting research for the war effort, in many ways imitating features of Rathenau's programme of military-industrial collaboration. Control of the Woolwich Ordnance Factory, where most shell filling took place, but was overloaded with empty shells as a result of the shortage of explosives, was transferred from the War Office to the Ministry of Munitions.

Moulton gathered together a group of capable scientists and administrators to maintain high production levels of both explosives and shells. The effort relied more on organizational ability than on completely new technical innovations, though shortages of aromatics stimulated development of novel chemical pathways. Imports of nitrates, including potassium nitrate from India, grew from 199,000 tons in 1914 to 388,000 tons in 1916, and reached 526,000 tons in 1918. However, the supply of ammonium nitrate from Norsk Hydro almost ceased, as was also the case for cyanamide, from both Norway and Sweden, as a result of the German threat at sea [23].

The role played by British engineers and scientists, particularly those associated with Brunner, Mond, in the production of munitions is particularly instructive to the understanding of post-war developments in high-pressure ammonia synthesis. The demand for the hasty design and erection of factories, the pressure to work out new routes to intermediate products, the exposure to poorly understood reactions, and the dangers involved, provided new and useful experiences in a variety of manufacturing processes and in solving problems in dissimilar environments. Brunner, Mond participants included chief chemist Francis Arthur Freeth (1884–1970; he joined Brunner, Mond in 1907), promptly called back from the trenches in early 1915, chemist Albert William Tangye (1866–1965), and the firm's erstwhile manager and consultant Herbert A. Humphrey. Their efforts would bring the leading alkali manufacturer to the first place in British production of strategic chemicals.

Humphrey was in continuous government service during 1914–1918 (Fig. 7.3). From August 1914 he was with the Admiralty. On 21 May 1915 he joined the Department of Explosives Supply as "an unpaid volunteer." He was soon after appointed an "Official of the Ministry of Munitions" [24]. In this capacity, Humphrey worked on ammonium nitrate, the design of a cyanamide factory, and the Häusser process for nitric acid. With colleagues from Brunner, Mond he participated in developments related to production and purification of aromatic and other chemicals required in manufacture of explosives. A number of the many wartime processes and products are discussed in the following. For sake of completeness, similar German, French and Austrian activities are included.

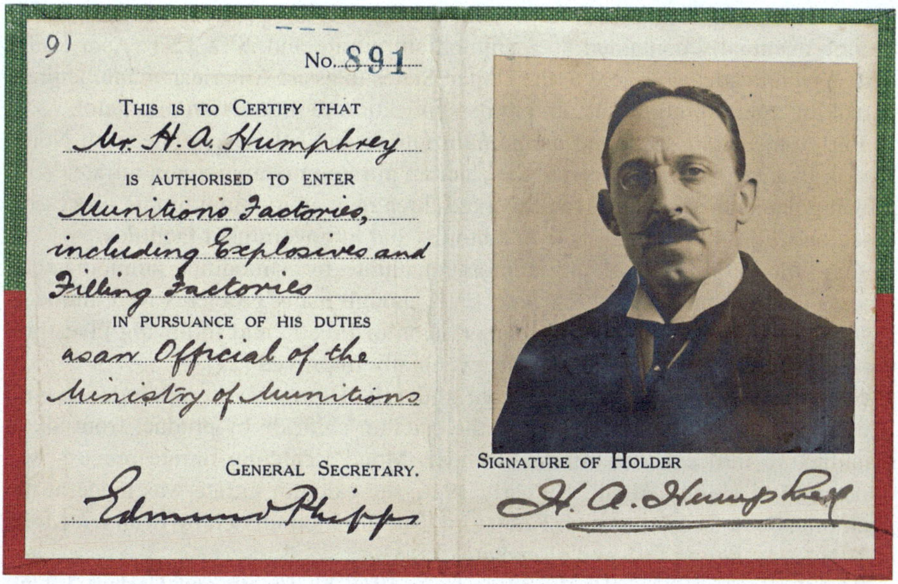

91

No. 891

THIS IS TO CERTIFY THAT

Mr. H. A. Humphrey

IS AUTHORISED TO ENTER

Munitions Factories, including Explosives and Filling Factories

IN PURSUANCE OF HIS DUTIES

as an Official of the Ministry of Munitions

GENERAL SECRETARY.

Edmund Phipps

SIGNATURE OF HOLDER

H. A. Humphrey

Fig. 7.3 A rather aristocratic-looking Herbert A. Humphrey appears on his British Ministry of Munitions identity card of 1917, no. 891, signed by Edmund Bampfylde Phipps, General Secretary, Ministry of Munitions. (Papers of Herbert Alfred Humphrey, Imperial College Archives and Corporate Records Unit, London)

Ammonium Nitrate and TNT Throughout World War I, military high explosives were based on aromatic nitro compounds, of which TNT, or trotyl, was more used than any other. TNT was manufactured from toluene, the aromatic coal tar hydrocarbon closely related to benzene; both were obtained from the distillation of coal tar [25]. Wartime demand for toluene greatly exceeded the supply—shortages were faced by all belligerents. One compound containing nitrogen and oxygen that could be safely mixed with TNT, in order to conserve and extend the use of toluene, was ammonium nitrate. Ammonium nitrate was a principal component of explosives used in mines. Under normal conditions it was made by the reaction between ammonia and nitric acid. This was more readily achieved in Germany than elsewhere as a result of the expansion in production of ammonia and its oxidation into nitric acid. The German mixture, known as Gestrecktes Füllpulver, contained 40% ammonium nitrate and 60% TNT [26]. Hoechst was the main German manufacturer, producing at one stage 500 metric tons per week, followed by Oppau at 200 metric tons per week. The equivalent Austrian product was called ammonal and in addition contained charcoal and aluminium powder.

In the absence of Haber-Bosch and cyanamide processes, the British had to develop novel alternative, and less direct, processes for obtaining ammonium nitrate. The main challenge was taken up by Brunner, Mond, working closely with Humphrey. Despite many technical difficulties several notable successes were achieved in Britain, where the mixed nitrate-TNT explosive product, called amatol, eventually contained 80% ammonium nitrate and 20% TNT. As a result, British technicians were sent to the United States to assist American manufacturers in making ammonium nitrate, and in its application to production of amatol.

Early on, Britain imported ammonium nitrate and calcium nitrate from Norsk Hydro. Brunner, Mond converted the calcium nitrate into ammonium nitrate. With growing demand, Brunner, Mond worked three new ammonium nitrate processes, based on Chile saltpetre, at its own factories and in government facilities.

First was production of the ammonium nitrate by a modified ammonia-soda process at Winnington, from 15 June 1915, though many problems were immediately encountered [27]. Under the supervision of Freeth, and following investigations by Humphrey, manufacture was eventually improved [28].

Second was a process that arose from studies by Humphrey in collaboration with Brunner, Mond on the conversion of the calcium chloride by-product from alkali manufacture into calcium nitrate. Brunner, Mond's calcium nitrate process was transferred to United Alkali. By early 1916, the calcium nitrate was made at the Ammonia Soda Company (near Knutsford, Cheshire; purchased by Brunner, Mond in 1916), Salt Union Ltd, and elsewhere.

The third process, as developed early in 1916 by Freeth and Herbert Edwin Cocksedge, produced ammonium nitrate from Chile saltpetre and ammonium sulphate—applying the phase rule and Freeth's pre-war study of the nitrate in the important crystallization step. Its scale up followed Humphrey's increasing role in the production of ammonium nitrate.

On 16 December 1915, Humphrey had been appointed the first civilian on the Ordnance Board. Less than a month later, on 11 January 1916, Lord Moulton "placed the direction and organisation of the ammonium nitrate works" under his charge. On August 28, Humphrey was appointed assistant director of the Department of Explosives Supply [29]. In this role he supervised erection of six new factories for production of ammonium nitrate, some of the work involving collaboration with United Alkali and Salt Union Ltd. The Freeth and Cocksedge process was used at a large facility near the town of Swindon, site of the works of the Great Western Railway in Wiltshire, west of London. On 28 November 1916, Humphrey chose the location, his department designed the works, and he supervised the installation of equipment; Brunner, Mond managed the works, opened in early 1917. The same process was also applied at Sandbach, Cheshire.

In 1916, Ludwig Mond's youngest son, the Liberal politician Alfred Moritz Mond (1868–1930), a director of Brunner, Mond since 1895 and later of Power-Gas, South Staffordshire Mond Gas, and Mond Nickel Company, resigned from these various industrial concerns to become First Commissioner of Works under David Lloyd George's coalition government (1916–1922), a post he held until 1921. Mond's responsibilities included oversight of the construction of munitions factories, including at Swindon. This further involved Brunner, Mond expertise in support of the war effort. Brunner, Mond was responsible for the manufacture of most of the ammonium nitrate produced in Britain.

Toluene Britain before the war and during its early stages had relied mainly on the explosive called lyddite, a mixture of picric acid (trinitrophenol) and guncotton (nitrocellulose). The British subsequently turned to the manufacture of TNT. Much of the effort to produce TNT was due to American-born chemical engineer Kenneth Bingham Quinan (1878–1948), who was brought to England late in 1914 by the Committee on Supply of High Explosives from the Nobel explosives factory near Cape Town [30]. He immediately implemented a programme of construction of new munitions factories, including the TNT factory at Oldbury.

The supply of toluene, normally obtained from coal tar, was, as indicated above, inadequate. As a result it was distilled out of Borneo petroleum spirit, which contained around 55% toluene, at Portishead, on the Severn Estuary, near Bristol. This was accomplished with the aid of the Asiatic Petroleum Co., Ltd of Rotterdam and Dutch chemists and chemical engineers. In 1915, Asiatic Petroleum laid down a plant at Oldbury for the production of mononitrotoluene from petroleum spirit. The process was also taken up at the Sandycroft state-owned undertaking near Chester, which later became part of the Queensferry TNT factory. The time for nitration at government factories was reduced from forty to ten hours [31]. The "difficult question" of purification of TNT was tackled by Freeth and Humphrey. Brunner, Mond was closely involved in new factories for the purification of TNT at Rainham, Essex, and Gathurst, Cheshire, and Humphrey at Silvertown, East London. Freeth's first process was dangerous in application. In January 1917, through lack of care, 69 deaths and many injuries resulted from a massive explosion at Silvertown. Six months later a TNT factory (of Hooley Hill Rubber and Chemical Works) at Ashton-under-Lyne blew up, with 43 deaths. In this case one of the nitrators went out of control, causing a fire that ignited stored TNT. The instability of TNT in warm weather contributed to an explosion at the National Shell Filling Station, Chilwell, Nottinghamshire, on 1 July 1918, killing 134 workers, mainly women.

In 1917 alone, the output of TNT in Britain was 76 million tons. TNT and amatol displaced lyddite for army use, though picric acid continued to be manufactured on a large scale since it was preferred by the Royal Navy. Modern war was driven by ammonia, nitric acid, toluene and phenol [32, 33].

Picric Acid The French relied extensively on the nitro compound picric acid, made from coal tar phenol, obtained by distillation of the "carbolic acid" fraction. This fraction was treated with sodium hydroxide, dissolved in soda solution and separated from oil; the crude phenol was liberated with sulphuric acid, or carbon dioxide. It was then fractionally distilled. Nitration of phenol commenced with sulphonation in sulphuric acid followed by addition, with stirring, to concentrated nitric acid, giving, in two steps, the picric acid, which was melted and poured into shells. The nitro compound mixed with guncotton, lyddite in Britain, was called melinite in France.

As with toluene, the supply of phenol was also often scarce. Demand was such that picric acid was manufactured by novel multi-step processes. In Britain, a picric acid process based on chlorobenzene was developed at United Alkali [34]. In another picric acid process, introduced by the dye intermediate firm L.B.

Holliday, benzene was converted into dinitrochlorobenzene, which was treated with caustic soda, to afford sodium dinitrophenolate. Acid treatment gave dinitrophenol; the third nitro group was introduced on nitration with concentrated nitric acid. Picric acid was also produced by both the South Metropolitan Gas Company and Brunner, Mond. British annual picric acid production reached a peak of 32,000 tons in 1917.

In Germany picric acid was made from benzene, in five steps: conversion to nitrobenzene, then, successively, to chlorobenzene, dinitrochlorobenzene, dinitrophenol and finally the trinitrophenol, picric acid. Bayer and Hoechst were the major suppliers, together producing around 450 metric tons per week out of a total of approximately 675 tons. While German firms had access to plentiful supplies of chlorine for the manufacture of organic compounds, that was not the case in France, which ordered 1000 tons from the United States. However, it was requisitioned on arrival by the army for use in gas warfare. The French then made synthetic phenol starting with the sulphonation of benzene. Treatments with lime and sodium sulphate gave calcium benzene-sulphonate sodium salt. The latter was separated, crystallized, fused with caustic soda and extracted with water to give sodium phenate and sodium sulphite. The sodium phenate was treated with sulphuric acid, setting free the phenol, which was decanted and distilled under vacuum. The phenol was then nitrated to picric acid. This was similar to the process used by Brunner, Mond. The initial sulphonation was carried out at Brunner, Mond's ammonia-soda plant at Lostock Gralam, Cheshire, near Northwich, and the caustic fusion at Warrington, based on a process perfected by A. W. Tangye in 1915 [35]. Lostock Gralam manufactured ammonium nitrate from both calcium nitrate and Chile saltpetre.

German nitro compounds employed in aerial bombs and torpedoes often contained such mixtures as trotyl and hexanitrodiphenylamine, on account of their great shattering power. There was also picryl sulphide, again a nitro compound, used for similar purposes by the Germans.

Two important high explosives did not require the valuable toluene or phenol; both were made from benzene. Tetryl (nitramine, tetranitromethylaniline, trinitrophenylmethylnitramine), a violent explosive, was used to transmit the explosion of the detonator in a shell to the less sensitive TNT or picric acid. The high explosive TNA (tetranitroaniline) was made in Germany, and also in the United States by the Calco Chemical Company of New Jersey [36].

The concentration of nitric acid required to produce explosives varied. Nitration takes place in a mixed acid, made up of nitric acid and sulphuric acid. As nitro groups are successively introduced into aromatic compounds, such as toluene, it becomes increasingly necessary to exclude water and as a consequence to employ more concentrated acids (see Chap. 8).

Propellants In addition to high explosives there were also the propellants, based on nitroglycerine and guncotton, the main constituents of the smokeless powders such as cordite. The nitrations were achieved with mixed acid. For nitroglycerine, nitric acid of 98–99% concentration was required. Nitrocellulose was made by the nitration of

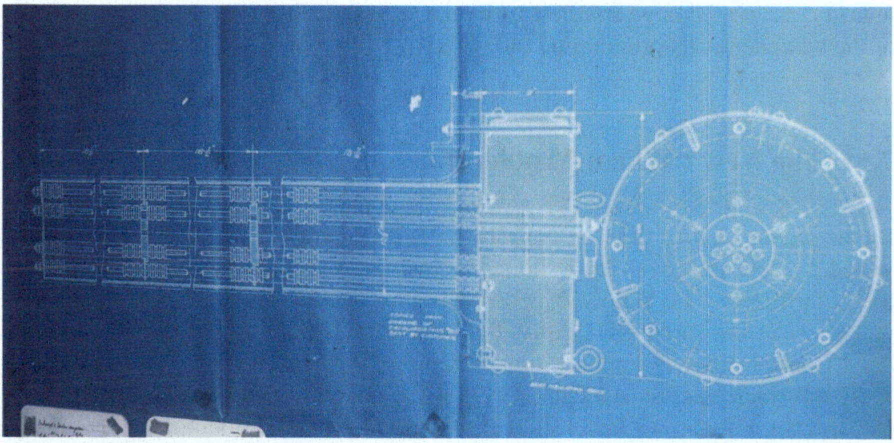

Fig. 7.4 Blueprint of vessel for Chaim Weizmann's fermentation process used in the production of acetone during World War I. (Photo by author, Weizmann Institute of Science, 1985)

dried cotton. The Austrian chemist Ernst Berl (1877–1946), as chief chemist at the Austro-Hungarian War Ministry in Vienna, worked on a process that converted textile factories, including the Tubize nitrocellulose facility, at Sarvar, Hungary, into producers of guncotton. He was in charge of the design of a large explosives factory near Vienna, which incorporated cyanamide and Pauling plants for nitrogen fixation, and apparatus for the oxidation of ammonia. The principal products were trinitrotoluene, trinitrophenol, nitroglycerine, and dynamite [37].

Tetryl and cordite required plentiful supplies of acetone, a solvent that was in short supply in Britain (it was previously imported from Germany), until the novel fermentation process of chemist Chaim Weizmann (1874–1852) was introduced on a large scale from 1915 (Fig. 7.4). Maize, rice, various cereals, and, less successfully, horse chestnuts were the starting materials [38, 39]. In the 1920s by-product hydrogen from the Weizmann process was used in the manufacture of synthetic ammonia and soon after methanol in the United States. In Germany, where wood cellulose was often used instead of cotton, Wacker Chemie manufactured acetone from wood from 1916 to 1917.

At the start of World War I, Russia was poorly prepared in chemical manufacture, notwithstanding lessons learnt earlier in the century in the Russo-Japanese War (1904–1905), and suffered more than other Allied nations from shortages of shells in 1915. Ipatieff was in charge of much of the wartime chemical effort, which proved to be impressive, including the large-scale production of by-product ammonia from coke ovens, though only from early 1917. The Russians had obtained benzene and toluene from Baku petroleum since 1916. As in Britain plans were drawn up for a cyanamide factory, but it never materialized. These various experiences were useful when Ipatieff turned to organizing the Soviet chemical industry, as was the large coterie of skilled workers that became available as a result of the war chemicals programme [40, 41].

7.5 The New "Wheat Problem"

By the second year of the war, concerns in Britain over the supply of nitrates were joined by concerns over food shortages, including productivity and the efficient use of fertilizers. This followed a poor wheat harvest and failure of the potato crop in 1916, causing rapid price rises. To address the situation, the first Minister of Food Control, Lord Devonport, took office on 22 December. The threat of shortages was increased in January 1917, after Germany announced the start of unrestricted U-boat warfare. Voluntary rationing was introduced in February. Devonport's successor, on 19 June 1917, was industrialist and Liberal politician Lord Rhondda. In order to enhance public awareness of the complex issues, Lord Rhondda underwrote the cost of a revised edition of William Crookes's *The Wheat Problem*. The chapters by the Americans Davis and Hyde were replaced by a chapter on the future of wheat supplies by Sir R. Henry Rew. Emphasis was placed on the increasing role of artificial nitrogen fertilizer in Germany and France, in contrast to Britain where farmers still preferred manure. Crookes added a lengthy review of the new nitrogen capture processes. In October 1917, a Royal Commission on Wheat Supplies was established to bring about state control of the supply and price of wheat [42, 43]. It was perhaps the incentive for considering manufacture of cyanamide in Britain.

7.6 Herbert A. Humphrey and Cyanamide

On 4 July 1917, Herbert Humphrey was informed by Lord Moulton that his services to the Department of Explosives Supply, Ministry of Munitions, would shortly be terminated [44]. Humphrey was deeply disappointed. It was not a matter of lost income, which was little enough, but of service to his country. He had thrived working alongside leading industrial managers in meeting the scientific and technological challenges posed by the wartime urgency. That was not uncommon for those who had enthusiastically taken on positions of responsibility among all belligerents. Humphrey had been satisfied to work for a "nominal salary," as both volunteer and consultant. Such arrangements, as acts of patriotism, were not unknown. Often they were expected. Sufficient additional financial support came from the three companies of which he was a director, Power-Gas, South Staffordshire Mond Gas, and the Humphrey Pump Company. Though nominally an independent consultant he remained on the payroll (perhaps on the basis of a retainer) of Brunner, Mond, which had come to a financial arrangement with the Department of Explosives Supply over his secondment to the department [45].

However, this was not the end of Humphrey's government wartime service. In August he was appointed chief engineer of the Munitions Inventions Department, and on November 8 he was instructed by Colonel Sir Henry E. F. Goold-Adams (1860–1935), Comptroller of Munitions Inventions, Munitions Inventions Department, to spend up to one year planning a "cyanamide scheme of some magnitude" in collaboration with J. H. Rider, who would deal with civil engineering and electrical issues. Humphrey would tackle the chemistry [46]. Oskar Carlson's continuous cyanamide process was the basis of their designs. Other processes investigated by Humphrey at this time included the Häusser process for nitric acid, which must have captured his interest since it involved an area of his expertise, gas combustion [47].

The cyanamide project never progressed beyond the drawing board, probably because by the time the designs were ready the war was drawing to a close. Also, and despite Humphrey's opining that technological know-how was severely lacking, the Ministry of Munitions had decided to embark on high-pressure ammonia synthesis.

For the engineer Humphrey, however, the wartime experiences in large-scale project management and development of processes for manufacture of nitrogen-containing compounds, and the acquisition of a network of contacts at the most senior levels in government and industry, would prove invaluable in the immediate post-war period. Along with his involvement with Brunner, Mond, consultancy work for the Mond gas enterprises, and fame associated with his pump, Humphrey's range of skills would serve to lead Britain, on a bumpy road, into the age of high-pressure industrial chemistry (Chap. 9).

References

1. H. A. Humphrey to Winston Churchill, 23 November 1914. Papers of Herbert Alfred Humphrey, Archives, Imperial College London, item B15.
2. Waeser B (1926) The atmospheric nitrogen industry: with special consideration of the production of ammonia and nitric acid (trans. Fyleman E), vol. I. P Blakiston's Son & Co., Philadelphia, p 22.
3. Stoltzenberg D (2004) Fritz Haber: chemist, Nobel laureate, German, Jew. Chemical Heritage Press, Philadelphia, pp 128–131.
4. Szöllösi-Janze M (2000) Losing the war, but gaining ground: the German chemical industry during World War I. In: Lesch JE (ed), The German chemical industry in the twentieth century. Kluwer, Dordrecht, pp 91–121, on 97–98.
5. Holdermann K (1954) Im Banne der Chemie. Carl Bosch: Leben und Werk. Econ-Verlag, Düsseldorf, pp 136–137.
6. Bäumler E (English version) (1968) A century of chemistry [Hoechst]. Econ Verlag, Düsseldorf, p 66.
7. Stoltzenberg D (1994). Fritz Haber: Chemiker, Nobelpreisträger, Deutscher, Jude: Eine Biographie. VCH, Weinheim, pp 234–236.
8. Moy T (1989) Emil Fischer as 'chemical mediator': science, industry and government in world war one. Ambix 36(2):109–120, on 110–113.

9. Szöllösi-Janze M (2000) Losing the war, but gaining ground: the German chemical industry during World War I. In: Lesch JE (ed), The German chemical industry in the twentieth century. Kluwer, Dordrecht, pp 91–121, on 100–103, 116–118.
10. Morris PJT (1982) The development of acetylene chemistry and synthetic rubber by I.G. Farbenindustrie Aktiengesellschaft: 1926–1945. DPhil thesis, University of Oxford, p 121.
11. Mantell CL (1940) Industrial electrochemistry, 2nd ed. McGraw-Hill, New York, pp 575–576.
12. Lefebure V (1923) The riddle of the Rhine: chemical strategy in peace and war. EP Dutton & Co., New York, p 38.
13. Haber F (1914) Liverpool section. The Hurter Memorial Lecture. Modern chemical industry. J Soc Chem Ind 33(2)(31 January):49–54, on 53.
14. Haber LF (1971) The chemical industry 1900–1930: international growth and technological change. Clarendon Press, Oxford, pp 200–201.
15. Beneke KOT (2006) Mitteldeutsche-Stickstoff-Werke AG, Piesteritz (heute Orsteil der Lutherstadt Wittenberg). Band 1. Aus dem Nachlass des ehemaligen Direktors Richard Beneke (Bilder von ca. 1920 bis 1925). http://www.unikiel.de/anorg/lagaly/group/klausSchiver/piesteritz1.pdf (accessed 20 September 2014).
16. Rathenau W (1918) Gesammelte Schriften, Band V. Reden und Schriften as Kriegszeit. S Fischer, Berlin.
17. Johnson JA (2006) Technological mobilization and munitions production: comparative perspectives on Germany and Austria. In: MacLeod R, Johnson JA (eds), Frontline and factory: comparative perspectives on the chemical industry at war, 1914–1924. Springer, Dordrecht, pp 1–20.
18. Szöllösi-Janze M (2000) Losing the war, but gaining ground: the German chemical industry during World War I. In: Lesch JE (ed), The German chemical industry in the twentieth century. Kluwer, Dordrecht, pp 91–121, on 105.
19. Lefebure V (1923) The riddle of the Rhine: chemical strategy in peace and war. For the Chemical Foundation. EP Dutton & Co., New York, quoting Sering, on p 211.
20. Sheppard D (2017) Robert Le Rossignol, 1884–1976: engineer of the 'Haber' process. Notes Rec R Soc London. DOI: https://doi.org/10.1098/rsnr.2016.0019 (accessed 18 March 2017).
21. Tamaru Oyama H (2015) Setsuro Tamaru and Fritz Haber: links between Japan and Germany in science and technology. Chem Rec 15(2)(April):535–549.
22. Reed P (2017) John Fletcher Moulton and the transforming aftermath of the chemists' war. International Journal for the History of Engineering and Technology 87(1):1–19.
23. Nitrates (1913–1919) (1920) J Soc Chem Ind 39(23):417R.
24. Authorisation for entry to 'Munitions Factories, including Explosives and Filling Factories', no. 891, signed by Edmund Bampfylde Phipps, General Secretary, Ministry of Munitions, 1917. Papers of Herbert Alfred Humphrey, Archives, Imperial College London.
25. Martin G, Barbour W (1915) Industrial nitrogen compounds and explosives. Crosby Lockwood & Son, London.
26. Marshall A (1925) Modern explosives. In: Sadtler SS, Lathrop EC, Mitchell CA (eds), Allens' commercial organic analysis, vol. III. P Blakiston's Son & Co, Philadelphia, pp 595–713.
27. H. A. Humphrey to Philip A. Henriques on 'supplies of nitrate of ammonia', 10 January 1916. Papers of Herbert Alfred Humphrey, Archives, Imperial College London, item G20.
28. Reader WJ (1970) Imperial Chemical Industries: a history, vol I. The forerunners, 1870–1926. Oxford University Press, London, pp 282–287.
29. Humphrey HA, History of the works at Billingham, nd [c. 1930]. Papers of Herbert Alfred Humphrey, Archives, Imperial College London, item F1.
30. McLeod RM (1998) Chemistry for King and Kaiser: revisiting chemical enterprise and the European war. In: Travis AS, Schröter HG, Homburg E, Morris PJT (eds), Determinants in the evolution of the European chemical industry, 1900–1939: new technologies, political frameworks, markets and companies. Dordrecht, Kluwer, pp 25–49, on 33.

31. Homburg E (2006) Operating on several fronts: the trans-national activities of Royal Dutch/Shell, 1914–1918. In: MacLeod R, Johnson JA (eds), Frontline and factory: comparative perspectives on the chemical industry at war, 1914–1924. Springer, Dordrecht, pp 123–144, esp. 131–132.
32. Macnab W (1922) Liverpool section. The Hurter Memorial Lecture. Some achievements of chemical industry during the war in this country and in France. J Soc Chem Ind 41(23)(15 December):353T–361T.
33. Johnson JA (2017) 28. Weltkriege. In: Sommer M, Müller-Wille S, Reinhardt C (eds), Handbuch Wissenschaftsgeschichte. J. B. Metzler Verlag: Stuttgart, pp 303–311.
34. Reed P (2015) Making war work for industry: the United Alkali Company's central laboratory during world war one. Ambix 62(1):72–93, on 83.
35. Reader WJ (1970) Imperial Chemical Industries: a history, vol. I. The forerunners, 1870–1926. Oxford University Press, London, p 286.
36. Travis AS (2004) Dyes made in America. The Calco Chemical Company, American Cyanamid and the Raritan River. Edelstein Center, Jerusalem, pp 47–50.
37. Weitensfelder H (2015) Dünger und Sprengmittel: Verwendung und Produktion von Stickstoffverbindungen in der Habsburgermonarchie im internationalen Kontext bis zum Ende des Ersten Weltkriege. Blätter für Technikgeschichte 77:37–54, on 49–51.
38. Freemantle M (2014) The chemists' war: 1914–1918. Royal Society of Chemistry, London, pp 95–119.
39. Travis AS (1992) Chaim Weizmann: colour chemist, biochemist and father of biotechnology. Biochemist (August–September), pp 15–19.
40. Brooks NM (2006) Munitions, the military and chemistry in Russia. In: MacLeod R, Johnson JA (eds), Frontline and factory: comparative perspectives on the chemical industry at war, 1914–1924. Springer, Dordrecht, pp 75–101.
41. Brooks NM (2007) The road to 'Big Chemistry' in the Soviet Union during the 1920s. Historia Scientiarum 16 (2007):264–274
42. Crookes W (1917) The wheat problem. Based on remarks made in the presidential address to the British Association at Bristol in 1898. Longmans, Green, London.
43. Brock WH (2008) William Crookes (1832–1919) and the commercialization of science. Ashgate, Aldershot, pp 366–388, on 387–388.
44. Lord Moulton to H. A. Humphrey, 4 July 1917. Papers of Herbert Alfred Humphrey, Archives, Imperial College London, item B21.
45. Brunner, Mond to Sir Keith Price, Department of Explosives Supplies, 2 July 1917. Papers of Herbert Alfred Humphrey, Archives, Imperial College London, item G88.
46. H. E. F. Goold-Adams to H. A. Humphrey, 8 November 1917. Papers of Herbert Alfred Humphrey, Archives, Imperial College London, item B23.
47. Ministry of Munitions, Nitrogen Products Committee, Processes Sub-Committee. 15 March 1918. Minute paper for use in connection with file no. P 1336 from R. T. G. French, secretary to the committee, 8 November 1917. Papers of Herbert Alfred Humphrey, Archives, Imperial College London, item B25.

Chapter 8
Wartime Expansion of the Nitrogen Industry

> The industry of atmospheric nitrogen has become a German industry, a world problem has been solved, and the most serious War danger of technical character had been prevented.
>
> Walther Rathenau, 20 December 1915, before the German Society of 1914. Bruno Waeser, *The Atmospheric Nitrogen Industry: With special consideration of the production of ammonia and nitric acid* (trans. E. Fyleman). Philadelphia: P. Blakiston's Son & Co., 1926, vol. I, p 23.

World War I was a long, brutal, protracted calamity carried out on vast scale. Its very nature, its total reliance on industry and the availability of raw materials, and especially the need to overcome shortages, disrupted the peacetime pace, style and scale of implementation of technological developments. Armies, and especially those who managed them, invariably trained to fight earlier wars, were often overwhelmed by situations that were completely unanticipated or unexpected. As described in the previous chapter this was particularly the case for the critical supply of modern explosives, for which the adaptability of industrial organizations engaged in the supply of nitrogen products was of paramount importance. Each nation faced a different challenge, according to its available resources or lack of access to them as a result of breakdowns in the usual lines of communication. This chapter provides an insight into the ways in which strategic industrial challenges were pursued in efforts to maintain unprecedented levels of nitrogen output, starting with the production of nitric acid from ammonia in Germany by a process that was little practiced before 1914. We then examine the steps leading to the massive growth of ammonia production in Germany and follow the tactics, not always successful, adopted elsewhere in Europe and in the United States. Some technical problems remained intractable until after 1918, none more so than the high-pressure synthesis of ammonia.

A.S. Travis, *Nitrogen Capture*, https://doi.org/10.1007/978-3-319-68963-0_8

8.1 Germany: Nitric Acid from Catalytic Oxidation of Ammonia

Around 1912–1913, laboratory studies into the manufacture of nitric acid from synthetic ammonia had been taken up by both Hoechst and BASF. The Bayerische factory at Trostberg, as noted, already produced nitric acid from cyanamide, while the coal industry used the Gerthe (Ostwald) process to make the acid from coke oven and coal gas works ammonia. Both used platinum as the catalyst, as did the important BASF contact process for sulphuric acid. The rare metal was in short supply after August 1914, since the principal source was the Ural Mountains in Russia.

The main research into developing an alternative, low cost, readily available catalyst for ammonia oxidation was undertaken at BASF, whose first patent, in 1913, made claim to a catalyst that might be "one or more oxides of metals in the iron group." Alwin Mittasch worked on improvements from early in 1914; he aimed at the efficient and economical production of nitric acid and, in turn, of nitrates. In collaboration with Christoph Beck (1887–1960), he came up with an iron oxide catalyst containing a promoter. An efficient iron-bismuth catalyst was also discovered. It was Mittasch's success that enabled Carl Bosch in October 1914 to promise that a new enlarged plant at Oppau would meet much of the needs of the military by the following spring [1]. For the oxidation units, BASF scientists and engineers introduced acid-resistant alloys containing silicon.

To start the exothermic ammonia oxidation, the catalyst was heated. Oxidation of ammonia was rapid, giving 96–98% nitric oxide. The temperature rose to around 900 °C; excess air prevented explosions from unwanted reactions.

$$4\,NH_3 \quad + \quad 5\,O_2 \quad \rightleftharpoons \quad 4\,NO \quad + \quad 6\,H_2O \quad \Delta H = -909\ \text{kJ mol}^{-1}$$

The nitric oxide combined with oxygen to give nitrogen dioxide.

$$2\,NO \quad + \quad O_2 \quad \rightleftharpoons \quad 2\,NO_2 \qquad \Delta H = -115\ \text{kJ mol}^{-1}$$

The dioxide, existing reversibly with its dimer, N_2O_4, was absorbed in water to afford dilute ("weak") nitric acid, not unlike a similar step in the electric arc factories. The absorption towers used at the latter did not provide the level of concentration required for most munitions grade nitric acid. Concentration by distillation was not possible since nitric acid forms an azeotropic mixture with water (a constant boiling mixture containing 68.5% acid by mass at 122 °C).

As a result, and well into 1915, concentrated nitric acid continued to be made from ammonia in four steps. In May 1915, however, BASF started production of nitric acid from ammonia by its new catalytic oxidation process. Concentration of the acid involved a counter-current operation in a tower packed with rings.[1]

[1]At this time BASF was already using Raschig rings, invented by Friedrich August Raschig (1863–1928).

A mixture of nitric acid of 50–60% concentration and sulphuric acid of 95% concentration, introduced at the top of the tower, met an upward flow of steam injected at the base. Vapour of concentrated nitric acid exited from the top of the tower. The sulphuric acid, now diluted, was collected at the bottom. This two-step process, involving oxidation and concentration, completely overcame the wasteful and long-winded four-step process and required less sulphuric acid. The resulting 99% nitric acid was ideal for use in several nitration processes.

Bayer's Leverkusen factory in 1915 adopted a similar process. Bayer's ammonia oxidation catalyst was made up of iron oxide with a heavy metal oxide, an innovation introduced by Mittasch and made available after Bosch agreed to share the results of BASF catalyst studies with other firms. Concentration of nitric acid up to 94%, was achieved by the addition of oil of vitriol and steam, and blowing into the mixture hot burner gases. At first, ammonia came from coke works.

In May 1915, Hoechst also became engaged in the manufacture of concentrated nitric acid. In this case, the ammonia was derived from cyanamide at Knapsack. Hoechst controlled almost half of this major cyanamide producer following acquisition during November and December 1914 of shares in Aktiengesellschaft für Stickstoffdünger from Berliner Handelsgesellschaft, Metallbank, and Metallgesellschaft. Hoechst, with its supply of cyanamide now ensured, expanded the Knapsack ammonia plant in 1915. The workforce was increased from 60 to 400, and new carbide furnaces were installed. New equipment for isolation of nitrogen from the atmosphere for the cyanamide process by the Linde process followed in 1916.

An improved process for the oxidation of ammonia was developed for the Hoechst Nitrogen Department by a small group of chemists headed by Martin Rohmer at Gersthofen, near Augsberg, a site that Hoechst had purchased around 1900 in connection with one of the indigo processes [2, 3]. This followed Rohmer's invention in 1913 of an ammonia oxidation process that gave mixed oxides [4].

Calcium cyanamide, though generally intended for agriculture, was now diverted to munitions production. In 1916, the War Ministry authorized further expansion at Knapsack, resulting in a facility capable of producing from cyanamide the ammonia that Hoechst converted into concentrated nitric acid on an increasing scale. In 1916, also, the Pauling process for the manufacture of concentrated nitric acid was installed by Hoechst.

As an ammonia oxidation catalyst, Hoechst first used platinized asbestos, diaphragms of platinum, or a narrow tube of platinum—none of which were entirely successful. In October 1917, a new acid plant in which platinum gauzes were used was installed. This enabled Hoechst to manufacture 16,500 tons per month of nitric acid using BASF ammonia liquor (20% ammonia) from Oppau and cyanamide ammonia liquor (up to 27% ammonia) from Knapsack [5]. Bayer at Leverkusen eventually relied on cyanamide ammonia liquor from Knapsack and gas and coke works ammonia, as well as ammonia released from ammonium sulphate by its reaction with lime [6].

A War Ministry report of a visit to the recently enlarged nitric acid facility at Hoechst in July 1918 observed: "...in its efforts to manufacture concentrated acid, Hoechst has been more successful than any other company. Hoechst has solved the problem of the nitric acid and sulphuric acid concentration in a manner that indicates that the required production output will be achieved also in future" [7]. By the war's end, Hoechst's newer ammonia oxidation plant had 252 converters, each fitted with a 20-in. gauze. The whole plant was supervised by one man per shift [8]. The daily output in 1918 was the equivalent of 375 metric tons of 100% nitric acid, considerably greater than at Oppau, where BASF produced 100 tons of 100% acid [9]. The Hoechst ammonia oxidation process, while important to the war effort, was far from economic, and much of it, as well as installations at Knapsack, was shut down after the cessation of hostilities. In 1923, a comparison of costs for the production of nitrogen compounds among the "IG" firms found that they were highest at Hoechst. The nitric acid equipment, in particular, was outmoded—much of it representing hasty wartime installations—and would remain so until well into the 1940s [10].

During the war there were significant improvements in platinum catalysed ammonia oxidation based on the independent investigations of Ostwald and Caro. Multiple platinum gauzes replaced single gauzes, and electrical heating was discontinued. "This type of plant supplied all the nitric acid required for explosives in Germany during the later years of the war... A much greater catalyst life was obtained in this design of plant, extending to six months, provided that conditions were uniform and that the gases were free from impurities that might have a poisoning effect" [11]. Supplies of platinum may in part have been restored after the armistice with Russia on 15 December 1917, following the Russian Revolution, and especially after the Treaty of Brest-Litovsk (3 March 1918). However Russian industry was in total disarray. Meantime, platinum production in Columbia increased considerably, though probably with little benefit to Germany.

The wartime emergency organization of a new sector of chemical industry in Germany based on ammonia and nitric acid was the outcome of outstanding research effort and commitment. It involved not only close integration between the activities of laboratory, pilot plant and construction departments, as well as manufacturers of specialized equipment—such as Krupp of Essen, and Bamag—but also on negotiations with the state and its military arm, and substantial support from the Reich Treasury [12].

8.2 France

The response in France to the nitrogen problem was not altogether different from that taken by Germany, apart from the absence of the Haber-Bosch process. At first, the French relied on imported Chile saltpetre, ammonium sulphate from British gas and coke works, ammonium nitrate produced from Odda cyanamide, and nitric acid and ammonium nitrate from Norsk Hydro. A small factory at

La Roche-de-Rame that worked the Pauling process, producing two tons of 50% acid daily, was taken over by the government, which late in 1914 had moved from Paris to Bordeaux.

Norsk Hydro held contracts with both German and Allied firms, and until 1915 supplied both sides with nitric acid and ammonium nitrate. In that year, the Rjukan Railway carried 250,000 tons of freight, including nitrogen compounds destined for the munitions factories of the opposing armies. Though Eyde was perceived by French investors in Norsk Hydro to be pro-German, the French armed forces received over half the output, mainly as nitrate, and the Germans just 13%, and perhaps less after the German contracts were completed in 1915 [13]. In the same year, as a result of the French investment in Norsk Hydro, and agreement with the French War Ministry, the Norwegian firm became engaged in the construction of a Birkeland-Eyde electric arc factory at Soulom, Hautes-Pyrénées. From 1916, the factory produced 300 tons of nitric acid per month, converted mainly into ammonium nitrate but also sodium nitrate. This was the only Birkeland-Eyde facility ever worked outside of Norway. Hydro-electricity came from a Midi Railway power station on the River Gave de Luz. In 1916, Rjukan produced 80,000 tons of ammonium nitrate for supply to Britain and France.

In 1917, German U-boat attacks on Allied shipping and the threat to supplies of Chilean nitrate encouraged the French to expand calcium cyanamide facilities at Notre-Dame-de-Briançon, and Bellegarde, Rhône-Alpes, both dependent on abundant hydro-electric power (Table 8.1). Ammonia derived from cyanamide was converted into nitric acid. Licenses were issued to the French government for working the cyanamide process of Nitrogen Products and Carbide, of London, which worked the Frank-Caro process in Scandinavia and held a license for the Ostwald process for the oxidation of ammonia. The intention was to produce both nitric acid and ammonium nitrate. For the catalytic oxidation of ammonia derived from cyanamide, a large plant was erected at Angoulême, in the Poitou-Chantres region. The platinum-catalysed ammonia oxidation process was named by the French the Kuhlmann-Ostwald process in honour of the Frenchman Frédéric Kuhlmann, who had undertaken ammonia oxidation experiments in 1838. Problems similar to those encountered in Germany had to be overcome, particularly corrosion of the reactors, until an improved stainless steel was introduced. Ammonia oxidation was in full production during 1917 at Angoulême, where the monthly output of nitric acid was 3500 tons [14].

French cyanamide production was increased from an annual output of just 7500 tons in 1913, to 100,000 tons in 1917. Among the new cyanamide works commissioned was Lannemezan, Hautes-Pyrénées, at the site of Le Société Française des Produits Azotés, in which Henri Gall's Société d'Electrochimie had a one-third share. It was supplied with hydro-electricity from stations at Bordères-Louron and Loudenvielle [15]. In November 1918, two enlarged cyanamide works were in operation and others, including Lannemezan, were nearly completed. One British observer remarked: "With all this plant in operation, France would have been independent of outside supplies of nitre. In this branch of chemical industry they got much ahead, and still remain far in advance of us" [16]. From 1917, the

Table 8.1 Calcium cyanamide factories at the close of 1917

Original affiliation, or licensing arrangement with:	Company	Place	Notes
Cyanid-Gesell-schaft (1899); La Società Generale per la Cianamide (1907)	Bayerische Stickstoff-Werke AG (Frank-Caro)	Trostberg, Bavaria	
La Società Generale per la Cianamide (1907)	Nitrogen Products and Carbide Company (Limited), London	Odda, Norway	Successor, with Alby United Carbide Company, to North-Western Cyanamide Company
	Nitrogen Products and Carbide Company (Limited), London	Meraker, Norway	Carbide manufacture
	Nitrogen Products and Carbide Company (Limited), London	Alby, Sweden	
	Società Italiana per il Carburo di Calcio/ Società Italiana Prodotti Azotati (SIPA)	Papigno, near Terni, Italy	
	Società Piemontese per il Carburo di Calcio	San Marcel, Aosta Valley, Italy	
	Le Société Française des Produits Azotés	Martigny, Switzerland	
	Gotthardwerke	Bodia, Switzerland	
	Le Société Française des Produits Azotés	Notre-Dame-de Briançon, Hautes-Alpes, France	
	Le Société Française des Produits Azotés	Bellegarde, Rhône-Alpes, France	
	Ost-Deutscher Stickstoffcalc und Chemische Werke	Bromberg, Province of Posen, Prussia (today Bydgoszcz, Republic of Poland)	
	Reichsstickstoffwerke (German government)	Piesteritz, near Wittenberg, Saxony	Managed by Bayerische Stickstoff-Werke
	Reichsstickstoffwerke (German government)	Chorzów/Königshutte, Upper Silesia (today Chorzów, Republic of Poland)	Managed by Bayerische Stickstoff-Werke
	Aussiger Verein	Falkenau, Bohemia, Austria-Hungary (today Sokolov, Czech Republic)	

(continued)

Table 8.1 (continued)

Original affiliation, or licensing arrangement with:	Company	Place	Notes
	Società per l'Utilizzazione delle Forze Idrauliche della Dalmazia	Cernica, near Sebenico, Dalmatia, Austria-Hungary (today Croatia)	
	Società per l'Utilizzazione delle Forze Idrauliche della Dalmazia	Dugi Rat, near Almissa, Dalmatia, Austria-Hungary (today Dugi Rat, Croatia)	
	Japan Nitrogenous Fertilizer Co. (Nitchitsu)	Minamata, Kagami, Kyūshū	
	Nitrogen Products and Carbide Company (Limited)	Dagenham Docks, near London	Ammonia from Odda cyanamide; then gasworks liquor. Late in the war joined with Alby United Carbide Company to enlarge the Odda works. Some work at Trafford Park, Manchester. Licensee of Ostwald process for oxidation of ammonia
	American Cyanamid Company	Niagara Falls, Ontario	
Deutsche Karbid Gesellschaft (1906); AG für Stickstoffdünger (1909, Polzenius)	AG für Stickstoffdünger	Knapsack, Westphalia	
	Electrizitäts-Werke Lonza	Visp (Viège), Switzerland	
	Österreichische Stickstoffwerke	Maria Rast, Steiermark (Lower Styria), Austria-Hungary (today Slovenia)	Commenced manufacture July 1918
	Ungarische Stickstoffdünger AG	Dicsöszentmárton, Austria-Hungary (today Tărnâveni, Romania)	Completed during the war, but not operational until 1919
	Mitteldeutsche Stickstoffwerke	Groß-Kayna, near Merseburg, Saxony	Subsidiary of AG für Stickstoffdünger. Limited production.
	Lonzawerke Elektrochemische Fabriken	Waldshut, Baden-Württemberg	

(continued)

Table 8.1 (continued)

Original affiliation, or licensing arrangement with:	Company	Place	Notes
Other			
Oscar Fredrik Svante Carlson's process	Stockholms Superfosfat Fabriks Aktiebolag/ Nitrogenium	Near Alby, Sweden	
Electro-Chemical Co. (Denka)		Tomakomai, Hokkaido; Ōmuta, Kyūshū	
Electro-Chemical Co. (Denka)		Fushun, Manchuria	
Frank-Caro; under license from Nitrogen Products and Carbide Company (Limited)	French government	Minor factories	
	French government	Lannemezan, Hautes-Pyrénées	Not operational at the Armistice

Sources include: Adolf Bräuer and J. D'Ans (editors), *Fortschritte in der anorganisch-chemischen Industrie an Hand der Deutschen Reichs-Patente*, Erster Band *1877–1917*, Zweiter Teil. Berlin: Verlag von Julius Springer, 1922, p. 2103. Air Nitrates Corporation (American Cyanamid Company) started construction of "US Nitrate Plant No. 2," at Muscle Shoals, Alabama, in December 1917. The Indre Ålvik works in Norway commenced carbide manufacture in 1919, but was then abandoned

French became increasingly dependent on explosives imported from the United States. Meantime, Georges Claude was undertaking experiments into high-pressure ammonia synthesis.

8.3 Italy

In 1916, Carlo Rossi commenced calcium cyanamide manufacture at Domodossola, in northern Italy, at the foot of the Alps. He had in 1911 been responsible for introduction of the Pauling electric arc process at the Legnano site and investigated improvements, probably at a small facility near Rome. An explosion almost destroyed the Legnano works, and it was not rebuilt.

Italian production of calcium cyanamide grew from 114,000 quintals in 1913 to 289,000 quintals in 1915. In an effort to improve the supply of food, increased use of gas and coke works ammonium sulphate and cyanamide as fertilizers was encouraged, including through decrees. One of these in 1917 offered agricultural societies and farmers in Puglia (Apulia), Calabria, and Sicily, as an inducement, reimbursement of up to one-quarter of the purchase cost. Another decree was aimed at combatting adulteration through the use of a special seal to close sacks of fertilizer [17].

As in France wartime investigations were undertaken into high-pressure ammonia synthesis, in this case by Luigi Casale.

8.4 Great Britain

In Britain by early 1916, losses to German submarine attacks stimulated efforts to introduce the new synthetic nitrogen processes. This was the remit of the Nitrogen Products Committee, an advisory panel of the Munitions Inventions Department, Ministry of Munitions, created in June 1916, following recommendations taken by the Council of the Faraday Society [18–20]. The committee was sanctioned by the then Minister of Munitions, David Lloyd George, and was supported by Sir William Crookes. The main interest was in the Haber-Bosch process. But how to cope with its severe working conditions was a complete mystery. This challenge was taken up by the Ministry of Munitions, but without success. However, the post-war outcome of the research, carried out from 1916 mainly at University College London, would lead, in the hands of a band of talented chemists and engineers, led by George Pollitt and Herbert Humphrey, to the establishment in Britain of a viable imitation of the Haber-Bosch process. This would have a lasting influence on the direction of the country's chemical industry. Its importance is such that British wartime and post-war research and development, mainly in industry, deserve special treatment and are brought together in a separate chapter (Chap. 9). Since nothing was achieved on an industrial scale before the early 1920s, with either synthetic ammonia or cyanamide, it is appropriate here to review wartime and post-war reactions by British observers and commentators to the impact of the German Haber-Bosch process. This is often closely linked to Haber's role in the development of gas warfare.

8.5 The Secret of Synthetic Ammonia

Interest in Haber's novel ammonia synthesis outside of Germany was considerable even before its introduction as the Haber-Bosch process, and it would continue to grow, particularly during the war years. However, despite the wartime efforts of individuals and groups of researchers in Europe, Japan, and the United States, it was not possible to bring about the industrial high-pressure combination of nitrogen with hydrogen. The practical constraints were enormous.

The Haber-Bosch process, an ingenious, elegant example of chemical engineering, and of the application of science to industry, is still viewed with awe today, even if not as much as in the past. It is perhaps that awe and the remarkable and unprecedented success of the industrial synthesis that exaggerated the role of the process, and in turn the role of Haber, in enabling Germany to wage extended warfare—as observed in the British parliament on 28 June 1917 by Liberal politician and chemical manufacturer Sir William Pearce (1853–1932) of Spencer, Chapman & Messel, Ltd [21].

> To-day Germany is making the whole of her high nitrogen explosives from nitrogen extracted from the air, and also a large portion of her fertilisers, and she is able to do this without any importation of nitrate of soda. If she had not been able to do this, the War would have been over more than eighteen months ago. That was an industrial achievement which was effected in Germany during this War. . .Germany has solved another pressing problem. German manufacturers had taken a very large interest, an almost controlling interest, in sources of supply in various parts of Europe and abroad. They proceeded to get rid of their interest and developed their chief power from low-class coal in their own country. . . To my mind, the German achievement is one of the most remarkable things in the whole War, and it is a lesson to us that it is by these ways that a nation can inherit the earth rather than by frightfulness and by cruelty. If Germany had been content to use her great brains and her great practical organisation in this way, I think that the countries would have had peace to-day and this awful War might have been avoided.

One thing was clear: the mystique associated with the Haber-Bosch process abroad contributed to BASF's unrivalled prestige, which it would retain for decades.

Not untypical of responses to the vital role played by the Haber-Bosch process in the German war effort was that of Major Victor Lefebure (1891–1947). Lefebure, who had studied chemistry at University College London, during 1915–1919, served with the British Army in France, specializing in gas warfare work. After the war he spent time in occupied Germany, including at Oppau and Ludwigshafen [22].[2] In 1921 he published *The Riddle of the Rhine: Chemical Strategy in Peace and War*, subsequently reworked for the United States Chemical Foundation, which lobbied for a powerful and strategic dye industry [23, 24]. The book was widely read and influential [25].

Lefebure opined, in connection with Article 172 of the Treaty of Versailles—signed by Germany on 28 June 1919—that: "probably the most important point in the clause is its interpretations with regard to the Haber process. Its critical importance in the manufacture of explosives is so great that our neglect to use the Treaty to remove the monopoly is a direct menace to peace. This process undoubtedly saved Germany in 1915 and is largely responsible for the three years of war agony which followed" [26].

Then there was British physical chemist Brigadier-General Harold Brewer Hartley (1878–1972), who before the war was at Balliol College, University of Oxford. He was an important post-war investigator of German developments, as the leading member of the "British Mission Appointed to Visit Enemy Chemical Factories in the Occupied Zone Engaged in the Production of Munitions of War,"

[2]Lefebure led a gas cylinder attack near Nieuport in 1916 and was subsequently rewarded with the rank of major. In 1917, he co-organized an inter-allied gas conference to which American observers were invited. He joined the British Dyestuffs Corporation at Blackley near Manchester in 1919.

sent out under the auspices of the Explosives Supply Department in late January 1919. Following the visit, which covered both British and French zones, Hartley reported—with the usual somewhat exaggerated reference to the role of the Haber-Bosch process, as well as with more than a hint of admiration—that [27],

> the key to Germany's war production of explosives was the Haber process for the production of ammonia from atmospheric nitrogen. It is significant that large scale production by this process only began at the end of 191[3], and that [late in] 1914 great pressure was put on the Badische Co., to increase its output. . . .Without such a process Germany could not have made the nitric acid required for her explosives programme, nor obtained fertilizers for food production after the supply of Chile saltpeter had been stopped by our blockade, and it is probable that she could not have continued the war after 1916.

During Hartley's on-site survey of the German chemical industry he was accompanied by seven British colleagues, in addition to delegates from the United States, France (four), Italy (five), and Belgium (one). The secretary of the mission was chemist Stanley Isaac Levy (1890–1968), whose letter drawing attention to the purported ignorance shown by a minister in the British war cabinet of the nitrate problem was published in *Chemistry and Industry* during March 1924. Levy had served as research assistant at the Ministry of Munitions during the war, which perhaps suggests that he had reasonable evidence of such ignorance [28].

Hartley opined, much in the same way that Ostwald had drawn attention in 1903 to the then critical role of Chilean nitrate: "In the event of another war we might be cut off from supplies of saltpetre." The implications were far reaching. Blocking the export of Chilean nitrate, as the recent war had demonstrated, threatened the ability to maintain a modern army. It was the German chemical industry's tradition of focusing on priority projects, and their successful outcomes, as so recently applied to synthetic ammonia, that emphasized the need to ensure independence in strategic nitrogen products. At the same time this particular endeavour had brought about a distinct shift to a new style of industrial research and development that was more akin to chemical engineering.

8.5.1 The "Haber" Myth

In February 1923, *Chemistry and Industry*, emphasizing the role of synthetic ammonia in agriculture rather than in war following a survey solicited among readers, could only add to the praises for this remarkable achievement [29]:

> THE FIXATION OF ATMOSPHERIC NITROGEN receives the first place in this week's letters. . .Pre-eminent, . . . [among chemical innovations], stands the synthetic ammonia process for the brilliance of the application of physical chemistry to its solution, and for the wonderful engineering skill displayed. But greatest of all we must recognise the profound influence which a cheap source of ammonia will exert on the most vital problem of the feeding of the rapidly growing population of the world by the intensive cultivation of the land with the aid of fertilizers.

That observation served to press home the importance of the Haber synthesis in both peace and war. Other observations, however, singled out, erroneously, the role of the Haber-Bosch process in enabling Germany to conduct the recent war from almost the first months. British chemist and industrialist Herbert Levinstein (1878–1956), who had managed to replicate the German synthetic indigo process during the war, had this to say of the ammonia synthesis: "Germany must have collapsed after the Marne but for Oppau. We [Britain] could not have carried on if the German blockade had cut off our Chile saltpetre" [30]. In a similar vein, Victor Lefebure observed [31]:

> It must be remembered that, after the first Battle of the Marne, the German Government turned to the I.G. [the community of interests of 1916] for a large part of its explosives and practically all its poison gas, and, as has been stated on many occasions, and with reason, Germany would not have been able to continue the war after the summer of 1915 but for the commercial development of the Haber process by the I.G.

Lord Moulton drew attention to the purported vital role of the "Haber factories" in Germany's preparations for war prior to 1914, and went so far as to claim that the Germans had deliberately waited for the factories to be got ready prior to embarking on war [32].

The myth was also perpetuated in Germany. In 1960, Nobel laureate Otto Hahn apparently remembered that on 29 January 1935, at a Kaiser Wilhelm Society gathering honouring the late Fritz Haber—a meeting from which civil servants were excluded by the National Socialists—the society's president, Max Planck, in the opening address pointed out that "had not Haber made his magnificent discovery, Germany would have collapsed, economically and militarily, in the first three months of World War I" [33].

These and similar statements continue to provide the basis for unsubstantiated claims that the Haber-Bosch process sustained the German munitions industry throughout the war, at least from 1915. Less controversial is the role of Fritz Haber in leading Germany's chemical warfare effort. This was a response to the stalemate from the latter part of 1914 and the shortage of nitrogen products for filling shells, both then and later. Toxic gases acted as both substitutes and supplements for conventional nitrogen explosives. Moreover, no account of Fritz Haber's life, particularly during 1914–1918, can omit a summary of this episode in "The Chemist's War."

8.6 Gas Warfare

For most people, Fritz Haber's name is synonymous with either synthetic ammonia or gas (chemical) warfare, and occasionally with both. In war, Haber was a loyal patriot of the Fatherland. On 4 October 1914, Haber, and Emil Fischer, were among 93 signatories, men of arts and sciences—including Carl Engler, Paul Ehrlich, Walther Nernst, Wilhelm Ostwald, Richard Willstätter, Max Planck, and Philip

Lenard—of Ludwig Fulda's "Der Aufruf der 93: 'An die Kulturwelt'" (Manifesto of the 93: To the Cultural World) that was published in German newspapers. Jointly they affirmed loyalty to the German Reich and rejected claims of German aggression and barbarism against Belgium and its citizens [34].

Haber's main role during the war was however in the mobilizing of science to the needs of the military [35, 36]. His outstanding organizational skills were increasingly harnessed in the service of the state through the laboratories of his own institute. This included research into gas warfare. However, he was not the initiator. In October 1914, the German military, by then fully aware of the danger posed by shortages of nitrogen products, began to take an interest in other products made by the synthetic dye industry. Carl Duisberg, head of Bayer, was encouraged to test toxic chemicals, particularly dye intermediates, in shells. On 27 October 1914, 3000 shells carrying a lachrymatory agent, the chlorosulphate salt of dianisidine, were fired at British and Indian troops near Neuve-Chapelle. In January 1915, shells filled with mixed xylyl bromide and explosive (T-shells) were fired at Russian troops. Both had little impact. The contents of the T-shells froze due to the cold weather.

By March 1915, there were severe shortages of explosives for shells on both sides, which encouraged the use of toxic gases. Haber, who originally suggested firing shells containing toxic chemicals from mortars, came up with a simpler method, release of compressed liquid chlorine from steel cylinders placed in trenches. He supervised the installation by mid-April 1915 of 6000 chlorine cylinders at Langemarck, near the Flemish town of Ypres. On 22 April 1915, the direction of the wind was in favour of the Germans; the valves of the cylinders were opened to release the yellow-green gas. It enveloped French and Algerian soldiers who suffered the effects of the first large-scale release of an effective war gas. Among the 5000 casualties, including 1000 dead, were a few German soldiers who found themselves in the way of the gas cloud in this Second Battle of Ypres. Just two days later a similar attack caused 10,000 casualties, with 4000 dead [37–39].

Though no breakthrough was achieved, in part because German commanders had no confidence in the tactic, a new type of warfare was introduced, particularly in frontal offensives.

On May 2, Haber's wife, Clara, committed suicide, though probably not altogether, as is often stated, in protest at her husband's new role in warfare, nor of his neglect for her (depression and suicidal tendencies were not unknown in her family; Hermann took his life in 1946, as did, later on, the oldest of his three daughters). Two-and-a-half years later Haber remarried, on 25 October 1917, to Miss Charlotte Nathan, who managed the German Society, a gathering place in Berlin frequented by influential men from government, industry, the arts, and sciences.

Fritz Haber's Kaiser Wilhelm Institute became the centre for war-related work involving toxic agents. His colleagues included Otto Hahn, James Frank, and Hans Geiger. Studies included the development of gas masks and suitable fillers, in which work organic chemist Richard Willstätter was involved. In February 1916 the institute's new role was formalized. A coordinating body was established, at the suggestion of Haber, who was backed by Koppel and Friedrich

Schmidt-Ott (1860–1956) of the Prussian Ministry of Education. This Kaiser Wilhelm Foundation for Military and Technical Sciences (Kaiser-Wilhelm-Stiftung für kriegstechnische Wissenschaften) became part of the War Ministry in 1917 [40]. The ministry's Department 8 was devoted to chemical warfare under the management of Haber's institute.

Despite the 1899 Hague Convention, in which the signatories agreed to refrain from using projectiles to disperse asphyxiating or deleterious gases in warfare, as ratified in 1907, at which time poisons were banned, retribution for the use of chlorine in war was axiomatic. The British (and French), though without facilities for compressing chlorine, which the Germans had taken into account while planning the use of the gas, were not slow in responding. A consignment of gas cylinders filled with chlorine was on its way to France by early July 1915 [41]. The gas was released on 25 September 1915, but the retaliation did not go well as the wind shifted direction, inflicting considerable casualties on both sides.

During the following years, the Allies and Central Powers developed ever more toxic and debilitating chemicals. The French called on the services of Victor Grignard, the British on Harold Hartley and William J. Pope at the University of Cambridge, and the Americans on Arthur B. Lamb, one of Haber's former assistants, at Harvard. Captain Hartley in 1915 had joined the Royal Engineers to work on gas warfare, taking on the role of chemical adviser to the Third Army in France. In 1917, as lieutenant-colonel, he was assistant director of Gas Services, General Head Quarters, France, and late in 1918, as brigadier-general, Controller of the Chemical Warfare Department, Ministry of Munitions. Toxic agents included phosgene, and mustard gas (yperite, introduced in summer 1917). Theories of meteorology guided the attacker. In the end the artillery barrages of toxic agents played hardly any role in determining the outcome of the war. They did however leave many casualties, suffering the long-term effects. There were also laboratory casualties among those engaged in gas warfare work, including Haber's colleagues at his Kaiser Wilhelm Institute. Gerhardt Just lost a hand, and Otto Sackur was killed as the result of an explosion when working with a bromine compound.

Notwithstanding the widespread condemnation of gas warfare, in public at least, and misgivings among some chemists, there were those in the military, including scientists, who believed in its potential for bringing about a quick end to battles. They included Victor Lefebure, Harold Hartley, and Major General Amos Fries of the US Chemical Warfare Service.

8.7 War Work: Ammonia Converters and Merseburg

The wartime Haber-Bosch facility at Oppau was "the last word on the subject," according to an American delegation from the Nitrate Division, Ordnance Department, U.S. Army, that visited Ludswigshafen-Oppau in June 1919. They were advised that "the German Government advanced something over 200,000,000 marks" towards its construction [42]. Another figure was 20 million pounds sterling. Carl Bosch, for his contribution to the nitrogen work and expansion of ammonia production, in addition to his skills as an organizer, was in 1916 appointed a director of BASF (and chairman in 1919).

Wartime developments had led to the introduction of a highly efficient, robust manufacturing facility. In 1913, the ammonia converters were 8 metres long. In 1915, Oppau began producing ammonia in ten 12-metre long converters, each one 75 tons in weight, and assembled from two 6-metre sections bolted together. The maximum size of forgings from solid ingots dictated the length of each section. Inner diameters of ammonia chambers were increased from an initial 500 to 800 mm. With the larger converters, pressures of 300 atmospheres were introduced. Outputs of converters increased from 20 tons a day in 1913 to 85 tons a day in 1915. Improvement of internal heat exchange closely followed the design of reactors devised by Rudolph Kneitsch for the BASF sulphuric acid contact process.

In order to maintain high production levels changes were made in the preparation of reactant gases. This included an improved process for preparing hydrogen from water gas, as developed by Wilhelm Wild during 1910–1912. It was introduced into the Oppau factory, in collaboration with Bosch, in 1915 [43]. Steam blown over red hot coke gave water gas, comprising equal volumes of hydrogen and carbon monoxide. The Bosch-Wild modification involved a further steam treatment at 350–450 °C over an iron-chromium catalyst. The result was a mixture of carbon dioxide and more hydrogen. This is called the "shift reaction."

$$H_2O + CO \rightarrow CO_2 + H_2 \quad \Delta H = -41\ \text{kJ mol}^{-1}$$

The reaction was also called the Bamag process after the engineering firm responsible for supplying the special equipment. The ammonia process steps were now modified, as follows: A nitrogen-hydrogen mixture was made up from producer gas and water gas. Sulphur compounds were removed, originally over activated carbon. The carbon dioxide formed in the shift reaction was absorbed, and subsequently recovered, by physical solution, or scrubbing, in water. Then, starting at 25 atmospheres pressure, the gas mixture was compressed in stages to 200 atmospheres, final traces of carbon monoxide were removed, and the hydrogen-nitrogen mixture was made up in the correct stoichiometric proportions, 75% hydrogen and 25% nitrogen.

Late in 1914, Oppau ceased the manufacture of ammonium sulphate fertilizer and produced instead ammonium nitrate for munitions manufacture, eventually at the rate of 60,000 tons a year. The Ammonia Laboratory continued to deal with design problems, including investigation of novel steel alloys. Bamag collaborated closely with Bosch in the development of ammonia oxidation plant [44]. By the end of the war, the 175-acre Oppau site was equipped with fifteen 12-metre ammonia converters, in addition to eight of the earlier type. Ten double-acting pumps continuously circulated the gas mixture. Gas engines drove all the compressors. There were 6000 workmen on the site, not counting prisoners of war brought in to undertake various manual tasks.

Until 1917, in the complete absence of Chile saltpetre, the production of nitrogen products relied on cyanamide, coal gas and coke works ammonia, and BASF synthetic ammonia. It was only then that the Haber-Bosch process came well to the fore with the opening of a new works on a green-field site in central Germany. Based on the success of Oppau, Bosch at the end of 1915 lobbied for the German government to fund construction of a new Haber-Bosch facility. In this he was successful. The site chosen was at Leuna, near Merseburg, on the River Saale, 20 km south-east of Halle, on the route of the railway connecting Berlin with Frankfurt. The location was in part chosen to ensure that the factory was out of the range of Allied aerial bombardment (the French had bombed Oppau on 27 May 1915, in what was the first ever strategic air raid on an industrial facility). Nearby were sources of water, brown coal, and gypsum (later used to make ammonium sulphate). Hydrogen was produced from water gas. Thirteen large boiler houses, for power and other utilities, each with a 120-metre tall chimney, extended almost the whole length of the factory. During this period Bosch had to deal with the AEG engineer Wichard von Möllendorff, then director of Kriegschemikalien AG, and subsequently head of the Reichsstickstoffwerke (August 1915–April 1916). There was another factor in the choice of location, in common with all major German nitrogen works constructed during World War I [45]. It was in Prussia, rather than Bavaria, which played a relatively small role in the war. The Merseburg facility, ultimately covering thirteen square kilometres, was later known as the Leunawerke (Figs. 8.1, 8.2, 8.3, 8.4, 8.5, 8.6 and 8.7).

An arrangement similar to that already in operation for the Reichsstickstoffwerke cyanamide factories was proposed by Karl Helfferich at the Reich Treasury. This would have meant government ownership of Merseburg, creating a state monopoly on nitrogen production. BASF, however, objected to this proposal. An agreement was finally reached on 10 April 1916 in which BASF accepted the cost of construction, with loans and subsidies from the Reich, that pledged to purchase the entire output. Construction at Merseburg, directed by Carl Krauch, commenced on 19 May 1916.

It was a critical time in the conduct of the war. The Germans had launched the long Verdun offensive against the French in February, leading to a series of attacks and counterattacks, with losses of around 70,000 or more men every month (by December the French managed to recover lost ground). On 1 July 1916, the British began their Somme attack, also with massive losses. The Germans were forced to move men and equipment from the Verdun sector (to face a Russian offensive). At the end of August, Paul von Hindenburg and Erich Ludendorff, following their successes since 1914 on the Eastern Front, were appointed joint chiefs of staff, replacing von Falkenhayn, whose Verdun offensive, suffering from the transfer of resources, and heavy casualties, was falling apart.

In August 1916, the Hindenburg programme for state controlled industrial reorganization, in which Carl Duisberg played a prominent role, was brought into operation. The industrialized war economy was conducted on an unprecedented scale, with centralized control of the production of armaments. It dictated urgent doubling and even tripling of war production [46, 47]. This required increased outputs of ammonia, cyanamide, nitric acid and ammonium nitrate, and toluene and other aromatics, as well as the war gases. Germany in 1916 produced 36 million shells, a fourfold increase on 1915.

Fig. 8.1 The BASF Merseburg factory, c. 1920. From a painting by Otto Bollhagen. Note the pipe and utility bridges, moving cranes, coal conveyors, extensive railway network, and gasometers. The sheer size of the facility symbolized the scale of the German wartime nitrogen programme from mid-1916. (BASF Unternehmensarchiv)

Fig. 8.2 Germany, including the provinces of Posen and Silesia, showing locations of operating Haber-Bosch BASF ammonia and calcium cyanamide factories, in 1917

Fig. 8.3 The BASF Merseburg factory, showing pipe bridges, and the gas plant, with the capacity for producing up to 80 million cubic feet every day. (Bild-Nr XIV/3 Haber, Archiv der Max-Planck-Gesellschaft, Berlin-Dahlem)

On August 18, the two groups of major German dye-making firms had created the IG (Interessengemeinschaft der deutschen Teerfarbenfabriken), which included Weiler-ter Meer and Griesheim-Elektron. Three areas excluded from their joint

Fig. 8.4 Machining a hollow forged ammonia converter at the Fried. Krupp AG works, Essen. The holes in the flanges enabled connections between vessels to be made with bolts and studs. (Edelstein Centre)

Fig. 8.5 Compressor room, BASF Merseburg. Several reciprocating compressors were required since there were technical restrictions on the volume of gas that each one could handle. (Edelstein Center)

interests were BASF synthetic ammonia, the Hoechst interest in acetylene chemistry, and the Griesheim-Elektron light metals business [48]. However, the merging of interests strengthened the financial standing of the chemical industry and enabled better working relationships, such as in the supply of BASF ammonia to the other partners (for production of nitric acid), and compliance with the needs of the Hindenburg programme.

On 15 September 1916, at the Battle of the Somme the British introduced the tank in an effort to clear the German trenches, something that gas warfare in 1915

Fig. 8.6 Ammonia synthesis, BASF Merseburg. (Harold Tongue, *The Design and Construction of High Pressure Chemical Plant*. London: Chapman & Hall, 1934, Fig. 221. IG Farben photograph)

had failed to achieve. A few days later, on September 20, Hindenburg's office wrote to the Reich Chancellor demanding that the Reich reduce the amount of nitrogen allocated to agriculture. This would release sufficient nitrogen to maintain the 20,000 tons per month required for preparing 14,000 tons of propellant powder (Treibpulver). The correspondence showed that monthly production of ammonia then amounted to: 8000 tons from coke ovens, 5000 tons from Oppau, and 4700 tons from the cyanamide factories. Included was a cyanamide facility at Waldshut (Lonzawerke), south-west Baden-Württemberg, with an annual capacity of 60,000 tons. No further increase in coke oven ammonia was anticipated, the planned Merseburg works would initially provide an additional 2500 tons, and ammonia output from cyanamide would be increased to 6500 tons. Hindenburg strongly recommended that the Reich Chancellor work with the Ministry of Agriculture, the KRA, and the Reich Treasury, to ensure the urgent reallocation of nitrogen products.

On the same day the nitrogen situation was discussed at a meeting in Berlin. The participants were Fritz Haber and Major Charles Garke (1860–1936), as representatives of the Prussian War Ministry; Nikodem Caro, representing the Reichsstickstoffwerke (managed by the Bayerische); Julius Bueb, commissioner

Fig. 8.7 Catalytic hydrogen plant at Merseburg. This process was based on the Bosch-Wild modification in which steam treatment of water gas at 350–450 °C over an iron-chromium catalyst converted carbon monoxide into carbon dioxide and increased the hydrogen content (the "shift reaction"). (Harold Tongue, *The Design and Construction of High Pressure Chemical Plant*. London: Chapman & Hall, 1934, Fig. 73. IG Farben photograph)

for the nitrogen economy; Eberhard Ramm, of the Prussian Ministry of Agriculture; and representatives of the Treasury, the War Food Office, the KRA, the Reich Interior Ministry, and the Bavarian Interior Ministry. The meeting was charged with a sense of considerable urgency. After allowing for additional needs of the military, shortfalls in the agricultural sector were anticipated for the autumn of 1917 and spring of 1918, unless (and probably even if) the new capacities became available. Agriculture required 50,000 tons of nitrogen per year, whereas only 30,000 tons were possible under the new arrangement. There was, overall, the need for an additional 6000 tons per month and, it was observed, "This cannot be achieved by *one* process alone" [49]. Suggestions for making up the shortfall included increasing cyanamide capacity by 1500 tons per month through the new carbide and power plant near Trostberg, and by 800 tons per month through expanding Chorzów from 25,000 to 75,000 tons per year. Haber anticipated that a further 2000 tons a month would become available from BASF by the end of 1917. This had to come from Merseburg, since, as the representatives of the Royal Bavarian Government pointed out, additional ammonia capacity could not be achieved at Oppau. They did however encourage the increased cyanamide capacity at Trostberg.

The figures given in the letter from Hindenburg's office demonstrate the extent to which even late in 1916 the Haber-Bosch process was by no means an overriding factor in the production of essential nitrogen products required for both the military and agriculture. Cyanamide-derived ammonia was still important, as demonstrated by the new carbide and hydro-electric facilities close to Trostberg and the threefold expansion of production at Chorzów. However preference was increasingly being given to the Haber-Bosch process. This, as it was argued, required less labour for construction and operation, power requirements were considerably lower than for the cyanamide process, a critical aspect when there was reliance on coal-fired power stations, and operation was less complicated than for cyanamide, insofar as working up the product was involved. Cyanamide production, while it used far less sophisticated equipment, required considerable manual work, including the sizing of lime and coke with sledge hammers, and breaking up by hand of carbide pigs and cyanamide ingots (until mechanical hammers were introduced). Coke oven ammonia was assigned, as ammonium sulphate, to the farmers.

Though Merseburg suited the demands of the Hindenburg programme, there were delays in construction caused by the general war situation, including shortages of coal and an unusually cold winter. The severity of the weather affected the whole chemical industry, as well as the population. Thus, on 7 January 1917, with the temperature at −21 °C, the Bayer Flittard TNT plant was completely wrecked by an explosion when a foreman attempted to free a frozen pipe from ice [50]. The situation was aggravated by severe shortages of food, particularly potatoes and meat, arising from a poor harvest. There had already been extensive food riots in Prussia early in 1916. The turnip became a substitute for other staple foods during what was long remembered as the Turnip Winter of 1916–1917.

On the evening of 27 April 1917, the Merseburg facility first produced ammonia. The next day it dispatched a train of tank wagons containing ammonia to Oppau (where it was converted into nitric acid, for production of ammonium nitrate and nitro compounds). Chalked on the tank of the first wagon were the words "Death to the French" (Fig. 8.8). That was just three weeks after the United States declared war on Germany. And just under three months before the British offensive in Flanders began.

The original planned ammonia capacity of what was referred to as Merseburg I was 30,000 tons per year, soon increased to 75,000 tons, then, in December 1916, in line with the demands of the Hindenburg programme to 160,000 tons (Merseburg II, but not in service at the Armistice) [51]. An Imperial Trade inspection of the Merseburg facility on 15 December 1917 noted that: "The ammonia water is 20 to 22 per cent. It is of purity suitable for nitric acid production" [52]. In July 1918, plans were drawn up for further expansion, to 240,000 tons per year (Merseburg III).

German ammonium nitrate production in 1917, described by one source as "sulphate of ammonia (really ammonium nitrate)," was 700,000 tons, equal to that of Great Britain and the United States combined [53]. In response to the severe food shortages experienced in Germany, and considerable social unrest, in January 1918 the Prussian Minister of Trade and Business was advised that some Haber-Bosch synthetic ammonia could be diverted to agriculture, as a result of expanding output at both Merseburg and Oppau. Oppau alone had increased production of ammonium nitrate from approximately 300,000 tons in 1916 to 500,000 tons in 1917, mainly relying on ammonia from Merseburg [54]. Expansion in Haber-Bosch ammonia enabled additional cyanamide to be released for agriculture. However, shortages of cyanamide—since carbide was also required for other

Fig. 8.8 The first trainload of ammonia awaiting departure from the BASF Merseburg works on 28 April 1917, bound for Oppau. Written by workers in chalk on the tank of the first wagon are the date, "Good Luck," and "Death to the French." Tree branches serve as decoration. (LHSA, MER, 1 525, Fotosammlung, Nr. 786. Landeshauptarchiv Sachsen-Anhalt, Abteilung Merseburg)

purposes and deliveries of coal were hampered by the overloaded railway network—and other fertilizers contributed to an erosion of confidence among Germans in the war effort [55].

Hartley's February 1919 report of visits to chemical factories in the occupied British and French zones engaged in the production of munitions indicates that in 1918 Germany produced 650 tons of synthetic ammonia per day: 250 tons at Oppau, and 400 tons at Merseburg (Fig. 8.9). Significantly, and again despite later views to the contrary, Hartley noted that [56]:

> No arrangements appear to have been made prior to the outbreak of war to utilize the resources of any of the dye factories for war purposes, and on mobilization their chemists were called up for military service. After the battle of the Marne the Government realized the need for expanding the output of explosives and most of the chemical works were producing small quantities by the end of 1914. The demands made on them increased during 1915, but it was not until 1916 that plant [at Merseburg] was laid down to assist in the enormous production of explosives required by the Hindenburg plan. Most of the big extensions of the synthetic ammonia and of the nitric and sulphuric acid plants date from this time, many chemists being released from the army and the scientific staff of some of the works being augmented.

From April 1918, Ludendorff, Supreme Commander of the German Army in Flanders, failed to hold back Allied forces along the coasts of Belgium and northern France. Apart from German military failures, particularly in early August, followed

Ammonia (metric tons NH_3 per day).

	1914	1918
Oppau	25	250
Merseburg	(1)	400
Total	25	650

1 Nil.

Nitric acid (metric tons 100 per cent acid per day).

	1914	1918
Leverkusen	56	180
Hochst	150	375
Oppau	?	100
Ludwigshafen	(?) 40	40
Weiler ter Meor	12	24
Total	258	719

Oppau has the power to produce now 500 tons HNO_3 daily, still retaining sufficient ammonia to supply the output at Hochst.

Fig. 8.9 Tables of German production of ammonia and nitric acid, 1914 and 1918. (*Dyestuffs: Hearings before the Committee on Ways and Means, House of Representatives, on H.R. 2706 and H.R. 6495, June 18–20 and July 14–18, 1919 [66th Cong., 1st Sess., 1919]*. Washington, DC: Government Printing Office, 1919, p. 211)

by the Allied push in September, there were severe shortages of food resulting in considerable part from the diversion of nitrogen products away from agriculture, as demanded by Hindenburg in September 1916, and the general slowdown in manufacture. The outcome of the war of attrition was increasingly dependent on developments far away from the trenches and battlefields. These included the British blockade; shortages of horses, farm labour, and coal; worn out machinery; and transport difficulties, particularly on the railways. The burden of the war was also taking a political toll, with strikes in factories, social unrest, riots, and other domestic problems. Vladimir Lenin's arrival at Finland Station, Petrograd (Saint Petersburg until 1914) in early April 1917, after a train journey from Zurich through Germany—the Germans ensured safe passage—across the Baltic Sea, and by train again via Sweden and Finland, was meant to hasten Russian withdrawal from the war, but the provisional government of Alexander Kerensky held out.

America's entry into the war, also in April 1917, forced the Germans into a series of major offences intended to bring the war to an end before the Allies could make full use of their expanded numerical advantage in manpower and resources. German chemist Hermann Staudinger (1881–1965), then at ETH, Zurich, predicted, on the basis of his study of available raw materials—including coal and iron ore, between 1860 and 1912—that Germany would be defeated. Staudinger opined that modern war was a question of technology rather than manpower [57]. However much German manufacturing technology had advanced, it would inevitably face a crisis brought on by severe shortages of essential raw materials. Now that the initiative had passed to the Allies, the defeat of Germany was inevitable.

Notwithstanding military failure, the shortage of food, and poor organization of factories and railways, the German nitrogen industry had expanded tremendously. In the final stages of the war, German calcium cyanamide production capacity, though not output, was close to an annual rate of 600,000 tons: at Reichsstickstoffwerke Piesteritz, operated by the Bayerische, 175,000 tons; at Chorzów, likewise operated by the Bayerische, 150,000 tons; at the Hoechst-controlled Knapsack, enlarged considerably during the war, 140,000 tons; at the Bayerische Trostberg works, 75,000 tons; and at Waldshut (Lonzawerke), unchanged at 60,000 tons [58]. Linde nitrogen liquefaction plant was installed at four sites—Piesteritz, Chorzów, Knapsack, and Waldshut. Immediate post-war expansion near the Bayerische Trostberg site included opening of the new Margarethenberg hydro-electric power station, close to Hirten (14.5 km upstream of the Schalchen carbide facility); started during the war, it was completed in 1919, and in 1924 named the Caro-Werk (Power station III). Also in 1919, a new carbide works began operating at nearby Hart (Figs. 8.10 and 8.11). Schalchen ceased manufacture of carbide in 1925 (Table 8.2).

From 1919 the state treasury administered the various industries that the Reich had sponsored during the war, particularly in the now important civilian areas of nitrogen, aluminium and energy production. Nitrogen meant effectively calcium cyanamide, and with it calcium carbide. In May the government established the Stickstoff-Syndikat, or Stickstoff-GmbH, a cartel whose original purpose was to support, through regulation of prices, the state's production of cyanamide. On 24 February 1920, the Piesteritz site became a state owned corporation, Mitteldeutsche Stickstoffe-Werk AG. This was followed on May 20 with formation of Bayerische Kraftwerke AG, incorporating the planned carbide factory at Hart and the nearby hydro-electric power station just to the north of the Bayerische Trostberg facilities. On 7 March 1923, the nitrogen, aluminium and energy concerns were merged into a new state corporation, the holding company Vereinigte Industrie-Unternehmungen AG (VIAG) of Berlin.

Cyanamide manufacture was cut back considerably in the early post-war period due to shortages of coal, which was exported in settlement of reparations claims. Once the cyanamide industry recovered the estimated annual output remained consistent, at around a quarter of a million metric tons, mainly from Piesteritz, Trostberg, and Knapsack. Chorzów was assigned to Poland, under Article 256 of the Treaty of Versailles [59]. Increasing demand, in part due to the loss of Chorzów, necessitated the completion of new carbide facilities at Piesteritz (1923), and at Bayerische Kraftwerke AG (Hart II, 1922–1924).

Poland also received over half the coal production facilities, and all the sulphuric acid plants, in Upper Silesia [60].

In 1920, the estimated potential in Germany for manufacture of fixed nitrogen was 500,000 tons, all as fertilizer: 60% from the Haber-Bosch process, and 20% each from the gas and coke works and cyanamide factories. From 1922, Merseburg was managed by Ammoniakwerke Merseburg-Oppau GmbH, a subsidiary of BASF. By the following year annual production of fixed nitrogen amounted to

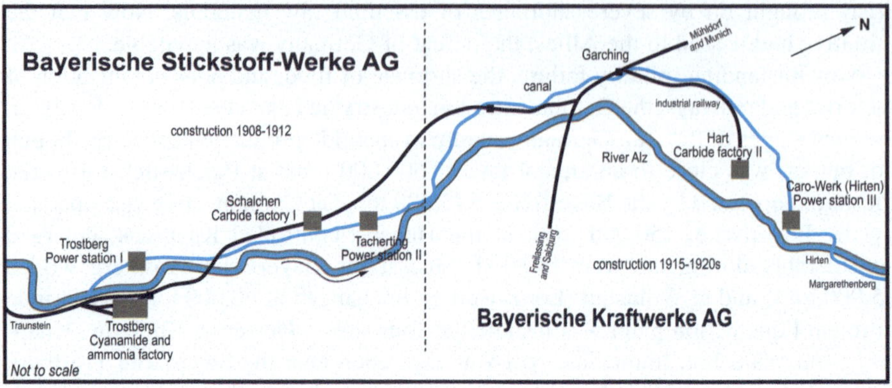

Fig. 8.10 Diagrammatic map showing the River Alz and canal excavated from Trostberg, Upper Bavaria, in the direction of Margarethenberg, showing locations of: the Trostberg factory, for calcium cyanamide and ammonia, the Schalchen carbide factory, and the Trostberg and Tacherting power stations of the Bayerische Stickstoff-Werke AG; and the Hart carbide factory (later known as Hart II), and the Caro-Werk (Margarethenberg/Hirten) power station of the Bayerische Kraftwerke AG. Railway connections are shown, including the 5-km long industrial railway from the Hart II carbide factory to Garching. The railway from Mühldorf to Freilassing, near Salzburg, was completed in December 1908. The community of Garching expanded considerably with the opening of the new facilities from 1915 (Based on a map included in an album of photographs of the Bayerische Stickstoff-Werke AG and Bayerische Kraftwerke AG dating from around 1925, held with the Adolph and Albert Frank Collection of papers and correspondence, Leo Baeck Institute, New York, AR 7176/MF772, series I: personal papers, 1857–1956, box 1, folder 13)

Fig. 8.11 The Bayerische Kraftwerke AG hydro-electric power station near Margarethenberg, close to the Hart carbide works. It was renamed the Caro-Werk in 1924, and was also known as power station III. (Evonik Industries AG, Corporate Archives, Hanau/Marl)

Table 8.2 The power stations of the Bayerische Stickstoff-Werke AG, and the Bayerische Kraftwerke AG, early 1920s

	Location	Construction began	Placed in operation	Owner
Power station I	Trostberg	1909	1911	Bayerische Stickstoff-Werke AG
Power station II	Tacherting	1909	1911	Bayerische Stickstoff-Werke AG
Power station III	Margarethenberg (Hirten)	1915	1919	Bayerische Kraftwerke AG (Caro Werk)

The Bayerische Kraftwerke AG was a separate entity, based on wartime plants, established in 1920, that included the facilities north of Tacherting. It had been sponsored by the Reich during the war; it worked closely with Bayerische Stickstoff-Werke

220,000 tons. The factory employed 11,000 labourers, 2500 craftsmen (over twice the number employed at Oppau), and 150 chemists and engineers. By 1930, there were 15,000 labourers, and 3000 technical and clerical employees. A quarter-of-a-million-tons of ammonium sulphate, one-quarter of the annual output, could be accommodated in a silo (Figs. 8.12 and 8.13). Every day, 12,000 tons of brown coal, delivered by the works railway, was consumed in producing 460,000 horsepower in 95 steam boilers. These drove 45 gas engines, dedicated to compressing the synthesis gas mixtures, made up from water gas (for hydrogen) and producer gas (for nitrogen) to 200 atmospheres. A Linde liquefaction plant supplied nitrogen to adjust the composition of the mixture so as to maintain the required proportions. The site was also equipped with its own electricity generating power station. Recently introduced, for the first time, was large-scale oxidation of ammonia under pressure of 5 atmospheres in the nitric acid process. Nitric acid was reacted with calcium carbonate to give calcium nitrate. The nearby "garden city Neu-Rössen" provided homes for workers [61].

8.8 The United States

In the United States, interest in the synthetic ammonia and other nitrogen fixation processes, as well as in oxidation of ammonia to nitric acid, was prioritized from 1916, following passage on June 3 of the National Defence Act. President Woodrow Wilson's remit was that investigations should be undertaken to "determine the best, cheapest and most viable means for the production of nitrates and other products for munitions of war, and useful in the manufacture of fertilizers and other products." Twenty million dollars were allocated to nitrogen fixation and oxidation of ammonia, under section 124, "Nitrate Supply." On June 19, Wilson formally requested that the National Academy of Science create the National Research Foundation (later the National Research Council) to encourage scientific

Fig. 8.12 Interior of BASF/IG Farben silo for storage of ammonium sulphate. The surrounding images indicate the advantages of various fertilizers. (*Erzeugnisse unserer Arbeit*. Frankfurt am Main: I.G. Farbenindustrie, 1938, opposite p. 157)

Fig. 8.13 BASF advertisement for ammonium sulphate fertilizer, c. 1918–1920. (BASF Unternehmensarchiv)

research in industry, specifically for purposes of strengthening the national defence. The formation of a Committee on Nitric Acid Supply was approved by the Secretary of War, Newton D. Baker [62]. Apart from increasing the import of Chilean nitrate, greater nitric acid capacity could be achieved only with the new nitrogen capture technologies. While there was great interest in synthetic ammonia, the Americans had little to draw on, excepting numerous BASF patents registered at the United States Patent Office, and the far from perfect General Chemical process.

The Bureau of Soils of the Department of Agriculture investigated high-pressure ammonia synthesis, as well as applications of the cyanamide process (Fig. 8.14).

Fig. 8.14 Barrels of military grade ammonium nitrate for mixing with trinitrotoluene to produce the explosive amatol, which contained up to 80% of the nitrate. Here seen stored at "US Nitrate Plant No. 2," Muscle Shoals, Alabama, a calcium cyanamide facility completed just as World War I came to an end. Though the plant did not contribute to the war effort, it demonstrated that calcium cyanamide was an important source of ammonia suited to the production of munitions, as widely adopted in Germany and France. Ammonia was released from the calcium cyanamide in autoclaves based on those developed at the Bayerische Stickstoff-Werke. The photograph was taken on 17 July 1919. (Library of Congress, Prints & Photographs Division, HAER [Historic American Engineering Record] ALA, 17-MUSHO, 1–70)

A catalyst test plant was installed in 1917 at Arlington Experimental Farms, Virginia [63]. However, little progress was made in developing suitable catalysts and equipment for the ammonia synthesis. The US Bureau of Mines from 1916 investigated conditions for the catalytic oxidation of ammonia. The American Cyanamid affiliate Ammo-Phos, at Warners, New Jersey, commenced the manufacture of nitric acid from cyanamide-derived ammonia, based on the autoclaves imported from Germany at the start of the war in 1914. (Obtaining the platinum catalyst, mainly from Russia, was less of a problem than for the Germans.) This contributed considerably to the financial position of Ammo-Phos, which until then had struggled to find a good market for its novel fertilizer [64].

Despite the emphasis on synthetic ammonia and its oxidation, American experts were divided in their opinions when it came to nitrogen capture. It was not simply a question of synthetic ammonia versus cyanamide. Thus, for example, there was the cyanide process of John E. Bucher of Brown University, Providence, Rhode Island, for which he filed various patents during 1914–1917. Bucher recognized the important role of iron in catalysing the reaction. He also undertook research into the synthesis of urea. Bucher described his cyanide process to the American Institute of Chemical Engineers in January 1917 [65–67].

For large-scale production, Bucher proposed to heat briquettes of soda ash, coke and iron in a vertical retort at around 1000 °C, under which conditions atmospheric nitrogen was captured as sodium cyanide. Ammonia could then be liberated from the cyanide. Bucher's process had been taken up by the Nitrogen Products Company, of Rhode Island. A pilot plant was installed at Greene, Rhode Island. A young chemist from MIT who joined Nitrogen Products in 1916 was Roger Williams (1890–1978). In 1918, he joined Du Pont, where he was to become the principal research leader in high-pressure chemistry, starting with synthetic ammonia, in the mid-1920s.

8.8.1 Three Opinions: Charles Parsons, the National Research Foundation, and the Ordnance Department

In 1916, Charles Lathrop Parsons (1867–1954), a Cornell graduate and chief chemist at the Bureau of Mines (1911–1919), was transferred to the War Department's Ordnance Department, as chief engineer. With fellow engineer Eysten Berg he visited Europe in October 1916 to investigate nitrogen processes, both for nitrogen fixation and the production of nitric acid. They were instructed not to enter Germany. Together they visited cyanamide and nitric acid facilities in France and Italy; Parsons then went alone to England, Norway, and Sweden to study cyanamide and electric arc processes. Parsons summarized his findings in two documents prepared for the Ordnance Department, a preliminary report of 17 January 1917 and a final report of 30 April 1917. His recommendations were for development of a modified Haber-Bosch process (as commenced by General Chemical) and, following a favourable report from the prominent chemist Charles Holmes Herty (1867–1938)—recently retired president of the American Chemical Society (1915–1916) and now editor of its *Journal of Industrial and Engineering Chemistry*—the Bucher cyanide process of the Nitrogen Products Company. Parsons expressed far greater hope in the latter, considering it decidedly superior, from a commercial point of view, to the Haber-Bosch process. Parsons had no time for the cyanamide process, which was far too expensive to work, even though it was the only viable process at the time [68].

Parsons's recommendations were somewhat at variance with those of the National Research Foundation's committee. The committee produced three reports, dated 2 June 1916 (preliminary), 6 January 1917 (summary), and 29 January 1917 (final) [69]. Its main recommendation was that the government support the production of hydro-electric power for fixation of nitrogen, and make the electricity available at preferential rates to companies engaged in this activity. The committee, unlike Parsons, clearly preferred the cyanamide process.

Early in 1917, the Chief of Ordnance of the United States Army, General William Crozier (1855–1942), created a Nitrate Division within the Ordnance

Department, directed from May 28 by Colonel John Warren Joyes. Now that the United States had entered the war the nitrogen situation had become a matter of some considerable urgency.

As a result of the quite different conclusions reached by Parsons and the National Research Foundation's committee, it was left to an Ordnance Department committee to decide on the best course of action. The initial emphasis of the committee was on: oxidation of ammonia; the General Chemical Company's synthetic ammonia process; the Bucher cyanide process; the use of by-product coke ovens for supply of ammonia and toluol (toluene); and maintaining a reserve of half a million tons of Chile saltpetre. There was no mention of calcium cyanamide.[3]

Joyes and colleagues examined the experimental synthetic ammonia apparatus of General Chemical in mid-August 1917. The process, based on the design of Frederick de Jahn, was experiencing severe technical difficulties. Meantime, the cyanide process was eliminated, at least from the manufacture of explosives. It received support for producing cyanide that could be adapted to use in war gases [70]. A plant with an intended daily capacity of 10 tons of cyanide, known as "US Chemical Plant No. 4," was laid down at Saltville, Virginia, in association with Mathieson Alkali Works. It began operating on 4 September 1918. Many difficulties were encountered, and after producing no more than a few thousand pounds of cyanide it was closed down in December 1918 [71].[4]

In the autumn of 1917, the Ordnance Department's committee authorized construction of modified Haber-Bosch and cyanamide factories. The government instructed General Chemical to build an ammonia plant at Sheffield, Alabama. This was based on the ailing process that Joyes and colleagues investigated. Partly completed in September 1918, what was called the "US Nitrate Plant No. 1" never became fully operational, and was abandoned; there were problems with the quality of the converters as well as with the catalyst. Wartime research on the process was conducted at Laurel Hill, Long Island; Arlington Experimental Farms, Virginia; and, from 1918, at Sheffield. Major Harry Curtis served at Arlington, where a high-pressure ammonia apparatus, working at 100 atmospheres, was put together, and at Sheffield.

Meantime, in September 1917, Crozier met in Washington with Frank Washburn, whose recommendations, based on American Cyanamid's offerings, had earlier been rejected. There had been suggestions that he had much to gain financially from a government contract, and that this was his main interest, particularly after Washburn claimed that the Haber-Bosch process was useless. Crozier realized, however, that American Cyanamid was the only firm with the requisite level of experience to bring about the rapid, large-scale production of nitrogen

[3]During the war, the National Research Foundation functioned as the Department of Science and Research of the Council of National Defense as well as the Science and Research Division of the United States Army Signal Corps.

[4]The Bucher process was investigated at the Solvay Process Company's laboratories, until around 1924, with particular reference to formation of briquettes.

products required for munitions. Contracts were now drawn up with American Cyanamid for the manufacture of calcium cyanamide, its conversion into ammonia, and in turn oxidation to nitric acid. Efforts were concentrated on a new cyanamide facility at Muscle Shoals, near Sheffield, "US Nitrate Plant No. 2," to be erected for the government by Air Nitrates Corporation, a subsidiary of American Cyanamid. The intention was to convert ammonia derived from cyanamide into ammonium nitrate, such that by 1919, producing ammonium nitrate at the rate of 110,000 tons per year, it would account for 13% of all Allied explosives [72–74].

Apart from nitrogen products, major shortages faced by the United States included of potash, previously imported from Germany. Solvay's Louis Jones and colleagues around 1910 had investigated Searles Lake in California as a possible source and, following further investigations after the outbreak of war, a substantial facility, based on collaboration with the Borax Company, extracted the mineral from Searles Lake, from 1917 until early 1921 [75].

8.8.2 Muscle Shoals

According to Williams Haynes, American Cyanamid had been called upon to undertake cyanamide manufacture at Muscle Shoals after difficulties were experienced with the synthetic ammonia factory at Sheffield. General William Crozier, apparently, "called Frank Washburn to Washington and said: 'Unless the Cyanamid Company can and will provide the plants and produce the materials within twelve months, the plans for placing a great fighting force on the battlefields in France for the spring offensive of 1919 cannot be carried out'."

For this enormous factory, encompassing 348 acres, American Cyanamid placed Walter Landis in charge of the design. Engineering work began on 17 December 1917. Ground was broken on 3 February 1918. The government constructed a "coal-powered electric generating plant to provide the electricity for the carbide furnaces." The logistics were nightmarish, including temporary, segregated, housing for a vast army of workers. Twenty-five thousand men had been brought into the four-square mile area. "And all the time the rain poured down. . .After the influenza epidemic struck, man-power became critical" [76].

Despite labour and other difficulties, "US Nitrate Plant No. 2" was completed just over eight months later, as the war came to an end. "The compressors in the [Linde] liquid air unit [were] started up on October 23 and three days later the electric current was turned on the carbide furnaces." After break-in trials (with nitric acid produced on November 19 and ammonium nitrate on November 25), and some work early in 1919, the facility was shut down through lack of demand (Figs. 8.15 and 8.16). Plans were drawn up for two other wartime cyanamide factories, at Toledo, Ohio, and Cincinnati, Ohio, on which, according to Haynes, work was started [77]. (In 1923, Landis, for his services to American Cyanamid, was appointed vice president of the corporation, a post he would occupy until his death in 1944.)

Fig. 8.15 Banks of cyanamide ovens, "US Nitrate Plant No. 2," Muscle Shoals, Alabama, 2 March 1919. (Library of Congress, Prints & Photographs Division, HAER [Historic American Engineering Record] ALA, 17-MUSHO, 1–65)

Also from 1917, the Nitrogen Products Company at La Grande, Washington, worked an electric arc oven developed in 1912 by the Norwegian inventor Franz Henrik Aubert Wielgolaski. He filed several patents, one in the United States jointly with Schönherr. Nitrogen Products was interested in producing sodium nitrite for dye manufacture, but the process was not economically viable. The factory was closed during 1921–1922, reopened in 1923, and abandoned after a few years, as was a facility at Lake Buntzen, British Columbia. Interest increasingly moved towards the high-pressure ammonia synthesis, particularly after more information became available about the BASF process.

American visitors to Germany in 1919, technical and non-technical, generally accompanied British and European experts on Allied missions, which helped to counter German complaints that they were engaged in industrial espionage [78]. Such was the case with the British-led mission in February 1919. An American account of the state of the German chemical industry supplemented those of the British and French. Major Theodore W. Sill, attached to the US Chemical Warfare Service, and though mainly interested in technical organic chemistry, in his report of the visit to Germany, declared: "At Oppau we saw what is probably the most phenomenal scientific development up to date, namely, the practical realization on an operating basis of the Haber process for ammonia production." Like his British colleagues, Sill was puzzled by the fact that "Haber, to whom Germany owes so much of her development in chemical products in warfare, had never

Fig. 8.16 Ammonia oxidation building, with ammonia spheres in the foreground, "US Nitrate Plant No. 2," Muscle Shoals, Alabama, 1919. (Library of Congress, Prints & Photographs Division, HAER ALA, 17-MUSHO, 1–4)

At Oppau we saw what is probably the most phenomenal scientific development up to date, namely, the practical realization on an operating basis of the Haber process for ammonia production. The buildings are all quite new and well constructed, and the vast amount of detail has been studiously and carefully worked out on a practical operating basis, producing upward of 100,000 tons of ammonia per year. This plant was a large factor in enabling Germany to stay in the war as long as she did, by means of producing large quantities of nitrates. The Germans have also another plant, a duplicate of this, which they are operating in the unoccupied area of Germany, so that it is really a great practical possibility at the present time.

Incidentally, it came to our attention that Haber, to whom Germany owes so much of her development in chemical products in warfare, had never attained a higher rank than captain in the German chemical warfare service, despite the responsibility and immensity of his job.

Fig. 8.17 Extract from the report of a visit to Oppau in February 1919 by Major Theodore W. Sill, US Chemical Warfare Service. (*Dyestuffs: Hearings before the Committee on Ways and Means, House of Representatives, on H.R. 2706 and H.R. 6495, June 18–20 and July 14–18, 1919 [66th Cong., 1st Sess., 1919]*. Washington, DC: Government Printing Office, 1919, p. 175)

attained a higher rank than captain in German chemical warfare service, despite the responsibility and immensity of his job" (Fig. 8.17) [79]. Sill, perhaps, had overlooked the fact that Haber's ethnic origins precluded him from attaining a higher rank in the Kaiser's army.

The vast Muscle Shoals cyanamide works, completed just as the war came to an end, remained a symbol of the uncertainties facing the immediate post-war

American nitrogen industry. Charles L. Parsons, in giving testimony, as one of the bidders for the Muscle Shoals facility, before the 67th US Congress on 8 May 1922 apparently believed that the end had come for the cyanamide industry [80].

> The cyanamide process is in a state of obsolescence. The plants in France, in Italy, in Dalmatia, in Silesia, in Norway, in Sweden are closed down. The plant in Canada is not producing cyanamide for fertilizer purposes...The plant is running only on a very small capacity. In the main the Cyanamide process is essentially already obsolescent.

He was, however, mistaken. New cyanamide facilities were opening in Europe and the Far East, while American Cyanamid would ensure the future of cyanamide in the United States through successful diversification into novel products. Maybe Parsons was downplaying cyanamide in order to reduce the value of the Muscle Shoals facility for his own benefit and those that he represented. In any case, in view of his earlier negative pronouncements, it is not clear what future plans they had in mind, apart from acquiring modern factory buildings and equipment, including an ammonia oxidation plant. More significantly, however, American investigators around 1920 were in general interested in acquiring, or mastering, the high-pressure ammonia synthesis.

8.9 Summary

After war broke out in 1914, and especially after the loss of Chilean nitrate, Germany at first relied on cyanamide for the production of ammonia, then increasingly on the Haber-Bosch process. The catalytic oxidation of ammonia to nitric acid was developed, as was a novel process for producing concentrated acid. By the last phase of the war, the Haber-Bosch process had become critical to the supply of nitrogen products. In contrast, the Allied nations continued to import and rely on Chilean nitrate; two million tons in 1915, and around three million tons in each of 1916 and 1917. By 1916, however, U-boats were causing considerable losses among Allied merchant ships, raising concerns over the reliability of Chilean supplies, quite apart from Scandinavian nitrogen products. One in four merchant ships that left Britain never returned. As a result, synthetic nitrogen processes attracted considerable attention outside of Germany. The French invested in cyanamide, with considerable success. As for the electric arc process, apart from in Norway and, for a time, the French Soulom plant, it was hardly viable.

The novel high-pressure ammonia technology was certainly realized to be a key to a new industrial sector, if not the key to a new industry, both in war and, especially, in peace. Therein lay its great significance. From 1916, considerable efforts among the Allied nations were invested in research aimed at replicating the Haber-Bosch process. The challenges were enormous, which alone attested to the level of sophistication achieved by BASF, which jealously guarded its hard-won advantage. No other major area of Allied wartime research was bedeviled by so many drawbacks. High pressure synthetic ammonia production would not be realized outside Germany until the 1920s.

References

1. Stoltzenberg D (2004) Fritz Haber: chemist, Nobel laureate, German, Jew. Chemical Heritage Press, Philadelphia, p 131.
2. Bäumler E (English version) (1968) A century of chemistry [Hoechst]. Econ Verlag, Düsseldorf, pp 62, 64.
3. Morris PJT (1982) The development of acetylene chemistry and synthetic rubber by I.G. Farbenindustrie Aktiengesellschaft: 1926–1945. DPhil thesis, University of Oxford, pp 85–86.
4. Lunge G (1916) Handbuch der Schwefelsäurefabrikation und ihrer Nebenzweige, Band I. Friedr. Vieweg & Sohn, Braunschweig, p 213.
5. Partington JR (1921) Ammonia oxidation at Höchst. J Soc Chem Ind 40(10)(31 May):185R–187R.
6. Allmand AJ, Williams ER (1919) Some chemical plants in the Cologne area. Part 1. J Soc Chem Ind 38(15)(15 August):285R–287R.
7. Bäumler E (English version) (1968) A century of chemistry [Hoechst]. Econ Verlag, Düsseldorf, p 68.
8. Evans EV (1920) The chemical industry of Germany. J Soc Chem Ind 39(3)(16 February):47-R–50R, on 49R.
9. Hartley H (1919) Report of the British mission appointed to visit enemy chemical factories in the occupied zone engaged in the production of munitions of war. In: Dyestuffs: hearings before the Committee on Ways and Means, House of Representatives, on H.R. 2706 and H.R. 6495, June 18–20 and July 14–18, 1919 [66th Cong., 1st Sess., 1919]. Committee on Ways and Means. Government Printing Office, Washington, DC, pp 210–214, on 211.
10. Lindner SH (2008) Inside IG Farben: Hoechst during the Third Reich. Cambridge University Press, Cambridge, pp 290–294.
11. Hunt LB (1958) The ammonia oxidation process for nitric acid manufacture. Platin Met Rev 2 (4):129–134, on 133.
12. Johnson JA (2003) The power of synthesis, 1900–1925. In: Abelshauser W, von Hippel W, Johnson JA, Stokes RG, German industry and global enterprise. BASF: the history of a company. Cambridge University Press, New York, pp 115–206.
13. Aftalion F (1991) A history of the international chemical industry (trans. Benfey OT). University of Pennsylvania Press, Philadelphia, pp 120–121.
14. Macnab W (1922) Liverpool section. The Hurter Memorial Leture. Some achievements of chemical industry during the war in this country and in France. J Soc Chem Ind 41(23)(15 December):353T–361T.
15. News and notes. France. The nitrogen industry (1921) J Soc Chem Ind 40(10)(31 May):193R.
16. Macnab W (1922) Liverpool section. The Hurter Memorial Leture. Some achievements of chemical industry during the war in this country and in France. J Soc Chem Ind 41(23) (15 December):353T–361T.
17. Italian decrees related to agriculture (1917) Commerce Reports (US Department of Commerce), no. 277 (26 November), p 778.
18. Pattison M (1983) Scientists, inventors and the military in Britain, 1915–19: the Munitions Inventions Department. Soc Stud Sci 13(4):521–568.
19. Nitrates (1913–1919) (1920) J Soc Chem Ind 39(23)(15 December):417R.
20. Van der Kloot W (2014) Lord Justice of Appeal John Fletcher Moulton and explosives production in World War I: 'The mathematical mind triumphant'. Notes Rec R Soc London 58(2)(20 June):171–186.
21. Pearce W (1917) Ministry of Munitions. Hansard HC Deb (28 June), 95, col. 605–col. 606.
22. Pratt JD (1948) Victor Lefebure, 1891–1949. J Chem Soc, pp 394–395.
23. Lefebure V (1923) The riddle of the Rhine: chemical strategy in peace and war. EP Dutton & Co., New York.

24. Lefebure V (1921) The riddle of the Rhine: chemical strategy in peace and war. Collins Sons & Co., London.
25. Steen K (2014) The American synthetic organic chemicals industry: war and politics, 1910–1930. The University of North Carolina Press, Chapel Hill, p 347, n. 53.
26. Lefebure V (1923) The riddle of the Rhine: chemical strategy in peace and war. EP Dutton & Co., New York, p 24.
27. Hartley H (1919) Report of the British mission appointed to visit enemy chemical factories in the occupied zone engaged in the production of munitions of war. In: Dyestuffs: hearings before the Committee on Ways and Means, House of Representatives, on H.R. 2706 and H.R. 6495, June 18–20 and July 14–18, 1919 [66th Cong., 1st Sess., 1919]. Committee on Ways and Means. Government Printing Office, Washington, DC, pp 210–214, on 214.
28. MacLeod R, Johnson JA (2006) The war the victors lost: the dilemmas of chemical disarmament, 1919–1926. In: MacLeod R, Johnson JA (eds), Frontline and factory: comparative perspectives on the chemical industry at war, 1914–1924. Springer, Dordrecht, pp 221–245.
29. Observations by 'Eye-Witness' (1923) J Soc Chem Ind 42(7)(16 February):143 (Chem Ind London).
30. Levinstein H (1923) The progress of the British chemical industry since 1914. J Soc Chem Ind (42)(11)(16 March):259–262, on 259 (Chem Ind London).
31. Lefebure V (1923) The riddle of the Rhine: chemical strategy in peace and war. EP Dutton & Co., New York, p 205.
32. Moulton JF (1919). Science and war. The Rede Lecture. Cambridge University Press, Cambridge, pp 11–12.
33. Hahn O (1960) Zur Erinnerung an die Haber-Gedächtnisfeier. Mitteilungen aus der Max-Planck-Gesellschaft 1:3–13.
34. Johnson JA (2004). The Kaiser's chemists: science and modernization in Imperial Germany. The University of North Carolina Press, Chapel Hill, NC, pp 182–188.
35. Stoltzenberg D (1994) Fritz Haber: Chemiker, Nobelpreisträger, Deutscher, Jude: Eine Biographie. VCH, Weinheim, pp 238–309.
36. Rathenau W (1925) Gesammelte Schriften, Band V. Wirtschaft, Staat und Gesellschaft. S Fischer, Berlin, pp 23–58.
37. Harris R, Paxman J (2002) A higher form of killing: the secret history of chemical and biological warfare. Random House, New York.
38. Brown FJ (1968) Chemical warfare: a study in restraints. Princeton University Press, Princeton, NJ.
39. Preston D (2015) A higher form of killing: six weeks in the first world war that forever changed the nature of warfare. Bloomsbury, New York.
40. Johnson JA (2017) Military-industrial interactions in the development of chemical warfare, 1914–1918: comparing national cases within the technological system of the great war. In: Friedrich B, Hoffmann D, Renn J, Schmaltz F, Wolf M (eds), One hundred years of chemical warfare: research, deployment, consequences. Springer: Berlin (DOI https://doi.org/10.1007/978-3-319-51664-6, Springer Open, 2017), pp 135–149.
41. Haber LF (1986) The poisonous cloud: chemical warfare in the first world war. Oxford University Press, Oxford.
42. Drinkner P (1919) Dyestuffs: hearings before the Committee on Ways and Means, House of Representatives, on H.R. 2706 and H.R. 6495, June 18–20 and July 14–18, 1919 [66th Cong., 1st Sess., 1919]. Committee on Ways and Means. Government Printing Office, Washington, DC, pp 533–534.
43. Teed PL (1919) The chemistry and manufacture of hydrogen. Longmans, Green & Co., New York, pp 101–106.
44. Appl M (1982) The Haber-Bosch process and the development of chemical engineering. In: Furter WF (ed), A century of chemical engineering. Plenum, New York, pp 29–53.

45. Johnson JA (2003) The power of synthesis, 1900–1925. In: Abelshauser W, von Hippel W, Johnson JA, Stokes RG, German industry and global enterprise. BASF: the history of a company. Cambridge University Press, New York, pp 115–205.
46. Haber LF (1971) The chemical industry 1900–1930: international growth and technological change. Clarendon Press, Oxford, p 281.
47. Tooley TH (1999) The Hindenburg program of 1916: a central experiment in wartime planning. Q J Austr Econ 2(2):51–62.
48. Morris PJT (1982) The development of acetylene chemistry and synthetic rubber by I.G. Farbenindustrie Aktiengesellschaft: 1926–1945. DPhil thesis, University of Oxford, p 15.
49. Translation and quote courtesy of Peter J. T. Morris, November 2014.
50. Verg E, Plumpe G, Schultheis H (1988) Milestones: the Bayer story, 1863–1968. Bayer, Leverkusen, p 202
51. Wille HH (1969) Der Januskopf: Leben und Wirken des Physikochemikers und Nobelpreisträgers Fritz Haber. Verlag Neues Leben, Berlin, p 229.
52. Translation and quote courtesy of Peter J. T. Morris, November 2014.
53. Chemical fertilisers in 1917 (1918) J Soc Chem Ind 37(14)(31 July):285R–286R.
54. Tongue H (1934) The design and construction of high pressure chemical plant. Chapman & Hall, London, p 4.
55. Peter J. T. Morris, personal communication, November 2014.
56. Hartley H (1919) Report of the British mission appointed to visit enemy chemical factories in the occupied zone engaged in the production of munitions of war. In: Dyestuffs: hearings before the Committee on Ways and Means, House of Representatives, on H.R. 2706 and H.R. 6495, June 18–20 and July 14–18, 1919 [66th Cong., 1st Sess., 1919]. Committee on Ways and Means. Government Printing Office, Washington, DC, pp 210–214, on 212.
57. Deichmann U (2001) Flüchten, Mitmachen, Vergessen. Chemiker und Biochemiker im Nationalsozialismus. Wiley/VCH, Weinheim, p 396.
58. Cyanamide production in Germany (1920) J Soc Chem Ind 39(21)(15 November):382R.
59. Nitrogenous fertilisers in Germany (1920) J Soc Chem Ind 39(3)(16 February):59R.
60. Effect of Upper Silesian plebiscite on chemical and metallurgical industries (1922). Chem Met Eng 26(2)(11 January):76.
61. Visit to the Ammoniakwerk Merseburg G.m.b.H. in connection with the World Power Conference at Berlin. Some technical data given by Direktor Dr Schneider, when addressing the visitors. Leunawerke, June 25th 1930. Papers of Herbert Alfred Humphrey, Archives, Imperial College London, Item M7, p 1.
62. Cochrane RC (1978) The National Academy of Sciences: the first hundred years, 1863–1963. National Academy of Sciences, Washington, DC, pp 209–211.
63. Tour RS (1922) Contributions to the study of ammonia catalysts, III. Chem Met Eng 26(13) (29 March):588–593.
64. Haynes W (1957) On the chemical frontier: the Cyanamid story. Copy held at Sidney M. Edelstein Library for the History and Philosophy of Science, Technology and Medicine, National Library of Israel, pp 40–41.
65. Bucher JE (1917) The fixation of nitrogen. Ind Eng Chem 9(3) (March):233–253
66. Haynes W (1945) American chemical industry: a history, vol. II. The World War I period, 1912–1922. D Van Nostrand, New York, p 84.
67. Hounshell DA, Smith JK (1988) Science and corporate strategy. Du Pont R and D, 1902–1980. Cambridge University Press, Cambridge, pp 184–188.
68. Haynes W (1945) American chemical industry: a history, vol. II. The World War I period, 1912–1922. D Van Nostrand, New York, p 94.
69. Haynes W (1945) American chemical industry: a history, vol. II. The World War I period, 1912–1922. D Van Nostrand, New York, pp 95–96.
70. DeKay Thompson M (1922) The Bucher process for the fixation of nitrogen. Chem Met Eng 26(3)(18 January):124–127.

71. Haynes W (1945) American chemical industry: a history, vol. II. The World War I period, 1912–1922. D Van Nostrand, New York, p 99.
72. Haynes W (1948) American chemical industry: a history, vol. IV. The merger era. D Van Nostrand, New York, pp 85–95.
73. Haynes W (1948) American chemical industry: a history, vol. V. Decade of new products. D Van Nostrand, New York, pp 101–118.
74. Coffey BF (1997) Fertilizers to the front: HAER and U.S. nitrate plant no 2. J Soc Industrial Archeology 23(1):25–42.
75. Mumford RW (1938) Potassium chloride from the brine of Searles Lake. Ind Eng Chem 30(8) (August):872–878.
76. Haynes W (1957) On the chemical frontier: the Cyanamid story. Copy held at Sidney M. Edelstein Library for the History and Philosophy of Science, Technology and Medicine, National Library of Israel, pp 43, 45.
77. Haynes W (1945) American chemical industry: a history, vol. II. The World War I period, 1912–1922. D Van Nostrand, New York, pp 104–107.
78. Steen K (2014) The American synthetic organic chemicals industry: war and politics, 1910–1930. The University of North Carolina Press, Chapel Hill, p 180.
79. Sill TW (1919) American officers in German chemical plants. The dyestuff plants and their war activities. In: Dyestuffs: hearings before the Committee on Ways and Means, House of Representatives, on H.R. 2706 and H.R. 6495, June 18–20 and July 14–18, 1919 [66th Cong., 1st Sess., 1919]. Committee on Ways and Means. Government Printing Office, Washington, DC, pp 173–177, on 175.
80. Muscle Shoals propositions, hearings, 67th US Congress, 8 May 1922, Committee on Military Affairs. Second session, p 446.

Chapter 9
Billingham: "The Synthetic"

The President of the British Nitrogen and Carbide Company expressed the opinion that it was hardly probable that the Haber-Bosch process could be extended much outside Germany because the operation of its costly and complicated plants presumes a high technical capacity.

From Arturo Miolati, *Synthetic Ammonia and the Casale Process. Amplified edition of a lecture delivered the 27th February 1927 at the Institute of Chemistry of the Polytechnic School of Prague*. Rome: Ammonia Casale SA, 1927, p. 13.

9.1 Nitrogen in Britain

Anecdotes like the above, as we have seen, may have been apocryphal but they certainly encapsulated the attitude of the time. The cutting off of nitrates to Germany and German submarine attacks on Allied shipping taught industrialized nations an invaluable lesson about independence in strategic materials. However, without exception the Allies made little if any headway in their attempts to imitate the Haber-Bosch process and gain complete security in supply of nitrates before the end of 1918. Interest in Britain, certainly more than in France, was stimulated by the fact that cheap and abundant hydro-electricity for the manufacture of cyanamide was not available, except for limited facilities in Scotland associated with the production of aluminium. In contrast, the French hydro-electric power stations in the Pyrenees and Alps enabled cyanamide production to be taken up and expanded during the war.

Solving the ammonia problem in Britain became the remit of the Munitions Inventions Department's Nitrogen Products Committee [1]. The committee, appointed in June 1916, was directed by the physicist John A. Harker. Research at the associated nitrogen products research laboratory into a high-pressure ammonia process, and the closely related oxidation of ammonia, was carried out in three sections led by Frederick G. Donnan (1870–1956) of University College London, who had received his PhD under Ostwald. Donnan, previously at Liverpool, was a

A.S. Travis, *Nitrogen Capture*, https://doi.org/10.1007/978-3-319-68963-0_9

consultant to Brunner, Mond, which after the war found his services of considerable value.

Each section under Donnan's direction included men who, in common with Donnan, had received training in Germany and were well matched with their tasks:

- Ammonia synthesis, under Lieut. Harold Cecil Greenwood (1887–1919), who undertook research for his PhD at Manchester University's Electro-chemical Laboratory and worked with Haber on a uranium-based catalyst just before the war;
- Ammonia oxidation, under James Riddick Partington (1886–1965), later a noted historian of chemistry, who had worked with Nernst in Berlin during 1911–1913 on the specific heats of gases. While there, Partington prepared a draft for *A Text-Book of Thermodynamics with Special Reference to Chemistry*, published in 1913 [2]; and
- Supply of gases, particularly pure hydrogen for ammonia synthesis, under physical chemists Captain Eric Keightley Rideal (1890–1974; knighted 1951), who received his PhD under Richard Anschütz in Bonn (1912), and Hugh Stott Taylor (1870–1974), a catalyst expert, who had worked with pioneer electro-chemist, and sometime assistant of Ostwald, Max Bodenstein, at the Technische Hochschule Hanover (and Berlin, from 1906).

Most of the British research was undertaken at University College, with some conducted at King's College. Donnan was assisted by chemical engineer Kenneth Quinan, who as director of factories in the Department of Explosives Supply oversaw the construction of new works for the Ministry of Munitions. Sir William Ramsay, who retired from University College in 1913 and had studied Haber's method and the oxidation of ammonia—and no doubt gained much useful information from discussions with Tamaru—provided some assistance to the Munitions Inventions Department, though for only a short time. He died in July 1916.

Following unrestricted U-boat attacks on Allied shipping from 1 February 1917, which threatened to cut off the supply of Chilean nitrate and wheat from the Dominions, the Nitrogen Products Committee recommended the construction of cyanamide and Haber-type plants, with priority for the latter. In August 1917, Herbert A. Humphrey was appointed chief engineer of the Munitions Inventions Department, and, in addition to his work on a cyanamide facility, became associated with "the little band of brilliant students working at University College, who were carrying out experimental work on the Synthesis of Ammonia and the problems of Ammonia Oxidation" [3]. An arrangement was made for Winston Churchill, Minister of Munitions from July 1917, to inspect the nitrogen laboratories at University College. Perhaps the odour of ammonia suggested that things were moving forward. That, however, was far from the case. There were even disagreements over where the research was leading. Humphrey recorded: "Dr Harker and I tried to dissuade the Department of Explosives Supplies from this project on the ground that there was an enormous advance needed from the small experimental scale to that of a large works, and that the information was not sufficient to warrant any attempt to bridge the extensive gulf" [4]. It was this situation that no doubt led to interest in the construction of a large wartime cyanamide factory in Britain.

Cyanamide in Britain turned out to be a problem. A small cyanamide facility was erected by Nitrogen Products and Carbide Company, Ltd, British licensee of Ostwald's ammonia oxidation process, at Dagenham Docks, Essex (another facility was installed at Trafford Park, Manchester). However, there were numerous technical difficulties. As a result, Nitrogen Products and Carbide relied on Odda and Alby, which it controlled through affiliated British companies, for cyanamide as the source of ammonia. When the Scandinavian supply of cyanamide became severely restricted, in part due to shortages of coal and raw materials, as well as demand for carbide from other industries and the U-boat threat, ammonia was obtained from gas works liquor. Despite the close working with Scandinavian factories and licenses issued by the British to the French, calcium cyanamide was never produced on an industrial scale in Britain. Oxidation of ammonia to nitric acid also met with difficulties.

Nitrogen Products and Carbide had anticipated strong and more profitable sales, far more so than from supplying the needs of the military, after the war. Experiments on the efficacy of cyanamide on different soil types were conducted at Rothamsted and Odda, and the quality of the granulated product was improved. However, as soon as the war was over the demand for cyanamide collapsed. In 1919, Alby United Carbide Factories, Ltd, merged with and took over the assets of Nitrogen Products and Carbide. As a result of long-term and crippling contracts for both carbide and power supply—which resulted in heavy debts and a lack of working capital—and shortages of coal and a strike, Alby United was liquidated [5]. Operations at Odda started again in 1924, as part of the Hafslund-Meraker combine, founder of Odda Smelteverk A/S, specializing in carbide manufacture. Electricity was supplied by a local company [6].

During the war years Kilburn Scott had continued to promote his three-phase electric arc process, drawing favourable comparisons with the other arc processes and discussing perceived advantages over the cyanamide process [7]. Where they could not compete, however, was in relative energy costs; electric arc processes consumed four to five times more power than the cyanamide process. In turn, the cyanamide process consumed considerably more power than the Haber-Bosch process.

Apart from the lack of success with the synthesis of ammonia, by mid-1917, some progress had been made with the oxidation of ammonia to nitric acid, using platinum or platinum-rhodium gauze catalysts. Partington took a prominent role in this work. The Munitions Inventions Department's apparatus, based on Partington's research, was not unlike that devised by Kaiser, and Frank and Caro. Instructions for assembly of the apparatus were published for the benefit of manufacturers, who were left to make do and improvise as best they could, working with ammonia from gas works and coke ovens. "At some small acid plants in England the necessary parts have been purchased here and there and put together by the plant superintendent.... All of these home-made outfits use ammonia liquor as the raw material, and include a 'stripping column' for stripping the ammonia gas from the liquor by means of steam or hot air. They also use a multi-layer flat platinum gauze" [8]. A major user of the process was United Alkali, in 1918, at its Widnes, Newcastle, and Bristol facilities [9]. As elsewhere, there were difficulties with steel absorption towers, as a result of corrosion, that were not completely overcome until well into the 1920s with the introduction of acid-resistant stainless steels. Partington was not happy with the lack of response to his nitric acid studies at University College. They were not taken up on a large scale: "It may fairly be said that laboratory workers in England and Germany seem to have been led to the same conclusions, but whereas the German technologists translated the results into plant the English work finished in the research laboratory" [10].

Ammonia research at up to 150 atmospheres pressure continued until at least the spring of 1918; Greenwood, the principal worker in this area, studied heat exchange using gases under high pressure [11].

9.2 "Little More Than a Wilderness"

Alfred Mond, *Synthetic Fertilizers. Billingham and Nitram Limited.* Billingham: A Nitram Publication, 1927. (Prepared for members of the Imperial Agricultural Research Conference on the occasion of their visit to Billingham, 19 October 1927).

In favouring the synthesis of ammonia the Department of Explosives Supply went against the advice of Humphrey and Harker, who had little confidence in success under wartime conditions. The department requested £5 million from the Treasury for the construction of a factory in the north-east at Billingham, then described as a "one-church, one-pub village, in County Durham." The site, with good railway connections, was just north of the River Tees and a short distance north-west of the industrial town of Middlesbrough (Fig. 9.1). Billingham, which Alfred Mond later described as "Little more than a wilderness," was chosen largely as a result of the nearby coal-fired power station of the North-Eastern Electric Supply Company Ltd, built during 1917–1921. A further £50,000 was requested for additional experimental work. However, after the Department of Explosives Supply had spent £1,300,000 on the project it was realised that there was little hope of success. Abandonment was covered by the "excuse that all the Department of Explosives Supply's staff and energy were needed to hasten the production of poison gas" [12].

Some research into the synthesis of ammonia was also undertaken in the private sector, including at John Thompson (Wolverhampton) Ltd, of Walsall, and at Siemens Brothers in Stafford. The results were even less promising. Introduction of high-pressure ammonia technology in Britain was hampered by the inability to develop equipment capable of withstanding the enormous operating stresses. Details of the BASF process were still masked behind a veil of secrecy. Moulton had "reluctantly decided that our scientific arrears of knowledge in this respect were too great to enable us to rely on such manufacture during the war" [13]. Moreover, there was another factor that held back progress: researchers were frustrated in their

Fig. 9.1 The Tees Valley region, comprised of south-eastern County Durham and north-eastern Yorkshire (the boundary is the River Tees), including Billingham, Stockton (site of the Mond Power-Gas Corporation works), West Hartlepool (source of anhydrite), and railway connections. From section of a map of the North Eastern Railway, dating from the early 1900s, at Whitby Station

efforts by frequent changes of management at the Munitions Inventions Department [14].

Although both the synthetic dye industry (source of aromatics) and the synthetic nitrogen industry by 1918 were recognized to be key sectors, no statistics on the various nitrogen products and their uses in Britain prior to 1914 were available. Technical and commercial information concerning existing facilities was certainly wanting.

Investigations into synthetic ammonia did not cease following the end of hostilities; they were quickly revived. After the signing of the Armistice on 11 November 1918, chemists and engineers from among the Allied nations believed that through direct access to the BASF Oppau site they could develop processes enabling the large-scale manufacture of ammonia [15]. Within 24 hours of the signing of the Armistice, Freeth, acting for Brunner, Mond & Co., which had played a major role in war-related production, proposed to Lord Moulton that a commission be sent out at the earliest opportunity to examine the ammonia works at Oppau and draw up a report for the Ministry of Munitions [16].

As for Billingham, it was transferred from the Department of Explosives Supply to the Munitions Invention Department, whose last comptroller, and chair of the Nitrogen Products Committee, Colonel Henry E. F. Goold-Adams, on 13 January

1919, placed Humphrey in charge of planning the future of the site. Three days later, Humphrey, still in government service, visited Billingham, and by the end of the month had drawn up an "Interim Report on the Billingham Scheme for the Production of Synthetic Ammonia and Derivative Products" [17]. Early in February, Humphrey suggested that the factory at Billingham—then hardly developed apart from a few sheds—should be taken over by a consortium of private firms, with Brunner, Mond as the principal component. On February 27, Lord Moulton "requested that B.M.&.Co., should either themselves undertake [completion of the factory] or should form a Syndicate to carry out the work" [18].

Meantime, arrangements had been made for official Allied inspections of German chemical factories, the first of which, in February 1919, was the "British Mission," headed by Harold Hartley. This mission, the first of two appointed to investigate chemical factories in the "Occupied Zone of Germany," included Greenwood as the government fixed nitrogen expert and A. W. Tangye from Brunner Mond. It was the Haber-Bosch process at Oppau that attracted most attention, and admiration. French officers, who had occupied the Ludwigshafen-Oppau factories on 6 December 1918, were on hand to meet the mission. The reception received from the Germans was considered cold and arrogant. Similar receptions awaited more or less independent visitors, as well as representatives of other Allied countries and their corporations. According to electrical engineer John Francis Crowley, writing in the introduction to the 1926 English edition of Waeser's 1922 study [19]: "when the members of the British and French Commission attached to the Allied Control Mission visited Oppau during 1919 they were told by the German chemists that even if they could erect plant identical with that at Oppau or Merseburg, they certainly could not operate it 'for lack of the highly trained workers, who were absolutely essential to successful operation'." Certainly BASF believed that this was the case.

In January 1920, the Treaty of Versailles came into force. In March, French troops and the remaining officers engaged in supervision and inspections at Oppau-Ludwigshafen vacated the two factories. France would reoccupy them on 15 May 1923, as a result of the Ruhr conflict. This was ostensibly caused by delays in the reparations deliveries of coal and coke, and was the pretext for Belgian and French troops to march into the Ruhr district, mainly on the right bank of the Rhine, on 11 January 1923. However, by then foreign firms were less interested in visits to observe the German chemical technology.

9.3 "Colonel Pollitt, Like Dr Mond, . . . Created Another Large Industry"

"Presentation of the Society's Medal," *Journal of the Society of Chemical Industry* (Transactions), vol. 46 (15 July 1927):290T–291T.

Physical chemist George Paton Pollitt (1878–1964) during the early 1920s would play the leading administrative role in the development of British-made synthetic ammonia (Fig. 9.2). Born at Mellor, near Blackburn, he spent his school

Fig. 9.2 George P. Pollitt. (Kenneth P. Magee/Edelstein Center)

years abroad, gaining fluency in French and German. After studying chemistry at the University of Manchester, he moved to Switzerland to embark on research at Zurich's ETH. He received his PhD from the University of Basel for a thesis on the contact process for sulphuric acid. In 1903, Pollitt was appointed assistant chemist at the Woolwich Ordnance Factory (Woolwich Arsenal), before taking up a similar position at Kynochs Ltd of Kynochtown, Essex, manufacturer of high explosives. During 1905–1912 he was a departmental manager at Brunner, Mond, which in 1911 and 1912 absorbed the soap makers Joseph Crosfield & Sons, Ltd, and William Gossage & Sons, of Warrington and Widnes, respectively. Pollitt acted as liaison officer between these firms. (They were sold to Lever Brothers Ltd in 1919.) In 1913, he became general manager of Hydrogenators, Ltd, and was involved in the opening of a subsidiary in Marseille. Pollitt joined the British Army on the outbreak of war in August 1914, and following distinguished service (he was taken prisoner in June 1918) was granted the rank of lieutenant-colonel on demobilization in 1919, when he returned to Brunner, Mond [20].

Much of Pollitt's subsequent career in chemical industry, until after the mid-1920s almost exclusively concerned with synthetic nitrogen, is detailed in a monograph published by Imperial Chemical Industries in 1957. *Billingham: The First Ten Years*, written by Brunner Mond, and later ICI, engineer Victor Emmanuel Parke, was dedicated to "Lieut. Col. G. P. Pollitt, D.S.O., M.Sc. Ph.D., first managing director of Billingham, as a mark of affection and admiration of the staff whom he led through the early years" [21]. Based on recollections and archival sources, it was a comprehensive account of the successful development of the manufacture of synthetic ammonia in Great Britain, including an overview of the many technical difficulties and biographical sketches of the main participants. The sense of danger while working with high-pressure equipment, the commitment and the excitement are palpable throughout, as is a strong hint of patriotic duty. The Germans had to be beaten at their own game. After things went wrong, which happened frequently, the team of chemists and engineers undertook extensive periods revising calculations and devising improvements. The result, one might judge from the monograph, was a triumph in both design and execution. In a sense that was true. The British had to confront the same trials and tribulations as had the Germans. Their accomplishment was well deserved. But to overcome certain major problems they relied on industrial espionage, something that was absent from the 1957 history, where a straightforward acknowledgement of intelligence activities was not to be expected.

Here Parke's account is combined with information drawn from documents earlier prepared by Herbert Humphrey and correspondence that, fortunately, he retained. While one or two reminiscences might give the impression of self-credit, something that was in many respects fully justified, this material adds considerably to the story. Humphrey's contribution to the historical record begins with the transfer of the government's site at Billingham to Brunner, Mond. Those involved at this stage were, representing the government: Lord Moulton, Colonel Goold-Adams, Humphrey, and Harold Hartley; and, representing Brunner, Mond: Roscoe Brunner (son of John Brunner), chairman; Colonel Pollitt; Major Francis Freeth, research chemist; Captain Amos H. Cowap, engineer; Leonard Harper; and A. W. Tangye.

On 7 March 1919, in the belief that Brunner, Mond, based on its wartime and pre-war activities, was the one firm most likely to succeed in bringing about the synthesis of ammonia, a "peace proposal" for the future of Billingham was placed by Humphrey before Roscoe Brunner, Tangye, and Freeth [22]. Shortly after, Moulton, Goold Adams, Humphrey, and Hartley met with Pollitt at London's Savoy Hotel to discuss the proposal, but little came out of the meeting. Nevertheless, on March 20, Humphrey showed Brunner, Mond representatives Pollitt, Tangye, and Harper around Billingham. They were, purportedly, neither inspired nor impressed. The next day, new Brunner, Mond director J. H. Gold (a solicitor who joined the firm in 1907) proposed to Moulton the setting up of a syndicate of firms, comprised of Brunner, Mond as technical adviser, and "the explosives makers and the large fertilizer makers and the companies who have contributed to the information now in possession of the Government and bearing on the subject" of synthetic ammonia [23].

Despite their apparent disinterest in Billingham, the Brunner, Mond scientific people were certainly anxious to receive a professional opinion on the technical manufacture of synthetic ammonia. To provide this, the comptroller of the Munitions Inventions Department loaned the services of its two experts, Captain Greenwood and Herbert Humphrey, to Brunner, Mond, both at the expense of the latter. Brunner, Mond now became interested in a site survey of Oppau—occupied by the French military—which was encouraged by both Gold and Freeth. Humphrey, through his association with Moulton, who received approval from Marshal Foch, succeeded in arranging a visit to the ammonia factory. On April 25, Moulton's secretary telegraphed: "All right mission may leave any time now and report to Major Lefebure" [24]. Their contact at Oppau was British officer Major Victor Lefebure, who was fluent in French and had worked closely with the French military in gas warfare work. On April 27, Humphrey, Greenwood, Pollitt, Cowap, and chemist Frank Fox, from Joseph Crosfield & Sons, travelled to Paris where they met with Lord Moulton.

Though the French military authorities were against non-government missions entering the BASF factory, permission was given in this instance for the Brunner, Mond party to "merely observe, with no right to ask questions, or to make any notes, sketches or measurements". On the day of arrival at Oppau, May 2, however, and perhaps through Lefebure's intervention, they were met by a very sympathetic French officer, physical chemist and nitrogen expert Captain Louis Hackspill [25].

The group set up camp at Oppau, curious to find out about every feature of the synthetic ammonia process. Their investigations continued, according to Humphrey, until June 5, but under far from ideal conditions. (Humphrey, on another occasion, noted that their stay at Oppau lasted from May 6 to May 20.) Many hindrances were placed in their way. The British were no doubt stunned by the audacity of BASF: "The Germans made this study very difficult and did all they could to impede our getting any information of value. At times they even became menacing." The workers had, "painted over the fronts of gauges, took down the lowest flights of works' ladders, disconnected pipes, covered up stores, and chipped the makers' names off their machines. The commission was forbidden to sketch or photograph and whenever they entered a building, all work at once stopped" [26]. These actions had been ordered by Carl Bosch, chairman of BASF. The ammonia plant, if it operated at all from November 1918 until June 1919, in any case suffered from a lack of coal. The Brunner, Mond people committed whatever they could to memory, each would later discretely make sketches, compare with others, and then in the evenings prepare detailed drawings. The daily diet during this time, Humphrey claimed, was restricted to bread and wine, since the fare in the French officers' mess blunted the appetites of the English visitors. Life at Oppau was not altogether miserable for the Brunner, Mond contingent. Cowap, an easy-going fellow, with a penchant for Gilbert and Sullivan operas, befriended French occupation troops from Madagascar, "who performed a war dance for his entertainment" [27].

At the end of the trip, the notes and drawings made by the visitors and packed in Humphrey's baggage disappeared from a railway wagon at Oppau in which they had been locked ready for onward movement (the intruders had sawn through the wooden floorboards). Cowap, however, had retained his notes in a kitbag; they were used to draw up plans and a report [28].

Notwithstanding the impression given of a lack of contact with staff at Oppau described by Humphrey there is circumstantial evidence that, in addition to surveillance, there were interrogations and clandestine activities in order to gain access to materials. It is likely that some level of coercion was involved. Between February and August, samples of catalysts had been spirited out of Oppau and were analysed at Northwich by H. E. Jones [29, 30]. Observations, and probably local sources of information, had revealed that earlier Allied wartime estimates of Haber-Bosch production costs were far too high.

In May 1919, around the time of the Oppau visit, the Nitrogen Products Committee published a report on the nitrogen industry [31]. It was a survey of the information that had been brought together during wartime investigations. Humphrey shortly after opined that a "No more important scientific report has ever been published by the Government than that of the over 350 pages presented in May last to the Minister of Munitions by the Nitrogen Products Committee of the Munitions Inventions Department" [32]. The report covered all aspects of the production and use of nitrogen products, and included—in addition to high-pressure synthesis of ammonia—cyanamide, the gasification of peat, and the Serpek process.

On July 1, the Ministry of Munitions became a department of the War Office. Pressure was placed on Brunner, Mond to take over the Billingham site, for which the Nitrogen Syndicate Ltd was created. Pollitt's visit to Oppau had given him a taste for the ammonia enterprise. On 10 July 1919, at the Carlton Hotel London he offered Humphrey, then still representing government interests, the post of engineer in charge of Brunner, Mond ammonia research. On July 24, Moulton reviewed the Oppau report and drawings with Pollitt, Roscoe Brunner, Cowap, Greenwood, and Humphrey. Moulton, according to Humphrey, "marvelled at the completeness of the information" [33].[1]

On August 27, Humphrey was appointed consulting engineer to a "syndicate" made up of Brunner, Mond, Explosives Trades Ltd (from 1920 Nobel Industries), and the heavy engineering firm Vickers Ltd that specialized in warships and armaments. According to Humphrey, it was he who had suggested that they combine in the ammonia venture, drawing on his experience of cooperation between industrial firms during the war.

Two syndicates suggested by Roscoe Brunner involved Brunner, Mond and Vickers, jointly with either the newly formed British Dyestuffs Corporation, in which the state had a major interest, or Johnson Matthey & Company, wartime supplier of platinum catalyst for use in ammonia oxidation. In 1919, Vickers was tied up with the takeover of British Westinghouse, an electrical concern that soon

[1]This differs from Parke's account, in which he states that the Oppau report and plans of the Brunner, Mond group were kept secret, even from Moulton's colleagues.

specialized in railway equipment (as Metropolitan-Vickers Electrical Company, founded in September), which was perhaps why another combination considered was Brunner, Mond, Armstrong Whitworth, whose activities were similar to Vickers, and Explosives Trades Ltd. Also in 1919, Pollitt and Tangye were appointed directors of Brunner, Mond.

Brunner, Mond, however, would go it alone. The man charged with control over the project was Pollitt. A main incentive, based on pre-war trade, was the potential for the export of ammonium sulphate, particularly to Japan and South East Asia. This was critical, since British gas works and coke ovens produced sufficient to satisfy the home market.

Following consultations with the Ministry of Munitions in October 1919 and much greater interest, if not enthusiasm, shown on the part of Brunner, Mond, terms and the price for acquiring Billingham were next on the agenda. A meeting was arranged in London, again at the Carlton Hotel, on November 20 between Brunner, Mond representatives and the director-general of lands at the War Office, Ministry of Munitions and Air Ministry, and of the Disposal Board. This was estate agent Sir Howard George Frank (1871–1932), partner in Knight, Frank & Rutley (founded in 1896), who was appointed the government negotiator on 1 November 1919. Frank was out to get the best terms for the state. His language and sales style—more in keeping with his usual line of business—were apparently not in accord with that of the assembled scientists from the chemical firm. According to Humphrey: "Sir Howard Frank took such a high-handed standing and demanded such impossible terms and conditions that the meeting abruptly terminated and Messers. Brunner, Mond & Co. decided to find another site for the proposed works." Humphrey and Harker were left to sort things out. Humphrey continued: "They felt that Sir Howard Frank's action in the matter would not be approved by the Ministry [of Munitions] and wrote to Sir [William] Graham Greene [1857–1950, Permanent Secretary, Ministry of Munitions] a joint letter on this subject. This led to Sir Howard Frank's writing to the Chairman of Brunner, Mond & Co. [Roscoe Brunner] that he would like to meet him, and later Captain Cowap and I were delegated to deal with the whole question of the purchase price as between Brunner, Mond & Co. and the Ministry. There followed an unforgettable struggle to obtain reasonable and fair terms, but in January 1920, I was able to draft a letter to Sir Howard Frank which was approved by the company, and a definite price was agreed by Col. Pollitt and accepted by the Minister on January 8th 1920" [34]. On 24 April 1920, Brunner, Mond concluded a contract for the purchase of the government's 266-acre Billingham site. Reflecting concerns over national security it was stipulated, as a condition of sale, that "By agreement with the Government the company will always be under British control, the directors are to be British-born, and the first directors are to be approved by the Government." There were no difficulties in fulfilling these conditions.

The principal contributors to the development of Billingham following the transfer, as identified by Parke, were Pollitt, Humphrey, Cowap, physical chemist Roland E. Slade, and administrator Phillip Charles Dickens, a qualified accountant. Roland Edgar Slade (1886–1968), graduated in 1909 at Manchester and then lectured at Liverpool, where he met Donnan. When in 1913 Donnan moved to

University College London, he took Slade with him, as lecturer. Slade had met Pollitt during military service. Amos Hayton Cowap (1878–1958) had joined Brunner, Mond in 1891, the same year as Humphrey.

9.4 Brunner, Mond Ammonia Research

Following the Oppau visit in 1919, investigations into the technology of ammonia synthesis gathered pace at Brunner, Mond. In the autumn, fifteen engineers were taken on by Humphrey and Cowap and four chemists by Pollitt to work in Brunner, Mond's new Nitrogen Fixation Department.[2]

On 13 October 1919 a private house called "Heyeswood" on the Brunner, Mond estate near the Winnington works in Cheshire was allocated to the ammonia project. Thirteen sketches and drawings, mainly based on information gleaned from the Oppau visit, were transformed into technical drawings and novel designs. There were a number of Brunner, Mond innovations, including the use of heat generated during the process in sustaining the reaction. "It was shown that with the ammonia pass conversion likely to be obtained a heat exchanger of about 80% efficiency would make such a converter autothermic. This was to be the outstanding English contribution to the process. It confined the reaction zone, which contained red hot catalyst, to a central, thin-walled cylinder, surrounded by a well-lagged heat-exchanger." With catalyst and heat exchanger together in the same vessel, "the design reduced the number of costly forgings for the plant and allowed each catalyst charge to run for a longer time. It provided an internal electric heater for keeping up temperature when the catalyst was losing activity, and for starting up" [35].

In the autumn of 1919, Harold Greenwood joined Brunner, Mond to work on the ammonia synthesis; but he died, suddenly, from blood poisoning at Winnington Hall on November 4. For the future of ammonia in Britain, this was an enormous loss, since not only had he been one of the two leaders in ammonia research at University College, but prior to 1914 he had worked with Fritz Haber on ammonia catalysts. A post mortem was considered, no doubt from concerns of deliberate poisoning by a foreign agent—and should doctors have had any suspicion of foul play "the Ministry are prepared to go to any length in the matter"—such was the strategic importance of the ammonia venture. R. E. Slade replaced Greenwood in February 1920. At this time, Cocksedge was investigating catalysts at Northwich.

In keeping with their original agreements, going back to the end of the previous century, information was exchanged between Brunner, Mond and Solvay Process in

[2]The engineers were N. Archer, G. R. Atkins, S. R. Boyce, F. H. Bramwell, C. W. Ellen, W. M. Gyles, A. G. Hinton, A. Johnson, G. S. Jones, F. S. Lundy, A. A. Munro, W. R. Tate, H. R. Tidswell, H. C. White, and T. M. Wilson. The chemists, all from the state-owned Gretna explosives factory, erected in response to the 1915 shell crisis, and as recommended by Harker of the Munitions Inventions Department, were: Victor E. Parke, F. M. Ray, D. H. B. Wride, and A. T. S. Zealley.

America. Solvay was then collaborating closely with General Chemical in ammonia research [36]. On 12 December 1919, Solvay and General Chemical agreed to join forces and establish the Atmospheric Nitrogen Corporation.[3]

In April 1920, Humphrey and Pollitt visited the United States to investigate progress in the General Chemical process, including the results of wartime work at Sheffield [37]. They brought back samples of catalysts and recipes, and tests were undertaken using the University College wartime synthetic ammonia apparatus. It had been in storage at Woolwich Arsenal and was delivered by horse and cart to the Walsall works of John Thompson (Wolverhampton) Ltd, to join that firm's experimental ammonia units, located in a ramshackle shed, "[t]he whole...watched over by a miserable, but faithful, one-eyed black tom cat" [38]. Thompson, through its engagement in fat hardening, produced the all-important hydrogen, which created an interest in synthetic ammonia. At Walsall, Brunner, Mond investigators found that the American catalyst, not unlike that used at (and taken from) Oppau, apart from a trace of vanadium in the latter, afforded 50% more ammonia than the Munitions Inventions Department catalyst. Moreover, the vanadium was found to be superfluous.

In 1919, Francis Freeth had visited the laboratory of physicist Kamerlingh Onnes (1853–1926; Nobel laureate 1913) at Leiden and was struck by the high quality of the research taking place there. From this time, Freeth became committed to introducing improved research facilities at Brunner, Mond. Among the chemists taken on in 1919 was Victor Parke, author of the 1957 Billingham history. He and a few of the other newcomers were assigned to laboratories at the Weston Point (Runcorn) works of Castner-Kellner Alkali Co., Ltd, in Cheshire—acquired early in 1920 by Brunner, Mond—and left to get on with it, with little or no supervision. A main reason for choosing Runcorn was, as with Thompson's, the ready availability of pure hydrogen, in this case from the electrolytic alkali process used to manufacture sodium hydroxide (caustic soda) and chlorine. Experiments were conducted on the collection and distribution of hydrogen, on a low pressure system for preparing the nitrogen-hydrogen gas mixture, and the high-pressure synthesis and circulation system, including the addition of fresh compressed gases as product ammonia was removed. The compressor was put together from a marine engine recovered from a sunken ship, and a cooling machine from the Liverpool Refrigeration Company; compression was created by running the engine in reverse.

A Brunner, Mond subsidiary, Synthetic Ammonia and Nitrates, Ltd, was registered on 3 June 1920. The capital of the company was £5 million [39]. Roscoe Brunner, chairman of Brunner, Mond, was chair of the new company. The engineer Herbert Humphrey was appointed to the board of directors. Captain Cowap was his chief engineer. Slade, at Runcorn from early 1920 when he replaced Greenwood as head chemist, moved to Billingham in June 1920, along with other Runcorn

[3]Early in 1920, Brunner, Mond considered bringing General Chemical, Solvay and Du Pont into a consortium to develop the ammonia process.

research workers (though ammonia work at Runcorn would continue). In the autumn Slade brought in another chemist, L. F. Knapp.

Cowap was concerned about the lack of understanding of chemical plant shown by the first group of engineering recruits—who appear to have had little or no knowledge of chemical industry—and made this known to Pollitt, who in turn declared that any faults lay with Cowap's methods and handling of his staff. Colonel Pollitt's stern and unforgiving approach to the training of manpower was no doubt moulded by his military experience. Apart from that, engineering staff, despite their growing importance, and even necessity, were not always given very much status, at least compared with chemists. When Humphrey received congratulations on his appointment from Walter H. Kniskern, chief engineer of the newly founded Atmospheric Nitrogen Corporation in New York, Kniskern remarked: "Engineers as such do not fare so well in this country, in fact, being looked upon in the light of a necessary evil" [40].

In December 1920, General Chemical and Solvay Process, in which Brunner, Mond had an interest, were merged with other firms to form Allied Chemical & Dye Corporation: Roscoe Brunner was appointed a director. Allied Chemical was directed by the highly successful but highly secretive businessman Orlando F. Weber (1879–1945). Atmospheric Nitrogen, Allied Chemical's subsidiary, began operating its synthetic ammonia apparatus at Syracuse in August 1921 [41].[4]

To gain as much information as was possible on European progress in ammonia production and power supply, Pollitt and Humphrey visited factories and laboratories in France, where they were introduced to the Claude process, and Belgium. Meantime, since September 1919, Carl Bosch had used BASF synthetic nitrogen to bargain with Solvay, and French and American firms, in attempts to gain access to and control markets in other chemical products and technologies through an international alliance of manufacturers that BASF would participate in, if not dominate. The perceived direct threat of the Haber-Bosch process to the Belgian Solvay & Cie ammonia-soda process (perhaps through use of synthetic ammonia in a competing product), led the corporation to acquire and peruse a copy of Camille Matignon's comprehensive report on the Oppau facility (Sect. 9.6). Synthetic ammonia continued to be the most powerful BASF bargaining chip, used until the mid-1920s to negotiate terms for access to products and markets of large foreign—and preferably multinational—chemical firms [42, 43].

Despite access to information, availability of equipment and a well-organized research effort, the gaps in technical knowledge at Brunner, Mond were not inconsiderable. On 29 August 1920, serious consideration was given by Pollitt, in a letter to Humphrey, towards reaching an arrangement with BASF—via "S & Cie"—for the release of know-how that could not be obtained during the Oppau site

[4]Freeth and Donnan during 1921–1922 undertook experiments on ammonia synthesis from nitrogen and hydrogen in high-frequency electromagnetic fields. This work was suggested by the research of a French inventor, Slatineau. Though of scientific interest it was of no use in industrial capture of nitrogen.

visit. This was Solvay & Cie, whose representatives had met with Pollitt and colleagues around ten days earlier to consider ways of dealing with BASF; its synthetic ammonia process was no less a potential threat to the Brunner, Mond ammonia-soda business. Brunner, Mond's target was to produce 100 tons of synthetic ammonia at Billingham every day.

Pollitt, in his clipped style, asked Humphrey [44]: "(1) Would we now pay anything for the remainder of the B.A.S.F. information; (2) Assuming we would, how much would you value it at now. (a) How much time wd it save in getting a 100 ton unit running; (b) Wht wd you be prepared to pay in perpetuity for all B.A.S.F. information, and assuming payment to be based on profits. That is wd you be prepared to say that it wd pay us now to make the agreement to hand over say 1% or 5% or 10% of the profits we are to make in the life of the Coy [?]." While requesting Humphrey's opinion on the value of acquiring such information, Pollitt's feeling was that we "shd be exceedingly loath to pay anything."

Humphrey's detailed response to Pollitt was not only telling in how far Brunner, Mond was removed from achieving successful manufacture. It also demonstrated the extent to which Humphrey understood the intricacies of all facets of the ammonia process. Moreover, in addition to revealing information arising from the sharing of knowledge arrangement with Atmospheric Nitrogen Corporation, through Solvay Process, it says much about the challenges encountered by all investigators into synthetic ammonia processes. For this reason the letter is worth quoting in full. Like Pollitt, Humphrey was not inclined towards contact with BASF [45]:

"I have your letter of the 29th inst., and my personal feeling is that I would prefer to carry on without further contact with the Germans.

Putting this feeling entirely on one side and having regard solely to the interests of the Company I still value the information we can obtain from the B. A. & S. F. at a figure of anything between £50,000 and £100,000, and anything on profits, up to 5% of our profits. [in hand "not of sale prices"].

My reasons are as follows and I will mention the plant in order:-

WATER GAS PLANT: we should obtain full information about mechanical grates on gas generators.

CO CONVERTER: We possess no information as to operating figures giving optimum conditions (temperatures, pressures, steam ratio etc.) and we have no knowledge of any detailed construction of this plant. In such absence of this knowledge we have had to assume that we must boost up our pressure to allow for the full 10 lbs. pressure drop, but it is conceivable that the Badische use the steam to inject the gas and so save much of this power.

The Americans [General Chemical/Atmospheric Nitrogen], after an attempt which has taken them some months to design a steel case converter, have given up the attempt as involving too many difficulties and have gone back to the small unit cast iron converter. They say they cannot overcome the difficulties of repeated contraction and expansion, but I think they are over-estimating such troubles.

We should like to know what type of hot valve they [BASF] use on the heat exchangers, also the pressure drops in each section of the plant and the life of their heat interchangers tubes and catalysts.

We have no knowledge of the actual manufacture of their catalyst especially in its preparation on a scale involving tons of material.

The various figures for horse power have all had to be guessed for this section, and we should obtain certainty on these matters; naturally our figures are calculated on a certain assumed basis.

COMPRESSORS: information was wanted about glands and packing and valves.

CO_2 REMOVAL: We have no detail as to the type of water injection pump, and the actual recovery of the pelton wheel, the ratio of water to gas, the actual absorption of other gases than CO_2 or the final CO_2 content.

We have no knowledge of the regulator which maintains the level of water in the CO_2 towers, the material of these towers i.e., whether cast iron or forgings, whether liners are used, or the type of packing which has been found best.

We should like to know the lives of the different parts of the plant and the troubles which they find most prominent.

CO REMOVAL: Our information from Oppau is very scanty on this section and we should get full knowledge of their liquor preparation and regeneration, together with losses, efficiency, final CO content, questions of attack on metal and value of refrigerating apparatus used in this section.

HEAT INTERCHANGERS: We should get definite data to scientifically design all heat interchangers.

AMMONIA CONVERTER: We probably have more detailed information about this piece of apparatus than any other at Oppau, but we would like to know a great deal more about the details, the life of the catalyst and the falling off in the efficiency of the catalyst.

We should learn the truth of the autothermic working and whether some air is always injected to maintain the requisite temperature (which the Americans declare to be the fact).

We want to know how they make their catalyst and the details of the apparatus employed in producing this material by the ton.

AMMONIA ABSORPTION: The apparatus is fairly well known to us, but we have to design every single detail which will mean hundreds of drawings, all this work will be saved and the same remark applies to a great deal of other apparatus apart from plant units.

GAS CIRCULATION: This is one of our chief difficulties, and the number of units installed at Oppau is, we believe, far in excess of the numbers of units used simultaneously. We should like to know the facts and something about the troubles over packing of glands and all there is to learn about autothermic control (oil-operated valves, etc., for sudden shut-down).

LABORATORY WORK: considering the number of chemists, and the length of time they have been engaged on the research work about catalysts and all other sections of the plant, there is no accumulated knowledge which can compare with that of the Badische Company. The ground covered is too wide to be mentioned in a letter.

PIPES: We have to proceed to make our own standard designs of all pipes, valves and other fittings.

We are also designing a good many measuring appliances; all this work could be saved.

BY-PRODUCTS: Much work must have been done on by-products and fertilisers generally and of which we should get full particulars.

TIME SAVED: I consider that if we were supplied with complete sets of drawings and full technical information we should save six months in getting the first 100 ton unit running on full output, and having regard to the larger plant we should save one year's work. We should also escape numerous infantile troubles and should, I take it, get the benefit of any further research[e]s and improvements which they may make."

Cowap added a note of agreement to Humphrey's detailed description.

It was clear that though data and measurements had become available, and descriptions were published of the BASF synthetic ammonia process, the endeavours of British researchers had met with limited success. Brunner, Mond certainly had to overcome severe shortcomings in bringing the ammonia synthesis on line. Humphrey's estimated value of the required information, at a then not inconsiderable sum of between 50,000 and 100,000 pounds sterling, was telling in how far there was to go before a viable working process could be achieved.

Brunner, Mond got what it needed through industrial espionage. This may have appealed to Pollitt who, as a result of his fluency in German, for a short time served with British intelligence during the war, when he enlisted a single agent, an employee of Brunner, Mond. Discreet feelers went out, indicating that the firm would pay for the information, but not through direct negotiations with BASF.

Within weeks, two Alsatian engineers, August Koebele and René Adler, based in Paris and claiming to have received close working experience with the high-pressure ammonia factories at Oppau and Merseburg, let it be known that they were ready to trade secrets with Brunner, Mond. A meeting was arranged, on which occasion they placed a layout drawing of Merseburg, and a list of suggestions for improvements, before Humphrey, Cowap, Slade, and Pollitt. The Brunner, Mond representatives, from their prior knowledge of the process, were convinced that the two informants were genuine. Koebele and Adler then went off to prepare a detailed flow sheet of Merseburg. On 20 December 1920, Brunner, Mond, satisfied that details of specifications and other requirements had been met, paid them 500,000 francs, with an agreement to pay a further 750,000 francs after the Billingham plant had been working successfully for six months. The conditions required Koebele and Adler to refrain from providing details to any other firm or country, apart from in France, for a period of ten years. Koebele and Adler had come to a similar arrangement with Etablissements Kuhlmann SA [46].

Brunner, Mond maintained complete silence over the negotiations with Koebele and Adler. At the same time, the firm wanted to verify the accuracy of the newly acquired information with a visit to Merseburg. For this purpose an approach was made to the War Office. There, Harker, at the Ministry of Munitions, had the impression that many details were still wanting (even if that was not entirely the case). On 28 January 1921, he passed on to Humphrey a classified report from the War Office dealing with the cost of producing ammonia in Germany. He advised Humphrey that a letter was to be sent to Brunner, Mond asking if the company "had all the information necessary for you to proceed at Billingham." Harker suggested: "The diplomatic answer for your people to give is undoubtedly to say you have not , and that you believe a mission to Merseburg would be really useful. I may tell you, for your private information, that the Premiers' Conference in Paris is going to put on the screw with regard to getting further information if we feel so disposed" [47]. Partington visited both Oppau and Merseburg, though whether or not to gather information for Brunner, Mond is not known. This enabled him to report early in 1921 on both factories [48].

It turned out that the information acquired from Koebele and Adler was sufficient—there was no need for a visit to Merseburg. Their reports contributed considerably towards bringing about the manufacture of synthetic ammonia in Britain. Brunner, Mond now moved forward rapidly in development of the ammonia synthesis. This included assigning responsibilities for management of the project. On 21 March 1921, Humphrey received confirmation that as director and consulting engineer to the new company he had as his chief assistant "Major [F. H.] Bramwell, with Captain [T. M.] Wilson and such further members of Staff (including Drawing Office) as may be free from time to time to be required." In keeping with the times, colleagues were addressed by their military ranks. Co-operation was emphasized, and "by means of the Executive Committee and the almost continual association between yourself and Colonel Pollitt, Capt. Cowap and Dr. Slade, there appears no reason why this close touch should not be continuously maintained." While Bramwell was called Humphrey's chief assistant his status as engineer was below that of Cowap [49].

At the end of 1920 the old University College apparatus at Walsall had been taken to Weston Pont (Runcorn). In the early months of 1921, a pilot plant, no. 1 unit, was constructed at Weston Point. Victor Parke was among the chemists assigned to this development work, carried out in a derelict cottage. The Ministry of Munitions provided nickel-chrome forgings, retained from wartime work, suited to high-pressure working at 150 atmospheres. These were ideal for making up a converter that could be operated on a semi-manufacturing scale [50]. The complete assembly, after modification, and a test run on May 21, was considered a success. The Oppau method of preparing the catalyst by fusion in a water-cooled steel pot was adopted. Output of ammonia soon reached two tons every day. No. 1 unit was worked continuously for several years.

Humphrey kept a file in which he noted the temperature rises over five stages of compression, up to 150 atmospheres, in a compressor obtained from the engineering firm Peter Brotherhood Ltd, of Peterborough, and installed at Runcorn [51]. This study was critical, since temperature in addition to pressure determined the yield of ammonia; also, for safe working it was necessary to prevent overheating in the catalyst area. Pure hydrogen for the pilot plant was obtained from the Castner-Kellner electrolytic plant. Work also began on the water gas (shift) process for producing hydrogen, the Bamag process, as developed by BASF. In May 1921, Brunner, Mond made known that it had been granted licenses from the British government to work processes outlined in sequestered BASF patents, with the Ministry of Munitions offering to acquire BASF's British patents that had been assigned to other firms. BASF had not been consulted with regard to these arrangements.

After the Atmospheric Nitrogen converter at Syracuse was put into operation, in August 1921, Slade and engineer W. R. Tate were despatched to America in order to undertake observations. They arrived in New York on 20 September 1921, and returned to Britain at the end of October, with information useful to both Runcorn and Billingham.

Brunner, Mond now set to work on confronting the technical challenges of large-scale working with a mixture of pragmatism and ingenuity, aided by an increased research and development budget, and the team building and leadership skills of Pollitt and Humphrey. Additional scientific and technical staff were hired to work at Billingham in a new laboratory that Donnan, the Brunner, Mond consultant, described as "perfectly designed and adapted to its works. . .For the first time in the History of this country, we have a technical and scientific laboratory which can bear comparison with anything to be found in Germany or the States" [52]. The research facilities had been designed by Slade, based on the laboratories at University College. Humphrey's five-stage compression experiments were now conducted at Billingham with the final pressure increased to 200 atmospheres. The new laboratory and workshops enabled not only the organization and management of know-how and the development of technical capabilities, but also the identification of gaps in high-pressure technology that needed to be addressed. The latter called for further expansion in the research staff.

From March to early July 1923, Pollitt was unwell, apparently as a result of blood poisoning, and resided in London, "forbidden any excitement whatever." Though "health forbids any technical discussion," he communicated with Humphrey on progress by exchange of letters. On May 16, Pollitt advised Humphrey that he was glad to learn that much needed additional engineering research personel were being taken on to assist the four leading experts at Billingham. After observing that Brunner, Mond director (William) Arthur Colgate (1883–1956; knighted 1951) "came and told me a lot of complimentary things about you and Billingham and progress there which cheers me mightily," Pollitt wrote: "May I emphasize one point. <u>These men sho[uld] be, must be, highbrows, nothing else is any use</u>. They sho. have a first class knowledge of physics, some considerable physical chemistry <u>as well as</u> the usual engineering subjects in their theoretical aspects. We do not want practical engineers for these posts. It is a chance of a lifetime to get some men of <u>your own type of training</u>. We have of this type—yourself—[F. H.] Bramwell—[T. M.] Wilson and perhaps to a lesser degree [W. R.] Tate but no others as far as I know. The men you take on now will be research men in the first instance but in the years to come will be our raw material for Chief Engineers and Engineers in

Charge. . .Don't turn a man down because he wants a decent screw. It will pay us to give it him" [53]. A few days later, on May 23, Pollitt advised the executive committee that while the ammonia process was viable, further mechanical and engineering studies were required. There were leaks, failures of joints, and as a result explosions. Catalyst deterioration was also a problem, as were the compressors [54]. The catalyst, however, was "highly efficient." On July 7, Humphrey informed Pollitt that a second converter was expected to arrive at Billingham within a few days. Manufactured at the Sheffield works of Cammell Laird, it had been assembled using electric welding, as with the first unit in order to avoid heat distortion [55].

Humphrey later described how, following laboratory and pilot plant studies, great emphasis was placed on detail and design. Frequent debriefing sessions involved the examination and reexamination of each step [56]: "Calculations were made to determine the cost of transporting various gases, liquids and solids which were involved in the processes; and the cost of distributing services (electric power, steam, and water). These calculations pointed the way to improve the lay-out so as to reduce these costs, and a second layout Plan was made and the calculations repeated. The above procedure was repeated 39 times before we were satisfied that, allowing for the necessary proximity of certain plant and buildings, the rest of the plant had been so disposed that the cost of operation would be a minimum. A series of workshops were erected to make all parts and instruments which we could not purchase and to carry out repairs. . .. Research continuously accompanied all designs, especially of the large cylindrical forging which had to contain red-hot gases at 4000 lbs/sq in. Especially regarding 'creep' of metal under these conditions."

Ammonia unit no. 2 was made ready for operation at Billingham. Among Cowap's contributions were specifications for the mild steel "forgings for the first ammonia converters. . .made from ingots of 180 tons, the largest ever poured at that time. The casting, trepanning and machining of these ingots constituted a first-class pioneering job" [57]. Billingham now had a complete, integrated synthetic ammonia apparatus. It offered a prototype for the future of the high-pressure manufacture of chemicals in Britain. All that remained was to make it produce the pungent ammonia.

During the last weeks of 1923 the technical staff worked with obsessive, feverish intensity, checking every component, valve, and gauge. Pollitt and Humphrey greatly appreciated and encouraged the camaraderie and efforts of their men who had now come so close to success.

Novel valves were invented, manufactured at the Britannia Works of J. Hopkinson & Co., in Huddersfield. A breakthrough in the large-scale synthesis of ammonia came, according to Humphrey, on Christmas Day 1923 [58]: "One day when we were trying out the Ammonia Converter Section, with a gradually increasing temperature of the circulating gases, the converter catalyst 'caught on' and began to make ammonia. We had not intended to start up the Works for several days but we took a sudden decision to put all sections of the plant into operation. So complete had been the organisation and training that this could be done at once. It was done with successful results." Slade, however, reported at the Executive meeting on 24 December that "the first traces of ammonia had appeared at 11.30 pm on 22 December." In any event it was an occasion for celebration, with Pollitt already drawing up plans for expansion, and a suitable social event, a fancy dress dance, at Elton Hall, where he was in residence at the time [59]. That was the return for their heroic endeavours. It was also a payoff for the culture of collaboration—inspired by Pollitt and Humphrey—that had developed between chemists and engineers intent on catching up with a new technology. It was just as much a major step in catching up with Germany.

This achievement enabled the manufacture of ammonia to commence at Billingham in 1924, with a daily output of 21 tons. Over a twenty-six day period, 150 tons of ammonia and 325 tons of ammonium sulphate were produced. The cost of the enterprise, as stated in 1925, was £3 million, paid for by the issue of preference shares in 1922 and profits from the alkali (soda ash) business, that had recovered after a depression in 1921. In 1924, Japan was by far the largest export market for Billingham's new fertilizer.

For a few more years, as undercover agents, August Koebele and René Adler continued to provide Brunner, Mond with information about BASF (later IG Farben) activities in the nitrogen industry. Though BASF/IG Farben were plagued with industrial espionage, and it was not unknown for the intelligence department to break up spy rings and catch infiltrators acting mainly for French and American interests, Koebele and Adler were not caught out, and would appear to have continued with their covert activities. On 28 February 1925, Humphrey wrote to Pollitt suggesting that Koebele and Adler be approached with a request to gather data on the costs incurred at Oppau and Merseburg in the manufacture of ammonium sulphate, including raw materials, freight, and rates of pay for labour. Humphrey considered, on the basis of reliable information that had already been procured, that Koebele and Adler would have no difficulty in penetrating the relevant departments: "I believe that...they could accomplish this quite well and in a reasonably short time. They have an intimate knowledge of the sources of supply and of all the raw materials, and would know exactly how to set about collecting the necessary data. I think the value of information would be so great as to repay the expense which would be involved."

Humphrey also discussed the implications of the recent Dawes Plan. Germany's early post-war economic and political crisis had culminated in hyperinflation that peaked late in 1923, around which time BASF paid off wartime loans from the state, particularly for the construction cost of Merseburg, in almost valueless marks. This dire situation was followed in August 1924 with the Dawes Plan—drawn up by Britain with support from the United States—designed to ease Germany's reparation's burden, and bring about improved relations with the former enemy. Humphrey opined that the plan promised to restore Germany to a state of prosperity, aided by price cutting, as well as improving working conditions for coal miners, who were working long hours and receiving low wages:

> I think the Germans have chosen a very good time in which to drop the price of Sulphate. Everything, including labour, is as cheap in Germany as it is likely to be. If the Dawes Scheme progresses steadily, Germany will become more prosperous fairly quickly, and one of the first effects would be that labour will demand higher wages. When I was in the Ruhr, the miners were working 12 hours a day, with only one hour off for meals. Ever since then, a Bill to establish an 8 hour day for the miners has been introduced, and this will lead to a general elevation of the labour position in other trades.

Humphrey also considered the longer term, including, surprisingly, the possibility of Germany once more controlling the coal and iron ore deposits in Alsace-Lorraine (Lorraine was previously Germany's chief source of iron). This was something that could be achieved only if Germany retained its monopoly in synthetic ammonia, and hence self-sufficiency in the manufacture of munitions, to the disadvantage of other nations. It was an issue with serious strategic consequences for Europe, according to Humphrey [60]:

> I see that the manufacture of Sulphate in Germany is now at its lowest cost and will tend to rise steadily. I also see that if Germany has any designs of getting back the valuable mining properties in Alsace and Lorraine, which are essential to make her iron and steel trade self-contained, she must start now to prevent other nations from putting up Synthetic Ammonia Plant; for, if such Plants become numerous, the materials for explosive manufacture will become too well established and distributed to make a shortage of explosives, such as occurred in the last War, impossible in the future. This would enormously decrease Germany's chance of success, and would remove her present great advantage in this respect.

The Armistice of November 1918 may have brought the battles of the Great War to an end, but it was only a partial victory for the Allies. Many outsiders still saw Germany as a threat. However, Alsace and Lorraine were safe from invasion after Germany in October 1925 participated in ratification of the Locarno Pact, and thereby guaranteed to respect the borders of its neighbours.

Table 9.1 Brunner, Mond/ICI ammonia units at Runcorn and Billingham

Unit number and location	Date of decision	Date operational	Capacity, tons/day
1 Runcorn	1920	21 May 1921	2
2 Billingham	11 July 1921	December 1923	20, later 30
	20 March 1924		extension to 52
		1926	extension to 65
3 Billingham	15 May 1925	27 February 1928	165
4 Billingham		28 April 1929	373
5 Billingham		September 1929	
		1932	415
		1936	486
		1941	642

Sources: Papers of Herbert Alfred Humphrey, Imperial College Archives and Corporate Records Unit, London; Victor E. Parke, *Billingham: The First Ten Years*. Billingham, County Durham: Imperial Chemical Industries Ltd, 1957. From 1932, the daily output is shown as "equivalent ammonia." In 1930, no. 2 unit was shut down. Manufacture of methanol using high-pressure converters began in 1928–1929

We can assume that Humphrey did not consider Billingham to pose a threat to peace in Europe. He and Pollitt were committed enthusiasts for the expansion of Billingham synthetic ammonia, as was Alfred Mond, who in 1923 returned as director of Brunner, Mond after serving as Minister of Health (1921–1922). Mond had several other interests. As a strong supporter of the Zionist cause, in 1923 he was asked by Chaim Weizmann to act as president of an association set up to support what became the Technion research institute in Mandate Palestine [61, 62].

In 1925, Humphrey filed a patent jointly with Synthetic Ammonia and Nitrates for a high-pressure catalyst tube in which the flow of gases was periodically reversed in order to provide a uniform temperature for the catalyst (BP 255,963).

Major expansion at Billingham involved introduction of the first of three large ammonia converters, units nos. 3, 4 and 5, that were installed by 1930 (Table 9.1). This required Brunner, Mond in 1925 to issue £2 million of 5% debenture stock, guaranteed under the government's Trade Facilities Act. As a result, the authorized share and loan capital stood at £7 million. In April there was a public outcry after the press, including the socialist *Daily Herald*, claimed that the British government had sold secret information concerning the ammonia process to Brunner, Mond. Meantime, the Dawes Plan encouraged export of coal from Germany, which adversely impacted on the British export trade. This was one of the factors contributing to the General Strike of May 1926, which delayed inauguration of Billingham's 165-ton ammonia converter, unit no. 3, until 1928. Victor Parke moved from Runcorn to Billingham, and on October 1, title of the Runcorn plant was transferred from Synthetic Nitrates to Castner-Kellner (both within ICI). Runcorn increased ammonia output from 1193 tons per annum in 1926 to 3362 tons in 1940.

In October 1926, Humphrey took over the day-to-day management at Billingham, in place of Pollitt, who moved to London to take charge of the Brunner, Mond group of companies [63]. Humphrey, according to a former member of staff at Billingham, lived at the nearby, newly refurbished Norton Hall, rent free [64]. "He was the bossman who was developing the site. He was a civil engineer. He lived there with all his servants around him. The pair-and-trap picked him up and took him to work and back again in the evening. In 1926 the total salary for Billingham was £78,000, and Master Humphries got £5000. Can you imagine—and he lived free." If nothing else, this opinion confirmed the leading position that Humphrey held at the site. Norton Hall had also served as the management club, where Pollitt and Roland Slade organized field sports. (Billingham's ammonia factory was associated with the Billingham Synthonia [synthetic ammonia] Football Club, and Billingham Synthonia Cricket Club, both founded in 1923.)

Following Pollitt's departure, Slade was appointed managing director of Synthetic Ammonia and Nitrates. Alfred Mond and Harry Duncan McGowan (1874–1961), of Nobel Industries Ltd, were then finalizing details of an unprecedented merger within the British chemical industry. In December 1926, Brunner,

Mond, Nobel Industries, United Alkali, and the British Dyestuffs Corporation formed Imperial Chemical Industries (ICI) [65]. This was mainly a response to the formation of IG Farben in Germany during 1925, which followed the wartime union of interests among dye firms. In both Germany and Britain the mergers were propelled by the drain on resources arising from heavy investment in high-pressure ammonia synthesis and in the case of IG Farben interest in high pressure conversion of coal to oil (Chap. 16). By the end of 1928, Pollitt had moved to ICI's new, prestigious Millbank headquarters, overlooking the River Thames, in London.

9.5 Observations in America and Independence in Energy Supply

While the energy needs of synthetic ammonia were far less than for cyanamide, the reliable supply of electricity, and its cost, were major factors in ensuring success. Studies into the generation of electric power, including self-sufficiency at Billingham, were prompted especially by progress in the United States. Cowap and a colleague had inspected power stations there during August 1924. Soon after, plans were made for Billingham to become self-sufficient in the supply of power. In May 1927, Humphrey was in the United States where he visited 22 central power stations and 16 factories. He was impressed with recent growth in all sectors of the economy: "The railroads, after a slump period, are once more prosperous and enormous freight trains over a mile long are an inspiring sight" [66]. The widespread application of electric power had brought about "this phenomenal growth of prosperity." He conferred with experts on the rival merits of gas engines and steam turbines in generating electricity.

By-product coke ovens were also on the agenda. The US steel industry, the main consumer of coke, made from bituminous coal, was in the final stages of conversion from beehive coke ovens, originally introduced in the 1830s—which released ammonia and other chemicals to the atmosphere or as waste tars—to by-product coke ovens, from which the recovery of ammonia, particularly as ammonium sulphate, was readily accomplished. Around 80% of coke ovens permitted by-product recovery in the mid-1920s, compared with just 17% in 1910. The Koppers Company practically monopolized the by-product coke oven trade. Its ovens, including those of its consultant engineer J. Becker, were observed in action at the Clairton By-Product Coke Company, of the Carnegie Steel Company [67]. The Semet-Solvay Company, founded in the mid-1890s to manufacture coke ovens, ceased their production with the 1920 merger that resulted in the formation of Allied Chemical. Humphrey was told that the Semet-Solvay Engineering Corporation had recently been formed to revive the coke oven business.

There was news about progress in synthetic ammonia in America. Louis Jones, who after leaving the Solvay Process division of Allied Chemical co-founded the Nitrogen Engineering Corporation, advised Humphrey that the Commercial Solvents Corporation, which relied on hydrogen as a by-product of the Weizmann process, had converted its synthetic ammonia plant into a methanol plant, based on the reaction between hydrogen and carbon monoxide (later relying on hydrogen from natural gas); and Du Pont had purchased controlling stock in the Ammonia Casale patents in the United States and Canada [68]. "This consolidates the Synthetic Ammonia interest in America and takes over the interest of the Electric Bond & Share Company in the [Casale related] Hydro Electric Chemical Company and the Niagara Ammonia Company." Major Charles O. Brown, a Solvay chemist and Jones's part-time partner, described the new electrolytic plant of the Solvay Process Division of Allied, in Syracuse, in which the electrolysis of brine would afford chlorine for dye manufacture at the nearby factory of National Aniline (also part of Allied), caustic soda for artificial silk (viscose) manufacturers in the vicinity, and hydrogen for the Allied ammonia plant.

Jones allowed Humphrey "to take a glance at a general sectional view" of the Nitrogen Engineering Corporation's converter. It was no more than a glance [69]: "The drawing was laid out on the table for half a minute and then removed. It was not explained and was only shown to impress me that they had a thoroughly worked out converter design." Humphrey had to work from memory to describe the converter, including: "The catalyst basket situated above the heat interchanger is pierced by a ring of vertical pipes and by a central pipe." The converter "appears to have a good many features in common with the Syracuse Converters [of Allied], of which we have drawings, and it aims at keeping down the catalyst temperature by heat abstraction from the catalyst, since the ring of pipes acts as a catalyst cooler." The Nitrogen Engineering converter was stated to work autothermically at one-third load. "It looks a good deal cheaper to build than our own converter and it has no internal lagging." Humphrey was intrigued to find that in American factories aluminium paint covered all exposed surfaces.

Humphrey's visit proved invaluable, and contributed to "Perhaps his greatest achievement at Billingham. . .the 40,000 K.W. electric power station. . .which had the special complication of letting down a large part of the steam through primary generators from 80 to 250 lbs. per square inch for use in the steam engines of the factory" [70]. Tenders for the power station were invited in December 1927, contracts were signed in February 1928, work began in May 1928, and it began operating by February 1929. It was Britain's sixth largest power station [71, 72]. Relying on pulverized fuel, the high-pressure, high-temperature power plant for generating electricity and steam incorporated what was stated to be the world's largest high-pressure fuel boiler of its type, with a distillation plant claimed to be the first of its kind, and it was highly automated. The new English Steel Corporation Ltd (the 1928 merger of Vickers, Vickers Armstrong, and Cammell Laird) supplied 32 boiler drums. The power station fulfilled its intended purpose: it gave independence from outside sources in the supply of electricity.

In 1927, Colonel Pollitt, who "like Dr. Mond, ... [had] created another large industry," received the Society of Chemical Industry's medal for his efforts in inaugurating an era of high-pressure industrial chemistry in England [73]. In the same year, Alfred Mond and Pollitt persuaded the board of ICI to authorize new capital expenditures for expansion at Billingham [74]. Carbon dioxide from the purification of hydrogen was used in the manufacture of ammonium sulphate by the anhydrite process and compressed into the solid "Drikold" for which there was a substantial market in Britain.

The enlarged Billingham ammonia factory was divided into sections by two appropriately named parallel, broad avenues, running east-west: Ammonia Avenue, north of which was the ammonia synthesis section, including the Bamag hydrogen plant, and south of which, bordered by Nitrates Avenue, was the ammonium sulphate production area, water gas plant, and coke ovens. Electricity was distributed, along with the gases and liquids used in various reactions, by a network of elevated pipe bridges. Most installations were in place by 1928, in which year synthetic ammonia production had almost reached a peak, and not just in Europe. In the United States, both Du Pont and Allied had just completed synthetic ammonia factories, while half a world away work had started on Noguchi's vast ammonia works, claimed to be the third largest in the world, at Hûngnam in north-eastern Korea.

During a tour of the Billingham factory—what the locals called "The Synthetic"—around 1930, the writer Aldous Huxley was greatly struck by the layout and organization, with cathedral-like structures dedicated to ammonia—tall converters lined up in single-storey structures, compressors, pumps, and apparatus for supplying pure nitrogen and hydrogen, dwarfing workers and visitors alike—all of a massiveness that truly astonished him. It was at once mysterious and heroic, which is why it served as an inspiration for his *Brave New World* (1932). Huxley recorded that, when set against the realities of life outside, with unemployment and poverty kindling widespread social unrest, Billingham was "an ordered universe. . .in the midst of the larger world of planless incoherence" [75].

But no sooner had Billingham, this marvel of high technology chemistry, been expanded than the export market for nitrogen fertilizer collapsed. In 1929, the year of the Wall Street crash, the "large industry," at least the British branch, was suffering from 50% overproduction, with less than 5% return on capital. The outcome was that ICI decided to lay off much of its workforce at Billingham.

Heavy losses were incurred on ammonium sulphate, which until the late 1920s had been a staple export product. New ICI products were introduced: in 1929, Nitro-Chalk, the first British granular fertilizer, and in 1931 concentrated mixed fertilizers, also in granular form.

As with synthetic ammonia, the impact of the recession was felt by the British gas-making industry. Thus, for example, the Gas Light and Coke Company, Ltd, in 1932 lost £4000 on the ammonium sulphate business, in contrast to £200,000–£300,000 profit made in previous years. In 1931, after thousands of workers had been released from Billingham, Humphrey, who had seen through the many changes since 1920, retired but remained a consultant to ICI [76–78].

9.6 Nitric Acid, Until 1939

The inter-allied joint technical mission of February 1919, headed by Brigadier Harold Hartley, and under the aegis of the British Ministry of Munitions, recorded German developments and capabilities in industrial chemistry, and reported to Lord Moulton. There was more than a hint, at least judging from the participants, of potential applications in Britain, notably at Brunner, Mond. In addition, there was a parallel programme involved in the undertaking of similar investigations, but in this case more as part of the effort to restore industrial operations in Germany, thereby providing jobs for workers, and at the same time enabling production of goods in order to satisfy reparations demands. This followed formation in November 1918 of the Inter-Allied Armistice Commission, headed by Marshal Foch, after which the zones of occupation were established. The headquarters of the British zone were in Cologne, under the command of Field Marshal Lord Herbert Plumer, who in mid-January 1919 created an Industrial Department. Among its members was the electrochemist Major Arthur John Allmand (1885–1951), who before the war had spent one year in Haber's Karlsruhe laboratory. At the end of January, Allmand and E. R. Williams investigated the manufacturing sequence: calcium carbide-calcium cyanamide-ammonia, at Knapsack; and then ammonia-nitric acid-nitrate, at Bayer's Leverkusen factory. At Leverkusen, in contrast to Hoechst, production of nitric acid had been problematic due to corrosion and the need for frequent repairs. At least that is what they were told.

No sooner had Hartley's mission completed its work in mid-February 1919, than French experts began arriving at German factories within the French zone of occupation, to investigate processes employed in the synthesis of dyes, munitions, and war gases. Among the first groups was that of Camille Matignon and two colleagues, chemist Paul Pascal (1880–1968), and a M. Vannin, who took a special interest in nitrogen products. At Oppau, they drew up a description of the interior of one of the 12-metre converters. Their efforts to determine the ammonia oxidation catalyst led to the "discovery of certain stocks in the outhouses of the shop, [which] lead us to think that briquettes of iron mixed with oxide of chromium with perhaps an admixture of a little bismuth salt was used." The Oppau plant was "characterized by the enormous dimensions of the converters and [nitric acid] absorption towers." They observed that ammonia converters were closed with hydraulic presses, and then enclosed in special steel jackets. For reasons of safety, the converters were installed partly underground in large halls whose walls were open at the top of the sides. The compressors were manufactured in Berlin to the instructions of BASF by Sulzer and Schwarzkopf. The nitric acid absorption towers, three parallel series of seven "gigantic towers," were accompanied by "A gigantic travelling crane of more than 30 m height [that] serves the whole installation" (Fig. 9.3). They then visited Hoechst, where, as at BASF, "engineers at times treated us with scant courtesy." Of special interest was the newer ammonia oxidation shop that used mainly BASF synthetic ammonia. Aqueous ammonia arrived at the works from Oppau, Knapsack (produced from cyanamide), and coke oven sources. The report on ammonia

Fig. 9.3 Absorption towers for nitric acid, at Oppau, IG Farben, 1930s. (*Erzeugnisse unserer Arbeit*. Frankfurt am Main: I.G. Farbenindustrie, 1938, opposite p. 173)

oxidation was no doubt of special interest to the French, due to their reliance on cyanamide. It was prepared by Pascal at Angoulême, the site of the large ammonia oxidation facility, and dated 22 February 1919 [79].

After the war, Bamag (Bamag-Meguin from 1924), working with Uhde's consultancy firm, became a major supplier of nitric acid plants throughout Europe, as well as to the Soviet Union and Japan. The introduction of acid-resistant stainless steels overcame many of the earlier corrosion problems in nitric acid facilities. The most important feature of the ammonia oxidation plant, as introduced in Germany, was the special chrome-steel converter in which layers of catalyst were laid out

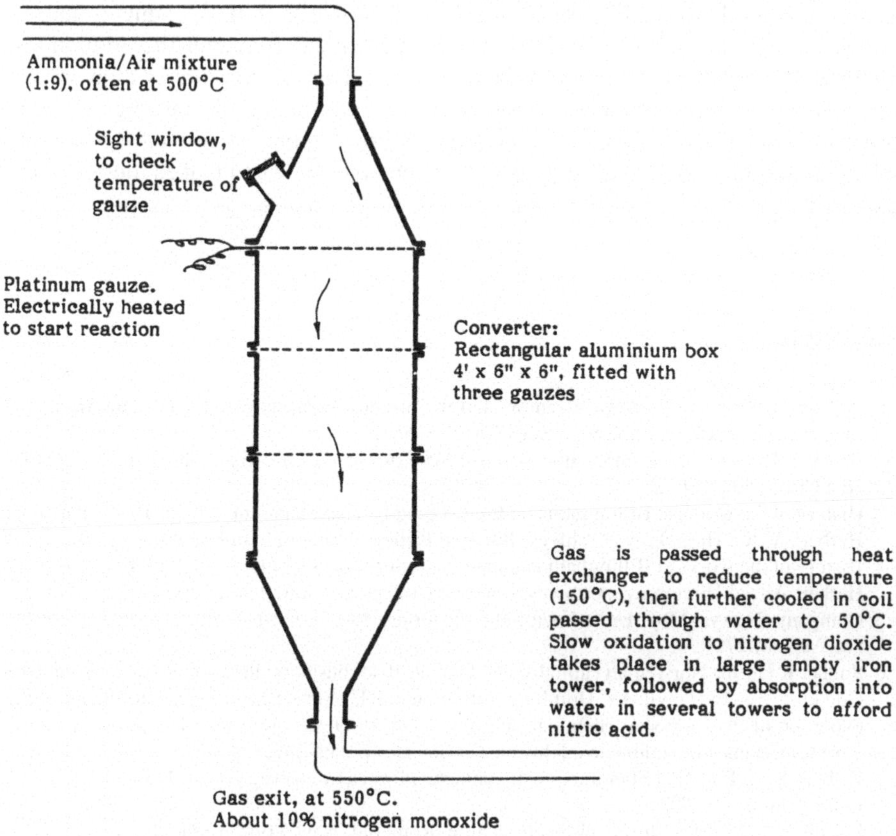

Fig. 9.4 Diagram of Imperial Chemical Industries (ICI) apparatus (ammonia burner) for the oxidation of ammonia to nitric acid. It is essentially the same as introduced in Germany by Nikodem Caro and others during World War I

horizontally. The catalyst, made up of a fine wire of platinum-rhodium alloy, was woven into a large circular gauze. The rhodium, representing 10% of the catalyst wire, gave strength and minimized the loss of platinum. Ammonia oxidation plants developed during World War I were to become the basis of all such installations in use until the 1950s and beyond. In 1926, ICI at Billingham constructed its first nitric acid plant for the oxidation of synthetic ammonia [80] (Fig. 9.4). This enabled the onsite manufacture of ammonium nitrate for blasting explosives and, by mixing with calcium carbonate, the fertilizer Nitro-Chalk. IG Farben's version of nitro-chalk was also called Leuna saltpetre. The use of saltpetre in the trade name was seen as a threat to its market by the Chilean nitrate industry. By 1930, Leuna-Merseburg had introduced ammonia oxidation equipment that operated under pressure.

In 1939, the British Ministry of Supply took delivery of 10-ft diameter gauzes, which became the norm for industrial use. In a typical converter the ammonia is vaporized over warm water and mixed with air. The gas mixture is then sucked through the converter or pushed through at low pressure and the gauze electrically heated to start the reaction. The reaction mixture is rapidly removed, cooled and taken up in dilute nitric acid, or water. The uptake is slow and makes the operation expensive.

References

1. Pattison M (1983) Scientists, inventors and the military in Britain, 1915–19: The Munitions Inventions Department. Soc Stud Sci 13(4):521–568.
2. Brock WH (2009) J. R. Partington (1886–1965): physical chemistry in deed and word. Bull Hist Chem 34(1):11–20.
3. History of the works at Billingham and their subsequent development, n.d. [c. 1930]. Papers of Herbert Alfred Humphrey, Archives, Imperial College London, item F1, p 4.
4. History of the works at Billingham and their subsequent development, n.d. [c. 1930]. Papers of Herbert Alfred Humphrey, Archives, Imperial College London, item F1, pp 4–5.
5. Company News. Alby United Carbide Factories, Ltd. (1920) J Soc Chem Ind 39(15) (16 August):279R.
6. Sogner K (1998) Norwegian capitalists and the fertiliser business: the case of Hafslund and the Odda process. In: Travis AS, Homburg E, Schröter HG, Morris PJT (eds), Determinants in the evolution of the European chemical industry, 1900–1939: new technologies, political frameworks, markets and companies. Kluwer, Dordrecht, pp 239–256.
7. Kilburn Scott E (1917) Manufacture of synthetic nitrates by electric power. J Soc Chem Ind 36 (14)(31 July):771–777.
8. Fairlie AM (1936) Sulfuric acid manufacture. Reinhold, New York, p 151.
9. Imison CS, Russell W (1922) The oxidation of ammonia. J Soc Chem Ind 41(4) (28 February):37T–45T.
10. Partington JR (1921) Ammonia oxidation at Höchst. J Soc Chem Ind 40(10)(31 May):185R–187R, on 187R.
11. Haber section-heat interchange-report on the rate of heat transfer in a simple type of heat interchanger using high pressure gases, viz., nitrogen, hydrogen, and mixtures of same. Lieutenant Greenwood and Charles Hobbs, 22 April 1918. Solvay Process Company archives, microfilm roll 17. Sidney M. Edelstein Library.
12. History of the works at Billingham and their subsequent development, n.d. [c. 1930]. Papers of Herbert Alfred Humphrey, Archives, Imperial College London, item F1, p 5.
13. Moulton HF (1922) The life of Lord Moulton. Nisbet & Co., London, p 207.
14. Parliamentary news. House of Commons (1918) J Soc Chem Ind 37(15 May):179R–180R, on 180R.
15. Travis AS (1998) High pressure industrial chemistry: the first steps, 1909–1913, and the impact. In: Travis AS, Schröter HG, Homburg E, Morris PJT (eds), Determinants in the evolution of the European chemical industry, 1900–1939: new technologies, political frameworks, markets and companies. Kluwer, Dordrecht, pp 3–21.
16. Reader WJ (1970) Imperial Chemical Industries: a history, vol I. The forerunners, 1870–1926. Oxford University Press, London, p 328.

17. History of the works at Billingham and their subsequent development, n.d. [c. 1930]. Interim report on the Billingham scheme for the production of synthetic ammonia and derivative products, NPC 1926. Papers of Herbert Alfred Humphrey, Archives, Imperial College London, item F1, pp 7a, 8.
18. Facts related to the Billingham factory, 2 June 1925. Papers of Herbert Alfred Humphrey, Archives, Imperial College London, item E31.
19. Waeser B (1926) The atmospheric nitrogen industry: with special consideration of the production of ammonia and nitric acid (trans. Fyleman E). P Blakiston's Son & Co., Philadelphia, vol. I, p xix.
20. Society medallist. George Paton Pollitt, 1927 (1931). Jubilee number, J Soc Chem Ind, p 99.
21. Parke VE (1957) Billingham: the first ten years. Imperial Chemical Industries Limited, Billingham Division, Billingham, County Durham.
22. History of the works at Billingham and their subsequent development, n.d. [c. 1930]. Papers of Herbert Alfred Humphrey, Archives, Imperial College London, item F1, pp 17–18.
23. Parke VE (1957) Billingham: the first ten years. Imperial Chemical Industries Limited, Billingham Division, Billingham, County Durham, p 9.
24. History of the works at Billingham and their subsequent development, n.d. [c. 1930]. Papers of Herbert Alfred Humphrey, Archives, Imperial College London, item F1, p 11.
25. Hackspill L (1922) L'azote. La fixation de l'azote atmosphérique et son avenir industriel. Masson & Cie/Gauthier, Villars et Cie, Paris.
26. Parke VE (1957) Billingham: the first ten years. Imperial Chemical Industries Limited, Billingham Division, Billingham, County Durham, p 11.
27. Parke VE (1957) Billingham: the first ten years. Imperial Chemical Industries Limited, Billingham Division, Billingham, County Durham, p 47.
28. Reader WJ (1970) Imperial Chemical Industries: a history, vol. I. The forerunners, 1870–1926. Oxford University Press, London, pp 354–355.
29. No. 77, Nitrogen prod. Synth amm. Haber proc. catalysts. Analysis of samples received from Germany between February and August 1919. H. E. Jones, Brunner, Mond & Co., Northwich, February–August 1919. Solvay Process Company archives, microfilm roll 2. Sidney M. Edelstein Library, Israel National Library.
30. No. 78, Nitrogen prod. Synth amm. Haber proc. catalysts and catalysers. Examination of Oppau catalysts from vanadium. H. E. Jones, Brunner, Mond & Co., Northwich, May 1920. Solvay Process Company archives, microfilm roll 2. Sidney M. Edelstein Library, Israel National Library.
31. Nitrogen Products Committee. Final Report (1919) Ministry of Munitions, Munitions Inventions Department, Cmnd 482. HMSO, London.
32. Humphrey HA (1920) The report of the Nitrogen Products Committee. J Soc Chem Ind 39 (31 January):25R–29R, on 25R, 26R, 27R.
33. History of the works at Billingham and their subsequent development, n.d. [c. 1930]. Papers of Herbert Alfred Humphrey, Archives, Imperial College London, item F1, p 12.
34. History of the works at Billingham and their subsequent development, n.d. [c. 1930]. Papers of Herbert Alfred Humphrey, Archives, Imperial College London, item F1, p 13.
35. Parke VE (1957) Billingham: the first ten years. Imperial Chemical Industries Limited, Billingham Division, Billingham, County Durham, p 21.
36. No. 317, Nitrogen prod. Synth amm G.C.Co. proc. Recovery of heat. Heat transfer in hot and cold interchangers in synthesis system at Sheffield, 1 August 1919. Solvay Process Company archives, microfilm roll 2. Sidney M. Edelstein Library, Israel National Library.
37. No. 319, Nitrogen prod. Synth Amm. G.C.Co. proc. Miscellaneous. Test of Cr-Va steel at temperatures ranging from 70 F to 1200 F. H. A. Humphrey, Brunner, Mond & Co., Northwich. Dec. 2, 1919. Solvay Process Company archives, microfilm roll 17. Sidney M. Edelstein Library, Israel National Library.
38. Parke VE (1957) Billingham: the first ten years. Imperial Chemical Industries Limited, Billingham Division, Billingham, County Durham, p 22.

39. Synthetic Ammonia and Nitrates, Ltd. (1920) J Soc Chem Ind 39(11)(15 June):209R.
40. W. H. Kniskern to H. A. Humphrey, 6 August 1920. Papers of Herbert Alfred Humphrey, Archives, Imperial College London, item E1.
41. Steen K (2014) The American synthetic organic chemicals industry: war and politics, 1910–1930. The University of North Carolina Press, Chapel Hill, pp 250–256.
42. Reader WJ (1970) Imperial Chemical Industries: a history, vol. I. The forerunners, 1870–1926. Oxford University Press, London, pp 360–362.
43. Bertrams K (2013) The years of crisis (1914–1950): the making and unmaking of international alliances. In: Bertrams K, Homburg E, Coupain N, Solvay: the history of a multinational family firm. Cambridge University Press, Cambridge/New York, pp 151–329, on 195–196, 201, etc.
44. G. P. Pollitt to H. A. Humphrey, 29 August 1920. Confidential. Papers of Herbert Alfred Humphrey, Archives, Imperial College London, item E2.
45. H. A. Humphrey to G. P. Pollitt, 30 August 1920. Confidential. Papers of Herbert Alfred Humphrey, Archives, Imperial College London, item E3.
46. Reader WJ (1970) Imperial Chemical Industries: a history, vol I. The forerunners, 1870–1926. Oxford University Press, London, pp 364–365.
47. J. A. Harker to H. A. Humphrey, 28 January 1921. Papers of Herbert Alfred Humphrey, Archives, Imperial College London, item E8.
48. Partington JR (1921) The Haber process at Oppau. J Soc Chem Ind 40(6)(31 March):99R–100R; The Haber process at Merseburg (Saxony). J Soc Chem Ind 40(8)(30 April):144R.
49. Abstract of letter from chairman, dated 21st March 1921, to H. A. Humphrey. Papers of Herbert Alfred Humphrey, Archives, Imperial College London, item E11.
50. Parke VE (1957) Billingham: the first ten years. Imperial Chemical Industries Limited, Billingham Division, Billingham, County Durham, p 23.
51. Runcorn compressors; Billingham circulators [nd]. Papers of Herbert Alfred Humphrey, Archives, Imperial College London, item M3.
52. F. G. Donnan to H. A. Humphrey, 8 March 1922. Papers of Herbert Alfred Humphrey, Archives, Imperial College London, item E13.
53. G. P. Pollitt to H. A. Humphrey, 16 May 1923. Papers of Herbert Alfred Humphrey, Archives, Imperial College London, item E18.
54. Parke VE (1957) Billingham: The first ten years. Imperial Chemical Industries Limited, Billingham Division, Billingham, County Durham, pp 26–28.
55. H. A. Humphrey to G. P. Pollitt, 7 July 1923. Papers of Herbert Alfred Humphrey, Archives, Imperial College London, item E2.
56. H. A. Humphrey to Professor Robert Salmon Hutton, 12 September 1948. Papers of Herbert Alfred Humphrey, Archives, Imperial College London, letter pages F52–53, item F51.
57. Parke VE (1957) Billingham: The first ten years. Imperial Chemical Industries Limited, Billingham Division, Billingham, County Durham, p 48.
58. H. A. Humphrey to Professor Robert Salmon Hutton, 12 September 1948. Papers of Herbert Alfred Humphrey, Archives, Imperial College London, Item F51.
59. Parke VE (1957) Billingham: the first ten years. Imperial Chemical Industries Limited, Billingham Division, Billingham, County Durham, p 80.
60. H. A. Humphrey to G. P. Pollitt, 28 February 1925. Papers of Herbert Alfred Humphrey, Archives, Imperial College London, item E27.
61. Wasserstein B (ed) (1977) The letters and papers of Chaim Weizmann. Series A, vol. XI, January 1922–July 1923. Transaction Books, New Brunswick/Rutgers University, pp 259–261.
62. Weintraub B (2003) Ludwig Mond: great chemist-industrialist; Alfred Mond (Lord Melchett): great Zionist leader. Bull Isr Chem Soc 14 (December):26–31.
63. H. A. Humphrey, Memorandum, 18 October 1926. Papers of Herbert Alfred Humphrey, Archives, Imperial College London, item E33.

64. Pettigrew AM (1985) The awakening giant: continuity and change in ICI. Basil Blackwell, London, p 129.
65. Reader WJ (1970) Imperial Chemical Industries: a history, vol I. The forerunners, 1870–1926. Oxford University Press, London, pp 451–466.
66. H. A. Humphrey, Preliminary report on visit to United States of America. 9th April to 28th May 1927. Papers of Herbert Alfred Humphrey, Archives, Imperial College London, item M6, p 3.
67. H. A. Humphrey, Preliminary report on visit to United States of America. 9th April to 28th May 1927. Papers of Herbert Alfred Humphrey, Archives, Imperial College London, Item M6, pp 45–51.
68. H. A. Humphrey, Preliminary report on visit to United States of America. 9th April to 28th May 1927. Papers of Herbert Alfred Humphrey, Archives, Imperial College London, item M6, p 94.
69. H. A. Humphrey, Preliminary report on visit to United States of America. 9th April to 28th May 1927. Papers of Herbert Alfred Humphrey, Archives, Imperial College London, item M6, pp 95–97.
70. Parke VE (1957) Billingham: The first ten years. Imperial Chemical Industries Limited, Billingham Division, Billingham, County Durham, p 45.
71. Humphrey HA, Buist DM, Bansall JW (1930) The steam and electric power plant of Imperial Chemical Industries, Ltd., at Billingham. J Inst Elec Eng 68(406)(October):1233–1275 (with discussion papers, 1276–1290).
72. Original drawings relating to the electrical equipment and power station are held with the Archive Centre, The Institution of Engineering and Technology, London, reference NAEST 236.
73. Pollitt GP (1927) The development of the synthetic nitrogen industry in Great Britain. J Soc Chem Ind 46(15 July):291T–295T.
74. Reader WJ (1970) Imperial Chemical Industries: A history, vol I. The forerunners, 1870–1926. Oxford University Press, London, pp 365–370.
75. Bradshaw D (2007) Introduction to Huxley A ([1932]) Brave new world. Vintage, London, p xxii.
76. For the later history of Billingham, until the late 1970s, see Pettigrew AM (1985), The awakening giant: continuity and change in ICI. Basil Blackwell, London, pp 120–155.
77. Williamson M (ed) (2008) Life at ICI: memories of working at ICI. Teesside industrial memories project, May 2008. Atkinson Print, Hartlepool.
78. In the 1990s ICI divested of various divisions, notably dyestuffs and pharmaceuticals (Zeneca, 1993), and acquired Unilever's specialty chemical business. The burden of debt forced the sale of the remaining divisions to the Dutch corporation AkzoNobel in 2008. Billingham continues to manufacture ammonia and fertilizer, since 2015 as part of CF Fertilizers UK, Ltd, under ownership of CF Industries Holdings Inc.
79. "On the visits of February 15, 16, 17, 1919, to the German plants of the French Zone of Occupation." Accompanying: "Visits to the German Plants Situated in the Occupied Zones," pp 62–85. Solvay Process Company archives, microfilm roll 28. Sidney M. Edelstein Library, Israel National Library.
80. Miles FD (1961) Nitric acid: manufacture and uses. Oxford University Press, London.

Chapter 10
Non-BASF Ammonia Technologies

Abbreviations Used in this Chapter

ASED	SA Ammoniaque Synthétique et Dérivés. (Montecatini-Fauser)
Azogeno	Società per la Fabbricazione dell'Ammoniaca Sintetica e Prodotti Derivati. (Claude)
Carburo	Società Italiana del Carburo di Calcio, Acetilene e altri Gas (Roma). In 1922, taken over by **SAFFAT**, at which time the **Terni** company was established.
DSM	Dutch State Mines (Nederlandse Staatsmijnen).
Montecatini	Montecatini Mining and Agriculture Company (Società Generale per l'Industria Mineraria e Agricola Montecatini). (Fauser)
Pechiney	Cie d'Alais, Froges et Camargue.
SAFFAT	Società degli Alti Forni Fonderie e Acciaiere di Terni. Also: Società Alti Forni Fonderie & Acciaiere; Società degli Alti Forni e Fonderie di Terni.
SIAS	Società Italiana Ammoniaca Sintetica-Processo Casale.
SIPA	Società Italiana per la Fabbricazione di Prodotti Azotati e di altri sostanze per l'Agricoltura. Also: Società Italiana Prodotti Azotati. (Montecatini)
SIRI	Società Italiana Ricerche Industriali. (Casale)
Terni	Terni-Società per Indùstria e l'Elettricità. Also: Società per l'Industrie e l'Elettricità 'Terni'. (Casale)

10.1 Monopoly Encouraging Innovation

On 24 January 1927, Nikodem Caro, co-inventor with Adolph Frank of the calcium cyanamide process for the capture of atmospheric nitrogen, lectured in Berlin on the nitrogen industry. He was particularly scathing of those countries that, he claimed, after World War I used "chemists and engineers in uniform" and "stolen"

A.S. Travis, *Nitrogen Capture*, https://doi.org/10.1007/978-3-319-68963-0_10

Haber-Bosch synthetic ammonia patents to develop nitrogen capture industries outside Germany. He could give credit only to Georges Claude in France and Luigi Casale in Italy for original innovations in ammonia synthesis. The great proliferation of nitrogen factories, he remarked, "makes one think and raises questions" about those countries that had without doubt stolen the ammonia industry to satisfy their strategic ambitions: "And now one understands that in most cases these new factories do not spring from economic but from political necessity, the strategic need for independence in nitrogen requirements in agriculture and for production of munitions" [1]. This resonated with Humphrey's scenario in July 1923, in which he foresaw the spread of nitrogen factories outside Germany's borders unless BASF maintained its monopoly on synthetic ammonia. But in 1927, with new processes available, the situation was very different. Nevertheless the strategy of controlling synthetic nitrogen technologies approached the kind of oversight reserved for industries serving the military.

Caro's allegiance at this time was to the German nitrogen producers, including IG Farben, the outcome of the 1925 merger of BASF with other leading German chemical companies. The cyanamide firm that Caro headed, Bayerische Stickstoff-Werke AG, was linked to BASF and then to IG Farben through the Stickstoff-Syndikat, the cartel established by the state in May 1919 to aid its nitrogen industry but that increasingly came under the control of BASF/IG Farben; the Stickstoff-Syndikat soon represented most of the German producers of nitrogen products.

In his talk, Caro had erred in remarking that foreigners had been able to use the BASF Haber-Bosch patents and visits to the Oppau ammonia facility to master the high pressure process. Observation and documentation had certainly been useful, as had samples of catalysts, but many features of the rival processes were certainly original, and circumvented the BASF patents. In any case, when BASF received enquiries concerning licensing from the United States, Holland, Italy, Japan, the Soviet Union, Germany, and elsewhere, it showed little or no interest in discussing terms. Occasionally a deliberately exorbitant price was quoted, along with conditions sufficiently stringent to turn away any would-be licensee. Before the late 1920s, there was just one exception, made under abnormal circumstances. In November 1919, the French, as military occupiers of Oppau, had come to an arrangement that provided access to Haber-Bosch technology, but even then they could not master the complexities.

For well over a decade following the introduction of the Haber-Bosch process, BASF's high asking price, or lack of interest in licensing, prevented the firm from capitalising internationally through the sale of technical know-how on the success of its ammonia process. Instead, BASF preferred to corner the world market in the supply of nitrogen products, particularly the fertilizer ammonium sulphate. As a result, outsiders took up research into high-pressure ammonia technology. The exigencies of war, with the urgent, effective bonding of science and industry, motivated the first research. In peacetime, prior to 1914, this urgency and merging had hardly been practiced, apart from in Germany, and then only in a half-hearted way. After 1918, through wartime experiences and the organization of science and technology, other nations began to catch up. It was their successes and the struggles over control of markets that brought about a state of over-production, particularly among the large producers—IG Farben more than any other—by around 1928. That was the year in which the original ammonia patents began to expire, following a 1923 change in German patent law that extended patent protection from 15 to 18 years. Only then did IG Farben, confronted with successful rival processes and competition in the ammonium sulphate export trade, consider sale of its much vaunted know-how.

On one matter, however, Caro was correct. Other nations had built up strengths in synthetic nitrogen, mainly in order to better defend themselves in times of conflict or shortages. The model was Germany, which had brought about the situation in the first place through developing an industry that made the nation self-sufficient in wartime. Outside Germany, even more than the synthetic dye industry, the synthetic nitrogen industry was perceived to be essential for national security, for the supply of fertilizer as well as of munitions. Wartime experiences suggested to many that military force, including preparations for its application, was a particularly useful tool in resolving international disputes and maintaining a balance of power. In that sense, self-sufficiency in nitrogen was important.

10.2 Casale, Claude, and Fauser

The BASF blocking strategy encouraged the rapid development of alternative high-pressure ammonia processes—in Italy, France, and soon after in Japan, Germany, and the United States—and competition among engineering and chemical companies eager to market and export their new technologies. The spur to inventors and entrepreneurs to use wartime experiences to devise synthetic ammonia processes is particularly associated with three individuals endowed with a hybrid of skills, in chemistry, electrotechnology, and engineering: Luigi Casale and Giacomo Fauser in Italy and Georges Claude in France. Casale and Claude worked with extreme pressures, 800 and 1000 atmospheres, respectively, that is, ultra-high-pressures, or what Claude called "hyperpressures." Important to all the new ammonia processes, once fully developed, was their widespread regional and international availability, and the means for producing hydrogen of adequate purity [2, 3].

This chapter reviews technical developments during the pivotal decade and a half that saw the emergence of not only close imitations of the Haber-Bosch process, as in Britain, but also rival processes. Given the heavy investment involved in constructing ammonia factories, nitrogen capture happened to fit in smoothly with the industrialization programmes of new political, strategic and economic systems that were impacting on vast expanses of the globe. Moreover, during the second half of the 1920s, this was indirectly aided by the failure of leading Western nitrogen manufacturers, particularly in Britain and Germany, that were dependent on export markets, to foresee the complications associated with overproduction and declining markets.

The story of the development of significant alternative high-pressure ammonia processes, particularly those worked at extremely high pressures, is one of the most exciting chapters in twentieth century chemical technology [4, 5]. These processes, though using the same basic chemistry as the Haber-Bosch process, incorporated a number of novel differences as well as different catalysts. Luigi Casale's process offered great simplicity as well as reliability and was ultimately the most successful in making high-pressure synthetic ammonia widely available. Apart from being the first original synthetic ammonia process to be successfully commercialized after the Haber-Bosch process, it provides the best example of the rapid global spread of a rival technology. The outcome was that by the mid-1930s the output of ammonia from Casale and other novel converters far exceeded that from BASF Haber-Bosch converters (BASF produced 90% of synthetic ammonia in 1928, but less than 50% in 1931). The following two sections provide backgrounds to the two Italian processes, those of Casale and Fauser.

10.3 Luigi Casale

Luigi Casale (1882–1927) was born in Longosco (Lomellina region, south-west Lombardy) on 22 November 1882 (Fig. 10.1). He studied chemistry under Arturo Miolati (1869–1956), professor of electrochemistry in charge of physical chemistry

Fig. 10.1 Luigi Casale. (Casale SA)

at the Royal Polytechnic of Turin (Reale Politecnico di Torino), who expressed a strong interest in the electrothermal nitrogen fixation processes. Casale graduated in 1908 and assisted Michele Piccini Fileti (1851–1914) at the University of Turin's Institute of Chemistry and Electrochemistry. During 1909–1912, he was with the Turin Institute of General Chemistry, where he received his postgraduate degree in 1910, after which he undertook research in physical chemistry under Walther Nernst in Berlin (1912–1913), which brought him in touch with the latest developments in the thermodynamics of gas reactions. This, and publicity from 1910 surrounding Haber's ammonia studies, no doubt stimulated Casale's interest in ammonia. On returning to Italy, Casale investigated the high-pressure synthesis of ammonia, as well as undertaking research in organic chemistry, as assistant of Miolati (1913–1915).

After Italy joined the Allies, in 1915, with declarations of war against Austria-Hungary on May 23 and Germany on August 28, there was for the first time close collaboration between the home-based chemical industry and the academic world. Casale was assigned to chemical studies related to the war effort at the University of Naples, including on asphyxiating gases and protective measures [6–8]. Following a laboratory accident and resulting illness (the cause of his death just over a decade later), he was forced to abandon this work, and in 1916 again took up the study of ammonia, encouraged by Miolati, a prominent member of the supply committees set up by the Italian Ministry of Arms and Munitions. Synthetic nitrogen products were of strategic importance—and not only for munitions, since Italy suffered shortages of grain supplies formerly obtained from Russia and Romania as a result of the closure of the Dardanelles. There was a dire need for fertilizer.

Casale's research was undertaken in the laboratories of two electrochemical firms. The first was Società Rumianca (Chimico Mineraria Rumianca), founded in 1915, in Genova, in the commune of Rumianca, north-western Italy, to manufacture alkali by electrolysis (1916); it later expanded into inorganic chemicals. At Rumianca, Casale undertook investigations into soil chemistry, retaining his strong interest in ammonia. The vice president of the firm was the engineer Lorenzo Allievi (1856–1941), Italian pioneer in aluminium production at Bussi, in Abruzzo in 1907, through his Società Italiana di Elettrochimica. He was also involved in the management of the calcium carbide manufacturer Carburo, located in the commune of Narni, province of Terni, Umbria, and elsewhere. Its acquisitions included the Narni carbide factory of Società Industriale Elettrica della Valnerina, established in 1908, and that had turned to manufacture of cyanamide. Carburo had a major interest in supporting cyanamide manufacture, which is why Allievi did not back Casale's interest in a potentially disruptive technology, that of synthetic ammonia [9]. The second connection was with the IDROS company of Terni (Soc. IDROS Terni), founded on 19 February 1916, for the production of hydrogen, oxygen, and nitrogen products using hydro-electric power. The availability of pure hydrogen was no doubt a major factor in merging the interests of Casale with IDROS. Casale subsequently resigned from Rumianca and moved full time to IDROS. There was a close connection with Società degli Alti Forni Fonderie e Acciaiere di Terni (SAFFAT), incorporated in 1884, which operated a large metallurgical works, specializing in naval armaments. The location of SAFFAT, originally Italy's largest industrial plant, was close to a source of hydro-electric power.

Patent applications indicate the type of ammonia research undertaken by Casale and co-workers at IDROS. First was a continuous, closed apparatus, with the recirculation of unreacted gases, for combining nitrogen with hydrogen, involving a novel catalyst held at 350–600 °C. This was devised during 1916–1917 in collaboration with inventor Carlo Andreucci and patented in 1917 jointly with lawyer Mario Santangelo, president of IDROS. Though the associated pipework was clearly stated to be capable of withstanding gases at high pressure, there was no discussion of high pressure as a requirement [10]. By 1919, Casale had developed a novel high-pressure process that, working at around 800 atmospheres, gave anhydrous ammonia simply by cooling. Casale's experimental converter, in the best tradition of converting "swords into ploughshares," was based on the redundant barrel of a cannon from the armaments of the battleship *Dante Alighieri*, as supplied by SAFFAT.

Casale placed inside the barrel two concentric steel tubes. The multi-tube converter was designed so that the barrel acted as an outer, pressure tube. This converter, with the catalyst tube centrally placed, as manufactured by SAFFAT, became the standard model in the early years (Figs. 10.2 and 10.3).

In 1921 and 1922, Casale filed patents for his ammonia synthesis in Italy, France, Britain, the United States and elsewhere [11]. Patents in the United States were lodged jointly with René Leprestre, of New York. The original high-pressure synthetic ammonia patent application referred to the synthesis of ammonia at pressures of between 100 and 1000 atmospheres, which suggests that Casale was aware of high pressure work undertaken in France from 1917 by Georges Claude. In both cases, the early experiments were beset with enormous dangers, far greater than those encountered at BASF around 1910 when working at 200 atmospheres. However, the very high pressure processes required less active, and less sophisticated, catalysts. Casale's catalyst was scrap iron.

The Haber-Bosch process, it should be emphasized, did not afford anhydrous ammonia unless expensive refrigeration plant was installed (usually ammonia was absorbed in circulating water, under pressure). At Casale's working pressure of 800 atmospheres, this disadvantage was completely done away with. It was possible to obtain liquid ammonia by simply cooling with water at ordinary temperatures. Moreover, production of anhydrous ammonia at higher pressures involved more compact units. As a result, and in addition to the absence of refrigeration equipment, the complete apparatus required less factory space [12].

Until around 1923, Casale jointly with Leprestre filed further patents related to high-pressure ammonia synthesis, including modifications to the electric heating coil, separation of ammonia and production of pure nitrogen. Casale also filed patents in his own name, including in 1921 for the production of a nitrogen–hydrogen mixture by first burning air in hydrogen[1]; for the recirculation of

[1]Casale, British patent, GB 148,885-A, filed 30 June 1921.

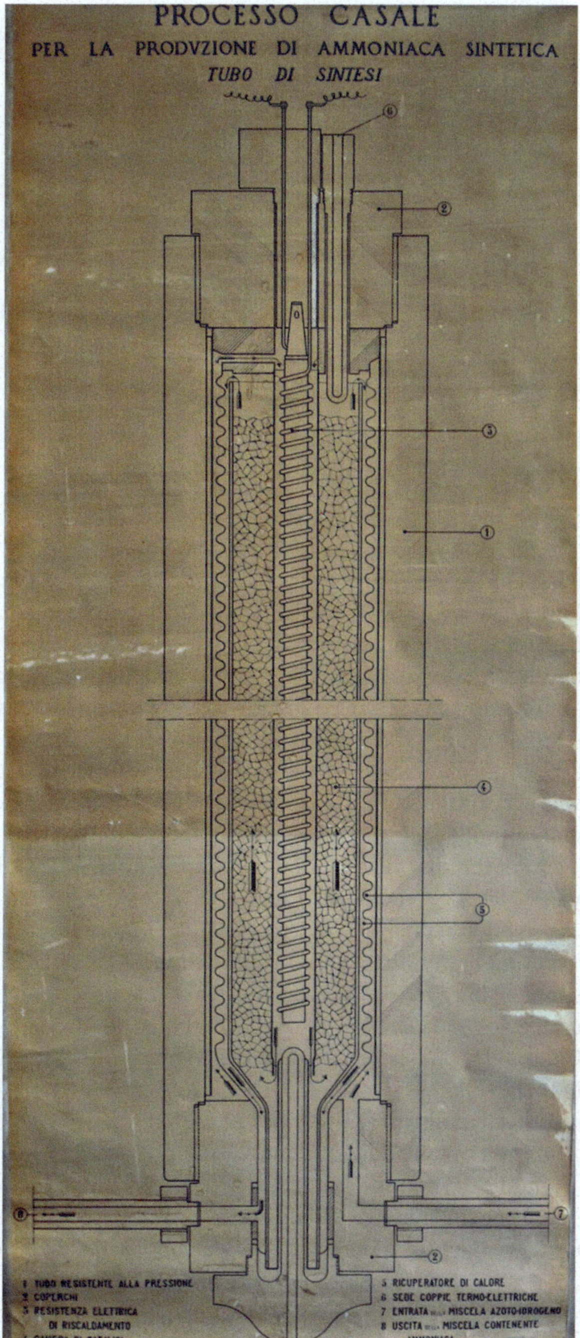

Fig. 10.2 Early Casale converter. (Casale SA)

Fig. 10.3 Assembling a Casale converter, 1923. The lid is open, showing the massive bolts and studs required to maintain the high pressure, of approximately 800 atmospheres. (Casale SA)

unreacted gases[2]; and for an improved ammonia synthesis apparatus.[3] Meantime Leprestre was active in the United States raising capital. The IDROS research facilities at Terni consisted of a modest workshop-laboratory and office. Pure hydrogen was available by electrolysis. For the separation of nitrogen from the air, oxygen was removed in a stream of hydrogen (Casale's burning of air process), to give steam. Mixed gases for the ammonia synthesis were compressed stepwise to 800 atmospheres. As in the Haber-Bosch process, a gas recycle circulation pump was incorporated into the early synthesis loop. Casale's pilot plant at Terni originally produced 100 kg of liquid ammonia daily [13].

Another important advantage of the use of very high pressures was the increased yield (15–18%, compared with 7–8% for the Haber-Bosch process), greater than expected by theory. Georges Claude reported yields of ammonia, as obtained experimentally, that agreed well with those calculated by Haber at up to 500 atmospheres but that at higher pressures exceeded theory. Alfred T. Larson, who undertook research at the US Fixed Nitrogen Research Laboratory (see Chap. 11), and later at Du Pont, found similar results, and developed a catalyst that, he claimed, gave a higher yield under Haber-Bosch conditions. From an industrial point of view what mattered was the considerably increased yield of ammonia obtained at extremely high pressures [14].

In the ammonia synthesis processes at the lower pressure range, at up to 350 atmospheres, the catalyst was specially formulated with the addition of promoters to ensure high activity, as in the Haber-Bosch process. At very high

---

[2]Casale, British patent, GB 194,740-A, filed 19 September 1921.

[3]Casale, Canadian patent, CA 230,616, filed 1 May 1923. See also Casale British patent, GB 193,789, 24 August 1922; addition to GB 185,179.

pressures, less active and inexpensive catalysts were used, as was the case in the Casale and Claude processes. These processes not only gave higher conversion but required around one-seventh the amount of catalyst as compared with the Haber-Bosch process. Also, less gas was pumped and circulated to produce the same volume of ammonia as in the Haber-Bosch process. The disadvantage of using pressures of 800 atmospheres and over was that the life of the catalyst was short. However, since Casale's iron catalyst, made from recycled catalyst or scrap iron, was inexpensive, it could be changed more often than the costly catalyst material used in the Haber-Bosch process. Casale's catalyst also suffered less from poisoning caused by the presence of carbon monoxide. Until the widespread introduction of coke oven hydrogen, from the mid-1920s, hydro-electricity was employed for the electrolytic production of hydrogen, in addition to the supply of power, in medium-size ammonia plants. Nitrogen was obtained by air liquefaction.

10.3.1 Controlling the Catalyst, and Casale's Ejector

As a result of the ultra-high pressure, great heat was generated in the Casale converter in the area of exposure of mixed gases to the catalyst. This became a critical problem, one that would cause the apparatus to explode if not controlled. The solution was to slow down the reaction. This was achieved by allowing some product ammonia to be retained in the recirculated gases [15]. In regulating the process, there were other ways of influencing the temperature of the catalyst: varying the temperature of the central electrical resistance, or varying the mass and speed of gases circulating in the synthesis tube. Flushing the inner wall of the outer tube, or shell, of the converter with synthesis gas kept it at a lower temperature, greatly reducing stress on the tube, enabling the use of ordinary steel rather than special alloys.

A converter giving an output of 20 tons of ammonia daily (as introduced by 1927) was 8.5 metres in length (with lids, 10 metres) and weighed around 40 tons. The inside diameter of the reaction tube was 500 mm, and outside diameter 850 mm. By 1927, a change in the method of closing converters involved substituting a screw closing for bolts. Not only was the apparatus safer but this permitted easier and more rapid assembly and dismounting. Loading and unloading of the catalyst was achieved with ease [16]. Once the synthesis apparatus was operating, the reaction continued normally for some time without internal heating, which, when required, could be adjusted by the operator according to the temperature reading.

The electrical heating element was placed within the innermost tube, and the catalyst in the annular space between this inner tube and the intermediate tube. Fixed between the intermediate and outer tube was a partition, an undulated metal surface, serving as heat exchanger, thereby creating a second annular space. Here, gases entering and moving up the vessel were preheated by the outgoing gas mixture, travelling in the opposite direction. The outer pressure tube was subjected to a minimum of stress by holding it at a temperature of around 200–220 °C, which was at least 100 °C below the temperature at which alteration of the structure of the steel was brought about by the action of hydrogen. The inner surface was cooled by the incoming cold gas mixture.

Gas circulation was normally by means of a piston (reciprocating) pump. However this involved lubrication of the moving parts, and required stuffing boxes and devices for separating out lubricating oil to prevent its entry into the catalyst tube, where, by carbonizing, it would poison the catalyst. These various drawbacks were overcome with a device that—borrowing from steam locomotive technology—resembled an injector but acted as an ejector. Invented in 1923, and in use by 1924, it replaced the circulation pump used in the earlier apparatus (Fig. 10.4) [17]. Its functioning was simple: A new charge of gas was forced into the device at a pressure of 50–100 atmospheres greater than the operating pressure. It thereby pulled through the system the unreacted gas mixture, returning it to the synthesis cycle. This static piece of equipment, replacing not only the circulation pump but also the oil separator and scrubber, offered distinct advantages over compressors then available, which were both costly and difficult to maintain. As a result the synthesis cycle required fewer moving parts, further reducing the apparatus space.

10.4 Rival of Casale: Fauser Ammonia

The engineer Giacomo Fauser (1892–1971), of Novara in the Piedmont region of north-western Italy, west of Milan, had at first worked more or less independently at his father's metal working factory. In 1913, Fauser invented an electrolysis cell, with the aim of producing oxygen for a welding machine. He investigated use of the burning in air of hydrogen obtained from the electrolysis cell for producing pure nitrogen. Fauser went on to study mechanical engineering at the Polytechnico di Milano, where he qualified on 23 December 1918. By then he was interested in the ammonia synthesis, no doubt stimulated by awareness of what the Haber-Bosch process had achieved and the potential for application to Italy's agricultural sector, including the fertile land in Piedmont. In 1919 he apparently turned up at Oppau, with the intention of acquiring a license from BASF for the Haber-Bosch process; but he was turned away. Subsequently he used an old howitzer as converter to synthesize ammonia. In 1920 he produced 100 kg daily working at around 500 °C, and 250 atmospheres in the open air. Fauser filed his main ammonia patent in Italy on 23 April 1921 [18]. Another patent described a novel apparatus for the electrolysis of water suited to the large-scale production of hydrogen (14 May 1921).

Interest in Fauser's process was shown by Ettore Conti (1871–1972), president of Impresse Elettriche Conti, founded in 1884 (and forerunner of the Italian Edison company), a major investor in hydro-electricity in Piedmont. As a result, Conti arranged for Fauser to meet with Guido Donegani, head of Montecatini. This led to the founding by Conti and Donegani in May 1921 of the Electrochemical Company of Novara (Società Elettrochimica Novarese), with a capital of three million lira. Its main objective was the development of Fauser's process. At the same time, Fauser was taken on by Montecatini, and a viable ammonia process that

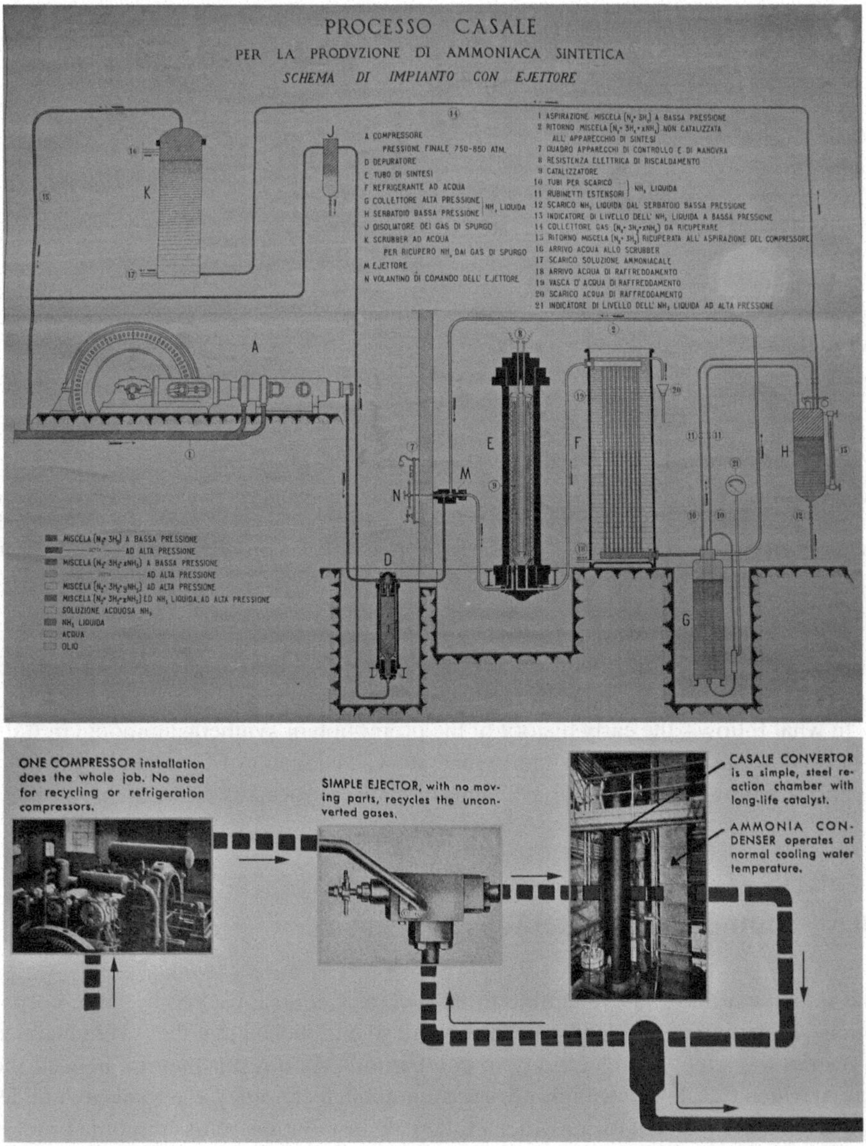

Fig. 10.4 Top: Scheme for Casale synthetic ammonia process, incorporating ejector (injector), M. The product was anhydrous ammonia, unlike the Haber-Bosch process in which the ammonia was absorbed in water. Bottom: The basic Ammonia Casale apparatus employing the ejector for gas circulation remained little changed after three decades, as shown in this diagram appearing in mid-1950s advertisements of the North American Casale representative Foster Wheeler Corporation of New York. (Casale SA)

Table 10.1 High-pressure synthetic ammonia processes in use by the mid-1920s

Process	Year of introduction	Pressure, atmospheres	Temperature, °C	Conversion, per cent, through single converter
Haber-Bosch (BASF/IG Farben, Germany)	1913	200	550	7–8
Casale (Italy)	1921	800	500	15–18
Claude (France)	1921–1922	1000	500–650	40 (first of a series of converters)
Fauser (Italy)	1921–1922	250–300	500	12–23
General Chemical/ Allied (United States)	1921	200	500	20–22
Nitrogen Engineering Corporation (United States)	1926	200–300	500	20–22
Mont Cenis (Germany)	1925–1926	100	400–425	9–20

Note: Catalysts were iron with added promoters; the Mont Cenis process employed an iron cyanide catalyst. An "American" process was developed after 1919 at the US Fixed Nitrogen Research Laboratory. However, "American" was variously applied to a number of processes. The Brunner, Mond ammonia facility at Billingham worked what was effectively a close imitation of the Haber-Bosch process

worked at 250–300 atmospheres was devised [19] (Table 10.1). Fauser's ammonia converter installed at the Novara facility soon had a capacity of one ton daily.

In what follows, the early history of the promotion of synthetic ammonia in Italy is necessarily presented as an intertwined story, as a result of Donegani's involvement, for a time, in both the Casale and the Fauser processes.

10.5 Ammonia Casale SA

The Italian carbide and cyanamide firms such as Carburo, backed by large corporations, monopolies, and leading banks, were so entrenched that they were enabled to restrict the supply of electricity to competitors. As a result they influenced the way in which Casale's new high-pressure ammonia technology was made available. In 1921, Casale planned to construct a full-scale installation of his ammonia process at the IDROS facility. However, Carburo, with its vested interest in carbide, the starting point for cyanamide, recognizing the very real threat of a new technology to its business, placed a number of obstacles in the way of IDROS ammonia production. In particular, the local electricity power company was persuaded against supplying electricity to the potential competitor.

Despite this drawback, and encouraged by the success of his pilot plant, Casale drew up plans for promoting his novel process. He decided on a course of action that involved founding a firm outside Italy to license his process worldwide rather than engage in large-scale ammonia manufacture in Italy. This independent firm would also attract international investors from nations with more stable currencies.

At the same time, with his backers, including SAFFAT, he would continue with technical developments.

On 21 April 1921, Casale set up Ammonia Casale SA in Lugano, Switzerland, in partnership with a banker, to license the process. The capital was 12 million Swiss francs, raised in the United States and soon after in Great Britain.

By establishing the new firm outside Italy, Casale avoided dependence on another short-term but influential backer, Donegani of Montecatini. This happened after chemist Giacomo Luigi Ciamician (1857–1922), in Bologna, convinced Santangelo, the president of IDROS, to undertake large-scale manufacture of ammonia based on the Casale process. To achieve this it was necessary for IDROS to increase capital and also overcome restrictions on the supply of electricity. This led to the involvement of Donegani—who in 1921 was prepared to follow up on any promising synthetic nitrogen process—as well as Carburo, no doubt for much the same reasons. On 22 May 1921, around the time that Donegani agreed to back Fauser, Luigi Casale, IDROS, Montecatini and Carburo merged their interests in Casale's process, through the Società Italiana Ammoniaca Sintetica-Processo Casale (SIAS), based at the IDROS company's works. The capital was 23 million lira, 5 million representing the Casale patents, 6 million IDROS, and 12 million Carburo and Montecatini. By the autumn of 1921, a converter with a daily capacity of two tons of ammonia was operating at Terni. At this time, Montecatini was pursuing an aggressive programme of controlling synthetic nitrogen in Italy; as a deliberate defensive policy it also invested in the other main nitrogen fixation product, calcium cyanamide, by acquiring a majority holding in SIPA (Società Italiana per la Fabbricazione di Prodotti Azotati e di altre sostance per l'Agricoltura) [20]. This organization managed most of the production and sales in Italy of cyanamide, including of the new Terni company, founded in 1922, following the takeover of Carburo by SAFFAT (see Sect. 10.7).

In 1923, the Domodossola cyanamide facility of Società Elettrochimica Dr Rossi, subsequently Società Agraria di Domodossola (1918), was taken over by the Montecatini subsidiary Società Anonima Prodotti Azotati, and capacity was increased to around 12,000 tons per annum. As a result of further expansion in the 1930s, the annual capacity was raised to 50,000 tons. Of the main cyanamide manufacturers this was the second largest, after the Terni company. Through SIPA, Montecatini controlled sales of cyanamide, Chilean nitrate, and gas works nitrogen products. (Distribution of fertilizers in general was through the Federazione Italiana dei Consorzi Agrari, founded in 1892.) High duties were imposed on imported Chilean nitrate in order to favour the domestic nitrogen industry.

10.6 Montecatini and Fauser

By 1923, Montecatini had lost interest in Casale and favoured development of Fauser's process. In 1924, Montecatini founded the Società Anonima Italiana Ammonia (SA Italiana; from 1929 Società Anonima Italiana Ammonia e Derivati), which had exclusive control over the Fauser process, soon known as the

Montecatini-Fauser process. Expansion in Italy was rapid. SA Italiana established the Società Sarda Ammonia e Prodotti Nitrici, and the Società Alto Adige Ammonia, for the manufacture of synthetic ammonia and fertilizer in, respectively, Coghinas, northern Sardinia, with a capacity of 10 tons daily, and Merano, South Tyrol (Venezia Tridentina; the present-day Trentino-Alto Adige region), the northernmost region of Italy [21, 22]. Coghinas received hydro-electricity for power and electrolytic hydrogen from a dam across the River Coghinas built in 1927. Merano (near Marlengo) had a capacity of 50 tons daily and relied on hydro-electric power, again for production of hydrogen by electrolysis, from the new Marlengo (Marling) power station (1924–1925). Merano was claimed to be the largest synthetic ammonia factory outside Germany. It was close to the newly created community of Sinigo, from which labour was available. Fauser ammonia factories were also erected at Mas, Sedico Bribano, in the Province of Belluno, Veneto (Venetia), in the north-east (1924); in the Province of Crotone, Calabria (1927); in the south (between Taranto and Messina); at San Giuseppe di Cairo, in the Province of Savona, Liguria (50 tons daily); and elsewhere in Italy [23]. At the end of the 1930s, the daily output of San Giuseppe di Cairo was 128 tons, of Merano 100 tons, and of Crotone 70 tons.

From 1926, Montecatini, began to license the Fauser process outside Italy. Jointly with the Belgian firm Evence Coppée, the specialist in coke ovens, established in 1850, Montecatini set up SA Ammoniaque Synthétique et Dérivés (ASED) in Brussels to license and manufacture ammonia and its products and construct Fauser equipment (through Société Centrale pour la Fabrication d'Ammoniaque de Synthèse). The Dutch State Mines (Nederlandse Staatsmijnen, DSM), after failing to gain access to BASF technology, in 1928 turned to engineering contractors, including ASED, for designs and cost estimates of ammonia plant. ASED was awarded a contract for gas preparation, and the supply and installation of Fauser ammonia apparatus and ammonium sulphate facilities. The outcome was that ASED, in collaboration with Coppée, constructed at Geleen (Province of Limburg) a synthetic ammonia factory with a capacity of 21,000 tons of fixed nitrogen per year.

Through this project Montecatini became interested in opening an ammonia factory to serve the Dutch-Belgian agricutural industries. As a result the Compagnie Néerlandaise de l'Azote was established in Belgium, in January 1929, with main offices at the Brussels headquarters of Coppée. In December 1930, an ammonia factory was opened in Dutch Flanders at Sluiskil, close to a coke oven facility, and on the Ghent-Terneuzen canal. Its capacity was 45,000 tons of nitrogen a year. It was claimed to be the largest synthetic ammonia plant in the world outside of Germany, overtaking in size the large facilities in Italy and Japan [24]. The Fauser ammonia process was also adopted in Canada, France, Poland, Japan, the Soviet Union and elsewhere. By 1936, fourteen synthetic ammonia plants based on the Fauser process were in operation outside Italy.[4] From the mid-1920s, Fauser

[4]Around 1930, Fauser invented a process for the oxidation of ammonia into liquid nitrogen tetroxide, which was then converted into concentrated nitric acid, suited to the manufacture of explosives. This was developed by Bamag-Meguin and used in many countries.

developed processes for ammonium sulphate, ammonium nitrate, nitric acid, methanol, and urea. Guilio Natta (1903–1979), at the University of Milan, contributed towards the high-pressure methanol process and gas production for use in the ammonia synthesis. In 1933, Fauser set up Montecatini's Central Research Laboratory [25].

Of the two Canadian synthetic ammonia plants, both opened in 1930, one relied on Fauser technology. This was owned by the Consolidated Mining and Smelting Company of Canada, Ltd, at Trail, British Columbia [26]. It drew electrical power from the Kootenay River, including for the electrolytic production of hydrogen, and by the end of the 1930s was producing 38,000 tons of ammonia annually. The other, smaller, plant belonged to Canadian Industries, Ltd, at Sandwich, Ontario. This 1929 merger of four companies, including Mond Nickel Co., used the Ammonia Casale process, again with electrolytic hydrogen. The annual output was 2500 tons of ammonia, mainly for the explosives industry. Canada's hydro-electric power facilities were greatly expanded during World War II.

10.7 Casale's First Licensing Arrangements

From 1921, Ammonia Casale widely promoted its process to would-be licensees outside Italy. In providing licenses and technical assistance Casale was following the strategy followed earlier by the Italian cyanamide licensing body that represented Frank-Caro interests. Moreover, just as with the cyanamide process before 1910, he found great interest from entrepreneurs in Japan. There, the very first large Ammonia Casale facility to open, in 1923, was at Nobeoka, in Miyazaki Prefecture, on the east coast of Kyūshū island (see Chap. 15). Prior to the opening of Montecatini's Merano factory, it was the largest ammonia factory outside Germany.

As mentioned earlier, Carburo was in 1922 absorbed by SAFFAT, which established Terni-Società per l'Indùstria e l'Elettricità (Terni). The Terni company, now allied with Casale, following Montecatini's preference for the Fauser process, gained control of SIAS. A new Casale plant was installed by Terni, at Néra Montoro (near the place named Terni), during 1923–1924 (Fig. 10.5). It produced synthetic ammonia, at 7–8 tons daily, as well as electrolytic hydrogen, oxygen, and distilled water. Later, it relied on coke oven gas for hydrogen. Terni now held interests in both cyanamide and synthetic ammonia.

The original Ammonia Casale converter, which by the autumn of 1921 was producing two tons of ammonia daily, was later set to work at Saint-Auban, in the Alpes-Maritimes department, south-east France, by Pechiney (Cie d'Alais, Froges et Camargue). As French licensee of the Casale process, Pechiney thus diversified into an activity that was less subject to price fluctuations than its main product, aluminium. It was reported in early 1927 that this very first Casale full-scale industrial apparatus had operated uninterruptedly and was still capable of producing two tons every day. Annual chemical and metallurgical analysis of the inside of the catalysis tube in the region subjected to the highest operating stresses showed that

Fig. 10.5 Ammonia factory of the Terni company, Casale process, at Néra Montoro, Italy. (Casale SA)

chemical and physical properties were unchanged [27]. In 1927, the Saint-Auban ammonia plant was undergoing expansion, with the intention of increasing daily output by 10 tons.

In 1925, the capital of SIAS was reduced and it was absorbed by the Terni company. On 8 July 1925, jointly with the Industrial Consortium of Velinao and Luigi Casale, it founded at Terni the Società Italiana Ricerche Industriali (SIRI), Italy's first modern research and development organization. In the high-pressure research, Casale was assisted by his wife, Maria Sacchi-Casale (1889–1950), who held a degree in chemistry and was an expert in analysis [28]. Also in 1925, Casale set up an International Consultative Committee. By early 1927, Néra Montoro was producing 30 tons daily. From this time, Terni declined in importance as a manufacturing location [29]. A new converter there produced 4 tons of ammonia daily, probably in part for test purposes. High-pressure reactions were developed for methanol and urea (the latter from reacting liquid ammonia with carbon dioxide). Other SIRI work included gas production for the ammonia process.

The range of Casale converters enabled production of 8, 16 and, by the mid-1920s, 20 tons of anhydrous ammonia daily. By 1927, converters of 20-ton daily capacity were used in France and Japan (Claude had also introduced 20-ton converters by this time). Casale, at a meeting of the International Consultative Committee a few days before his death in early 1927, advised his audience that he was ready to produce even larger units [30].

Casale equipment worked with coke oven hydrogen was introduced in Belgium and France. This was through Société Ammonia, an association of French mining firms and Pechiney that licensed the Casale process with special reference to the use of by-product coke ovens as the source of hydrogen. In December 1923, Belgian Solvay acquired a 13.3% stake in Société Ammonia. It was a deliberate move to get Solvay more closely connected with the French chemical industry and to counter, with others, Claude interests [31]. After obtaining a license for the Casale process in France, Kuhlmann established joint subsidiaries with some leaders in the French coal and coke industry. In June 1924, Kuhlmann announced the formation of Anzin-Kuhlmann Chemical Products Co. (Société des Produits Chimiques Anzin-Kuhlmann) at Anzin to produce ammonia by the Casale process. Production began on 4 December 1926, using hydrogen from the coke ovens of SA Mines d'Anzin. In 1925, the Casale process was introduced at the Soulom facility in France that until then had continued to work the Birkeland-Eyde electric arc process [32].

Casale's process was described by Arturo Miolati, Casale's teacher at Turin, in a detailed scientific and technical lecture he gave in Prague on 27 February 1927 [33]. The Ammonia Casale factory, as Miolati emphasized, resembled an electric power station insofar as a handful of workers maintained operation of the machinery and synthesis apparatus by means of measuring and controlling instruments, through observing meters and indicators on a switchboard. This was common to all synthetic ammonia processes. The published version of the lecture, printed in several languages, showed that theory, experiment and large-scale application had paid off handsomely. Miolati would continue to emphasize the superiority of the process developed by his former student.

Since Montecatini had placed a brake on growth of the Casale process in Italy, Luigi Casale had been wise to set up Ammonia Casale in Switzerland, from where he could promote synthetic ammonia elsewhere. As a result, the Casale process became the most widely adopted synthetic nitrogen process; by 1924, it was in use in Italy, Japan, France, Belgium, and the United States. In 1925, Casale converters were installed in Spain at Sabiñánigo, Aragón, of 5000 tons annual capacity (Energia e Industrias Aragonesas SA; Sociedad Española de Industrias Aragonesas), and a factory was under construction in Switzerland (Lonza at Visp, of 2500 tons annual capacity, opened in 1927) (Figs. 10.6, 10.7, and 10.8) [34].

In 1925, Casale-based ammonia facilities worldwide, working, under construction or planned, were claimed, according to their designs, to have a total annual capacity of 169,200 metric tons; actual production was just 30,200 tons. Factories reported as still under construction accounted for 103,500 tons, while those commisioned but not yet started represented 35,500 tons. In Japan, Noguchi's Nobeoka factory produced 10,000 tons of ammonia annually and was subsequently expanded; in 1926 a new factory was opened at Minamata with a capacity of 32,000 tons. With its rugged simplicity, reliability and high yield, the Casale process was rightly seen as a threat by the management at BASF (and later IG Farben), particularly in 1927, when three Casale plants were ordered for use in Germany.

Fig. 10.6 Electrolysis plant for hydrogen used in the Casale ammonia process, at Sabiñánigo, Spain. (Casale SA)

Fig. 10.7 Compressor room, Casale process, Union Chimique Belge, at Ostend, Belgium. (Casale SA)

Fig. 10.8 Synthesis gas compressors, Casale process, at Visp (Viège), Switzerland. (Casale SA)

Annual Casale ammonia process production reached 160,000 tons by 1927, and with new plants then in hand or ordered, was expected to achieve 350,000 tons (Table 10.2). While these figures were lower than for ammonia produced by IG Farben (in Germany), they were greater than corresponding figures for all other competing processes (Table 10.1). Ammonia Casale's growth arose from licensing arrangements with firms outside Italy, and developments at its research laboratories. As indicated earlier, Casale converter output invariably exceeded guaranteed figures. Thus by 1927, Nobeoka used converters that were warranted to give a production of 20 tons per day but that actually were greater. The first Swiss Lonza converters, at Visp, averaged, per unit, 4 tons per day, exceeding Ammonia Casale's guaranteed 3.5 tons a day.

By the early 1930s, the many Ammonia Casale converters were producing around half of total world production of ammonia. (At that time output from the large British and German nitrogen factories had declined due to their reliance on export markets). Around 200 factories eventually worked first-generation Ammonia Casale technology. Many of Miolati's students undertook research in the Casale research institute (SIRI). Of passing interest is the fact that closely involved with Ammonia Casale from the late 1920s until 1940 was engineer Carlo Emilio Gadda (1893–1973), who had studied at the Politecnico di Milano. He is best known as a leading Italian writer and poet [35].

In 1935, Ammonia Casale filed a patent for an internal combustion engine powered by ammonia and undertook road trials with FIAT 509 and 527 Ardit vehicles. Later, ammonia fuel was replaced by synthetic methanol, another of Casale's high-pressure products.

Table 10.2 Licensees of the Casale process for synthetic ammonia, early 1927

Country	Licensee	Location
Italy	Terni-Società per l'Indùstria e Elettricità	Terni
	Terni-Società per l'Indùstria e Elettricità	Néra Montoro
Belgium	SA des Fours à coke Semet-Solvay & Piette (merged into Union Chimique Belge in 1928)	Ostend
	Etablissements Kuhlmann	Selzaete, East Flanders
Yugoslavia	Società per l'Utilizzazione delle Forze Idrauliche della Dalmazia	Dugi Rat
Spain	Energia e Industrias Aragonesas SA	Sabiñánigo, Aragón
France	Cie d'Alais, Froges et Camargue	Saint-Auban
	Société des Mines de Dourges	Hénin Liétard
	Société des Mines de Lens	Pont à Vendin
	Société des Mines de Viçoigne, Noeux & Drocourt	Hénin Liétard
	Société des Produits Chimiques Anzin-Kuhlmann	Anzin
	Société Houillère de Sarre & Moselle	Carling
	Société des Engrais Azotés & Composés	Soulom
	French government	Toulouse (opened 1927)
	Société Marles Kuhlmann (Société des Mines de Marles)	Chocques
	Compagnie de Roche la Molière	Firminy
Germany	Kohlechemie AG	Duisburg
Japan	Nippon Chisso Hiryō Kabushiki Kaisha (Nitchitsu)	Nobeoka, Minamata
Soviet Union	Severny Chimichesky Trest (Severnii Khimicheskii Trest) (Northern Chemical Trust)	Chernorechenskii factory, Dzerzhinsk, Nijninovgorod (Nizhny Novgorod). Later Gorky
Switzerland	Usines Electriques de la Lonza SA	Visp (Viège)
United States	Ammonia Corporation/Niagara Ammonia Corporation	Niagara Falls (acquired by Du Pont in 1927)

Sources include: Arturo Miolati, *Synthetic Ammonia and the Casale Process. Amplified edition of a lecture delivered the 27th February 1927 at the Institute of Chemistry of the Polytechnic School of Prague*. Rome: Ammonia Casale SA, 1927, pp. 42–43; and Mirko Lamer, *The World Fertilizer Economy*, issue 10. Stanford: The Food Research Institute of Stanford University, 1957, pp. 397–402

Note: At the first 24 Casale facilities, hydrogen was prepared in 12 cases by electrolysis, in 9 cases from coke oven gas, and in 3 cases from water gas (Toulouse and Choques by the BASF modified Bamag process; and Dzerzhinsk by the Pintsch process). Nitrogen in 12 facilities was produced by the liquefaction of air, out of which two facilities also employed the Casale burner (burning hydrogen with atmospheric oxygen). Four facilities relied on the Casale burner alone

10.8 Claude Ammonia, and Casale in France

In 1902, Georges Claude (1870–1960) in France devised a system of air liquefaction that enabled, by fractional distillation, the separation of liquid oxygen, and later nitrogen and argon. In the same year, in partnership with Paul Delorme he founded Société L'Air Liquide. The Claude process, based on compression and then decompression of air, allowing it to do external work, competed with the earlier system of Linde. Claude also invented neon lighting. From 1917, as part of his secret government wartime research, Claude undertook experiments on high-pressure ammonia synthesis at around 1000 atmospheres and 500 °C. This was achieved prior to the Armistice in November 1918 [36]. Pilot plant tests carried out at Claude's "hyperpressures" were undertaken in 1919 at the Montereau (Île-de-France) facility of the new Air Liquide subsidiary—formed jointly with Compagnie de Saint-Gobain—Société Chimique de la Grande Paroisse. Hydrogen was generated from water gas. The Claude ammonia process was worked such that the gases were passed through a series of tubes without recirculation. After coke oven gas became the source of hydrogen, the first tube was the purifier, where carbon monoxide was removed. This was followed by passage through four ammonia converter tubes (all five were also referred to as bombs). At each stage the gas was cooled and liquid ammonia was collected. At 1000 atmospheres, the process gave 40% ammonia in the first converter, and 25% in the second [37]. In common with Casale's process, Claude's use of high pressures involved handling smaller gas volumes than in the Haber-Bosch process, and use of a less active catalyst, since the process was more efficient. As with the Casale process, the more compact equipment, with converters of decreased height and vessel diameter, compared with the Haber-Bosch process, suggested viable operation of a number of medium-size units. Because of the very high pressure the overall yield was high, and only 10% of unreacted nitrogen and hydrogen remained after passage through the four converters [38–40].

In 1923, Claude, backed by two Swiss banks, set up a partnership in Milan to license his process in Italy. This was the Società per la Fabbricazione dell'Ammoniaca Sintetica e Prodotti Derivati (Azogeno). Despite opposition from Donegani, including his use of political connections, Azogeno erected at Bussi a Claude ammonia plant of capacity 7 tons daily; the nearby works of Società Italiana di Elettrochimica supplied hydrogen [41, 42]. A second Claude factory in Italy was located at the coastal town of Vado Ligure, west of Genoa, close to the border with France. In this case, hydrogen was available from an adjacent coke oven plant.

Claude had been active elsewhere. In 1921, a Claude converter with an output of 5 tons daily had been installed at Montereau, and in 1922 two more were installed at Bully-les-Mines, in northern France. A proposal to produce ammonia using the Claude process by the Cumberland Coal Power & Chemicals, Ltd, in north-western England, ended with liquidation of the company in December 1922 [43]. Cumberland

Coal, with which inventor James West was associated, had intended to introduce the process in Norway.

Claude worked on producing hydrogen from coke oven gas at Béthune, in the Pas-de-Calais Department. He succeeded to isolate and purify hydrogen suitable for use in the ammonia synthesis. This greatly appealed to coke oven firms and enabled the setting up of ammonia factories in countries with plentiful supplies of coal and coke and steel industries. In part to exploit coke oven gas, Claude participated in formation of Société Belge de l'Azote in 1923. A small plant was installed at the works of the coal mining concern Ougrée-Marihaye. In 1930, two hyper-compressors manufactured by Sulzer of Winterthur were installed at Béthune and operated successfully in the 1000 atmosphere range for several years.

From around 1924, the Claude process using coke oven gas was taken up in Spain by a consortium, the Sociedad Ibérica del Nitrógeno, at Flix, Catalonia, and, by 1930, in Asturias, at a site close to the large coal mining and engineering firm Sociedad Metalúrgica Duro Felguera. However, technical difficulties were experienced, followed in the 1930s by disruption caused by the Spanish Civil War (1936–1939).

In a number of instances, Claude's ammonia process was beset with technical problems, until at least the mid-1920s, as experienced in the United States by Du Pont and in Japan. Often performances were poor and the process did not produce the claimed high yield. The great heat generated in the catalyst chamber, as in the Casale process, was a major problem. Converters made of normal steel corroded under the reaction conditions and exploded. As a result, there were delays in introducing Claude's process. The corrosion problem was at first overcome with special chrome-nickel steel converters made in Sheffield, England, by the engineering firm Vickers Ltd. For the smaller Claude converters, catalyst tubes of a chrome-nickel-tungsten alloy, trade name BTG, were manufactured by the large French steel and mining firm Société de Commentry-Fourchambault et Decazeville, of Imphy. Hadfields Ltd, also of Sheffield, manufactured a similar alloy (HR-1) and produced high-pressure catalyst tubes. Problems with the Claude converters meant that by early 1927 the most successful French ammonia factories worked the Casale process, mainly through license holder Pechiney. This included a large state-owned factory in the south.

After the Armistice the BASF Ludwigshafen-Oppau complex, situated on the west bank of the Rhine, was occupied by the French, who were keen to acquire details of Haber-Bosch know-how. They were in a far better position to do so than the British. They threatened to invoke the article of the Treaty of Versailles which related to facilities for the manufacture of munitions, and have the Oppau facility dismantled unless Haber-Bosch know-how was released and a license granted. Several Allied observers, in particular Major Victor Lefebure, had (in addition to Merseburg) associated Oppau with the production of armaments [44–46].

On 11 November 1919, a contract was signed between representatives of the French state and the BASF management whereby in return for technical assistance and a license the French agreed not to permit destruction of the Oppau works. It was claimed that French military and other officials at Oppau, mainly chemists, in collusion with Bosch and in order to protect French interests, restricted close access by representatives of other Allied countries to the converters and related ammonia apparatus, as experienced by the British contingent from Brunner, Mond, and American officials.

It took some years before a French synthetic ammonia plant based on foreign technology began operating, at Toulouse, in the south-west. This delay was partly as a result of the loss of interest on the part of potential investors and of competition from the French Claude process for ammonia. The French minister of industrial reconstruction, Louis Loucheur, who concluded arrangements with BASF in 1919, had brought together a consortium of industrialists headed by Joseph Frossard of Kuhlmann and Emile Lambert of Société Lambert, Rivière & Cie to acquire a license, through the Société d'Etudes, for operating the Haber-Bosch process. Probably with the Claude process in mind, Loucheur declared that even if BASF refused to co-operate, the French had the necessary technical data to go ahead with the production of ammonia. BASF had demanded 50 million francs for the license, but when the contract was signed it stipulated just five million francs [47]. Royalties were also far less than BASF demanded. The choice of Toulouse—the site of a former powder factory that produced nitrocellulose during the recent war—for the ammonia factory was made with national security in mind. Its remoteness from Germany was warranted; but its remoteness from French coalfields deterred investors from entering into what was intended to be either a private or a state-private partnership concern, which contributed to the delays. In February 1921, the Académie des Sciences was directed by the government to form a commission to investigate the merits of the rival Claude and Haber-Bosch processes.

However, there were strong and vociferous objections in France, both commercial—principally Claude interests—and nationalistic, to the German process [48]. Concerns were expressed in Germany, for similar reasons. According to *Chemistry and Industry*, "The German press angrily accuses the directors of the Badische Co. of having betrayed their country for 11,500 million marks (at present rates!) to the 'hereditary enemy'" [49].

On 11 April 1924 the French state, having written off the idea of attracting private funding, established the National Industrial Nitrogen Office (Office National Industriel de l'Azote), administered by the Ministry of Public Works, and authorized construction of the new national explosives factory at Toulouse [50]. Soon after, technical experts from BASF visited Toulouse, and later a French delegation travelled to Ludwigshafen. The state's negotiators also maintained contact with Pechiney, licensee of the Casale process. This time, Claude interests contested the validity of the Casale patents [51, 52]. In this they failed. The ammonia factory of the Pouderie Nationale de Toulouse opened in 1927, under the aegis of the Compagnie National de l'Azote. The converters were based on Casale technology and the hydrogen generators on BASF/Bamag water gas technology. The capacity, for the first of six units, was around 50 metric tons a day. IG Farben received licensing fees from this time, but only for gas preparation, based on the contract, which was to run for fifteen years from the time that ammonia output reached 30 metric tons daily. By around 1930, the daily output of Toulouse reached the planned for 120 metric tons.

Prior to and even well after the installation of synthetic ammonia factories, the French output of cyanamide factories was increased. In 1921, in which year 14,000 metric tons were produced, the French War Ministry disposed of its interest in the cyanamide factory at Lannemezan in the Pyrenees, returning the site to Le Société Française des Produits Azotés [53]. By 1925, French cyanamide output was 55,000 metric tons; almost 30,000 tons were imported. Lannemezan in 1932 changed over to synthetic ammonia, based on electrolytic hydrogen.

Fourteen French synthetic ammonia factories existed in the early 1930s, of which ten were geographically concentrated in or near the Pas-de-Calais and Nord departments. Three of these ten worked the Claude process, five the Casale process, and two the American Nitrogen Engineering Corporation process. The other four factories were at Roche-la-Molière/Firminy (Loire; Casale process), Grand-Quevilly (near Rouen, Normandy; Claude), Soulom (Hautes-Pyrénées, Casale), and the state powder factory at Toulouse. Output of ammonium sulphate from the Claude process was claimed to be 125,000 tons and from the Casale process 117,000 tons (apart from Toulouse). One important outcome of the availability of the various ammonia processes was that they clearly demonstrated to IG Farben that its monopoly was not inviolable, which in turn eroded its bargaining power when dealing with foreign rivals over various markets in chemicals.

10.9 Promoting Ammonia Casale in Britain

As a result of the difficulties encountered by Claude, Montecatini's refusal to license the Fauser process outside Italy before 1924, and BASF's extortionate terms and lack of interest in licensing, Casale had a four-year headstart in making ammonia technology an international enterprise.

Though Casale's process was never adopted in Britain, the following account provides an example of how his nitrogen technology was promoted through lobbying by foreign representatives before well-chosen audiences of businessmen, military leaders, and scientists. Here the representative of Casale in Britain was Irish-born electrical engineer John Francis Crowley (1884–1951).

During World War I, Crowley had investigated nitrogen fixation at the sequestered Siemens Brothers works in Stafford, England. In 1919 that works became the property of English Electric. Crowley left the firm soon after at a time of reorganization and cutbacks. He set up a consulting engineering partnership in London with John A. Harker (previously organiser of the Munitions Inventions Department's Nitrogen Products Committee). In the autumn of 1921, Harker visited the Casale workshop and expressed surprise at the apparently casual attitude shown towards conducting synthesis at ultra-high pressures: "One of the things which struck me most about the working of the Terni plant, which had been only two days in operation, was that none of the individuals concerned appeared to regard the operations they were conducting as anything out of the common" [54].

Crowley negotiated with Luigi Casale, with the intention of promoting and representing Ammonia Casale interests in Great Britain. Harker's death in 1923 did not dampen Crowley's enthusiasm. In 1924, with Eric M. Bergstrom of Balfour, Beatty & Co., Crowley emphasized the potential for installation of Casale synthetic ammonia factories in the Scottish Highlands, taking advantage of abundant water power. They drew attention to the ease of the electrolytic production of hydrogen as used in the Italian processes, and the advantages of the Casale process, including the quick start up time, as compared with the Haber-Bosch process [55].

On 30 April 1925, Crowley arranged a dinner to honour Casale and the "Conseil d'Administration" of Ammonia Casale at the Royal Automobile Club in London. Present from the Casale company were the president, H. Bouchayer, and administrator, Baron d'Ornellas. The audience included several British military men, industrialists, and scientists, including chemists Sir Arthur Cope, Professor J. R. Partington and Sir Robert Robinson, and Sir Charles Parsons (inventor of the eponymous turbine) [56]. Crowley drew attention to the strong financial participation of British interests in the Casale process. Dr. C. G. Clayton observed that the Casale process was easily adaptable to small-scale operation—in part because the ultra-high pressure apparatus required smaller plant than the Haber-Bosch-type process as used at Billingham—which made it more strategically advantageous, insofar as several compact plants could be distributed throughout the country. (Such a proposal was adopted prior to and during World War II, but with ICI's technology, following concerns over the vulnerability of Billingham to bombing by the Luftwaffe. This involved the construction of ammonia and synthetic fuel facilities well away from the north-east coast, and in Canada. The new ICI ammonia factory, at Mossend in Lancashire, opened in September 1939. Also in Lancashire, ICI jointly with Shell erected a large synthetic fuel factory at Heysham, opened in 1941, which later became an ICI ammonia facility.)

10.10 The Mont Cenis Process and Its Introduction into the Netherlands

The development of a second German synthetic ammonia process, based on coke oven hydrogen, was sponsored by a consortium of German mining companies that had been unable to acquire a license for the Haber-Bosch process. This Mont Cenis process was named after the Gewerkschaft der Steinkohlenzeche Mont Cenis, of Herne-Sodingen, Westphalia, which in 1924 commissioned chemical engineering contractor Friedrich Uhde of Dortmund to design a converter and ancillary equipment (Fig. 10.9). The Mont Cenis process operated at just 100 atmospheres, not unlike the conditions Uhde had investigated during World War I for the synthesis of liquid hydrocarbons (at 350–500 °C, over a nickel catalyst) (Fig. 10.10). Uhde's low-pressure process required an extremely potent catalyst, which was provided by Ivar Cederberg, who had undertaken catalyst investigations at the University of Stockholm (1916–1919, and from 1919 at Norsk Hydro) [57]. The Mont Cenis

Fig. 10.9 Friedrich Uhde. (ThyssenKrupp, Corporate Archives, Duisburg)

Fig. 10.10 The first Uhde-Mont Cenis ammonia plant, Herne-Sodingen, mid-1920s. (ThyssenKrupp, Corporate Archives, Duisburg)

ammonia process was promoted from late 1925. The patents were held by Gasverarbeitungs-Gesellschaft GmbH, also of Herne-Sodingen, where the first facility was erected in 1925 and began operating in 1926. At the time of full-scale operation, in 1928, there were four converters, each capable of producing 25 tons a day. As elsewhere, the converters incorporated novel heat exchangers, and each unit included two-stage refrigeration. Uhde also engaged in the design of nitric acid plant, and a process for the manufacture of calcium ammonium nitrate.

The Mont Cenis process was taken up in Holland during 1927 by Royal Dutch-Shell. Historian Ernst Homburg and colleagues have described the particular circumstances of this move by a large oil company into chemical manufacture. It is an early example of a counter trend to the chemical industry's involvement in the conversion of coal to oil, and the later refining and processing of oil. In mid-1927, Royal Dutch-Shell was at loggerheads with IG Farben (and Standard Oil of New Jersey) over the terms of use of the Bergius coal hydrogenation patents (see Chap. 16). In August, Carl Bosch, head of IG Farben, in a fit of anger, predicted the decline of the oil industry through production of synthetic gasoline by the Bergius high-pressure hydrogenation of coal process, in the same way that BASF had decimated the natural indigo industry in the early 1900s. In response, the board of Royal Dutch-Shell decided to attack IG Farben's fertilizer business, for which the Netherlands was a good market—the second largest market for German ammonium sulphate after Japan—by entry into the manufacture of synthetic ammonia. The Mont Cenis and Casale processes were investigated. In September, Royal Dutch-Shell opted for the Mont Cenis process. Contracts were signed in the following month. Royal Dutch-Shell's IJmuiden MEKOG facility was situated near the coke ovens of the Hoogovens steel works, the source of hydrogen. The facility, designed by Friedrich Uhde's engineering firm, began operating in September 1929. It was the first of a trio of large Dutch ammonia factories to open by the end of the following year. Despite initial difficulties involving poisoned catalyst, and problems with compressors, heat exchangers, gas losses, and refrigeration units, several improvements were made, including an increase of working pressure to 150, and later 300, atmospheres. These higher pressures were introduced into the Mont Cenis process following expiration of certain BASF patents. Royal Dutch-Shell's improvements enabled the firm to bring about a break with the German patent holder in February 1934. The two other Dutch ammonia facilities, those of the Dutch State Mines (DSM) and the Compagnie Néerlandaise de l'Azote (Montecatini), used the Fauser process. Around 1930, Dutch mines produced annually about 13 million tons of coal (compared with less than two million tons before 1914), ensuring an adequate supply of hydrogen from coke ovens. The aggregate annual capacity, as fixed nitrogen, of the Dutch nitrogen factories exceeded 90,000 tons, including by-products (production in the mid-1930s was reduced, partly as a result of international cartel arrangements (Tables 10.3 and 10.4). The emergence of the Dutch ammonia factories would have far-reaching impacts on engineering and plant design in the Netherlands [58–63].

Table 10.3 Nitrogen production in the Netherlands for the agricultural years 1929–1930 until 1935–1936 (30 June–1 July), in tons

1929–1930	18,201
1930–1931	55,461
1931–1932	89,069
1932–1933	90,405
1933–1934	85,077
1934–1935	81,051
1935–1936	85,231

Source: Ernst Homburg

Table 10.4 Nitrogen production in the Netherlands for the agricultural years 1933–1934 and 1934–1935, according to outputs of by-product and synthetic ammonia, in tons

1933–1934	By-product 10,678	Synthetic 74,399
1934–1935	By-product 11,028	Synthetic 70,023

Source: Ernst Homburg

10.11 Pure Gases: Mainly Hydrogen, and Nitrogen

The manufacture of cyanamide relied on the availability of pure nitrogen; and of synthetic ammonia on both pure nitrogen and hydrogen. Industrial gas production emerged in the 1880s, following the introduction of the Brins process for isolating oxygen from the atmosphere. This was adapted to the isolation of nitrogen in the United States at American Cyanamid. By 1910 the Linde and Claude (Société de L'Air Liquide) processes, also originally developed for oxygen, had likewise been adapted to production of nitrogen. Whereas the availability of oxygen depended on distribution networks to remote users, atmospheric nitrogen for the fixation processes was extracted and used at the sites of manufacture of nitrogen products. The Linde and Claude processes were costly to install in large manufacturing facilities. A less expensive source, used in manufacture of synthetic ammonia, was producer gas, as adopted at Oppau before 1914. However, Linde liquefaction for nitrogen was used to adjust the composition of the synthesis gas mixture.

The supply of hydrogen relied on electrolytic and coal-based processes. The lighter than air and highly reactive gas was originally required for airships (and later hydrogenation processes). Production on a large scale, however, was stimulated by the emergence of synthetic ammonia processes. BASF attempted to use the Linde-Frank-Caro process, for obtaining hydrogen from water gas, but found it both expensive and inefficient [64].

Hydrogen production was the most costly feature of ammonia plants. Charles O. Brown, vice chairman of American Cyanamid's Chemical Engineering Corporation, emphasized the critical role of hydrogen thus [65]: "[T]he differences [among the synthetic ammonia processes] are, in general, those required because of the different sources for obtaining hydrogen. . . The development and purification of the hydrogen is still the outstanding feature of cost and extensiveness of plant equipment. The synthesis equipment is obviously a smaller and less expensive part, this being one of the advantages of the use of high pressure synthesis."

The source of hydrogen for the early ammonia industry in large part determined its widespread regional and international availability. Water gas was associated with larger ammonia facilities (Germany, and later Britain and the United States); electrolysis with early experiments and pilot plant studies and, mainly, the availability of hydro-electric power (Italy, Japan, etc.); and coke oven gas originally with France and the Low Countries, before it became the most favoured process until the 1940s. In Europe, hydrogen from water gas and coke oven processes was freed of other components following liquefaction.

10.11.1 Electrolysis

Electrolysis of water gave pure hydrogen but required low cost electric power. This was the case, for example, at Montecatini's Merano factory. Traces of oxygen were removed by reaction with hydrogen over a platinized asbestos catalyst. Later, a copper deoxidiser was employed. Several Casale licensees generated hydrogen by electrolysis, notably the Japanese factories of Noguchi (Nitchitsu). Electrolysis remained an important source of hydrogen in locations where there was a plentiful supply of hydro-electricity. It was also available from the chor-alkali electrolytic processes.

10.11.2 Water Gas

The abundant supplies of coal in Germany, particularly lignite, was a deciding factor in its use as a source of hydrogen for the BASF ammonia synthesis. In terms of volumes produced, water gas was the main source of hydrogen. Several complex operations gave hydrogen of suitable purity for ammonia manufacture. At BASF, water gas was made by the action of steam on coal or lignite, with the catalytic conversion of carbon monoxide into dioxide. This was the basis of the water-gas shift reaction, designed to increase the yield of hydrogen. The shift reaction required a catalyst based on ferric oxide. Elimination of catalyst poisons was essential.

In the United States, catalyst and other studies related to the production of hydrogen from water gas were conducted at the Fixed Nitrogen Research Laboratory [66]. The water gas process was used by Du Pont until the 1940s, at Toulouse (by the BASF/Bamag process), at some Soviet ammonia factories, and at ICI (until the 1950s). A notable feature of several manufacturing processes was the way in which they were mixed, such as Casale for synthesis of ammonia and BASF for generation of hydrogen at Toulouse.

Table 10.5 Percentage of synthetic ammonia made from hydrogen produced from coke oven, electrolytic, and water gas processes for the agricultural years 1926–1927 until 1929–1930, according to George P. Pollitt (1930)

Agricultural year	1926–1927	1927–1928	1928–1929	1929–1930
Coke oven	3.0	5.8	11.1	14.2
Electrolytic	6.4	8.8	10.5	15.2
Water gas	89.0	84.1	77.3	69.7

Source: Harold Tongue, *The Design and Construction of High Pressure Chemical Plant*. London: Chapman & Hall, Ltd, 1934, p. 109

According to ICI's George Pollitt in 1930, during 1926–1927, 89% of hydrogen used in the manufacture of ammonia came from water gas (in part reflecting the great output of German synthetic ammonia), 3% from coke oven gas, 6.4% from electrolysis, and 1.6% from other processes (Table 10.5). The main disadvantage of the water gas process was in the handling of solids, in particular lignite, and in the purification of gases. The process required very costly plant. For this reason it was best suited to large facilities.

In the mid-1920s, Fritz Winkler (1888–1950) at IG Farben developed a fluidized-bed generator for producing water gas in which a six-foot layer of finely divided lignite (generally waste material) was kept in motion by an air or steam blast.

The German engineering firm Bamag studied the reaction in which iron decomposes water at a high temperature to afford water gas. The iron-steam reaction was investigated by Pechiney for its Ammonia Casale process. It required the careful control of temperature to prevent secondary reactions and decompositions. The carbon monoxide in the hydrogen was not injurious to the Casale process, in which it was rapidly reduced to methane, which could be removed from the circuit.

10.11.3 Coke Oven Gas

By the mid-1920s, firms in France and Belgium had started the shift to coke oven gas as a source of hydrogen for ammonia synthesis (Table 10.6). This brought them to the fore among the leading manufacturers of synthetic ammonia (Fig. 10.11). Coke oven gas contained around 50% hydrogen, in addition to methane, carbon monoxide and nitrogen. The hydrogen, once isolated, required careful purification, using processes developed by both Linde in Germany and Claude in France. This enabled coke oven hydrogen processes to be taken up on a large scale from the late 1920s, with enormous implications for the global spread of the synthetic ammonia industry. It encouraged the setting up of ammonia factories in all industrialized countries with extensive reserves of coal. Until 1930, however, the production of coke oven hydrogen was often restricted by the lack of ovens.

The importance of coke oven gas as a source of hydrogen is most clearly illustrated by developments in Belgium, where Casale converters were installed during 1923–1924 at Zanvoorde, Ostend, by SA des Fours à Coke Semet-Solvay &

Table 10.6 Source of hydrogen, location, ammonia process, operating company, and manufacturing capacities for 1927

Source of hydrogen, location, and ammonia process	Operating company	Annual operating capacity, tons of synthetic ammonia	Capacity under construction, tons of synthetic ammonia
Water gas			
Hopewell, Va., USA, ALLIED	Atmospheric Nitrogen Corp. (Allied)		35,000 (completed late 1928)
Charleston (Belle), W. Va., USA, CLAUDE	Lazote, Inc. (Du Pont)	6350	
Oppau, Germany, HABER-BOSCH	BASF/IG Farben	125,000	
Merseburg, Germany, HABER-BOSCH	BASF/IG Farben	375,000	
Billingham, England, Modified HABER-BOSCH	Synthetic Ammonia & Nitrates, Ltd (ICI)	20,500	52,000 (completed 1929)
Montereau, France, CLAUDE	Société Chimique de la Grande Paroisse (Air Liquide)	1500	
Toulouse, France, CASALE	French government: Pouderie National de Toulouse. L'Office National Industriel de l'Azote		37,000
Willebroek, Belgium 1926–1927, FAUSER	Evence Coppée et Cie; with Montecatini formed SA Ammoniaque Synthétiques et Dérivés		10,000
Hikoshima, Japan, CLAUDE	Suzuki Shōten		[3000]
Nijninovgorod (Nizhny Novgorod, later Gorky), Soviet Union, CASALE	Severny Chimichesky Trest (Severnii Khimicheskii Trest)		7560
Coke oven			
Béthune, France, CLAUDE	Compagnie des Mines de Béthune; *also* Bully Grenay	6000	12,000
St Etienne, France, CLAUDE	Société des Houillères de Sainte-Etienne	540	3260
Decazeville, France, CLAUDE	Société de Commentry, Fourchambault et Decazeville	3000	
Waziers, France, CLAUDE	Société l'Ammoniaque Synthetique		4350
Hénin-Liétard, France CASALE	Société des Mines de Dourges	4725	
Pont-à-Vendin, France, CASALE	Société des Mines de Lens [Ammonia]	7000	

(continued)

Table 10.6 (continued)

Source of hydrogen, location, and ammonia process	Operating company	Annual operating capacity, tons of synthetic ammonia	Capacity under construction, tons of synthetic ammonia
Hénin-Liétard, France, CASALE	Société des Mines de Vicoigne, Noeux & Drocourt	4725	
Anzin, France, CASALE	Société de Produits Chimiques Anzin (Kuhlmann)	7000	
Carling, France, CASALE	Société Houillière de Sarre-et-Moselle		4725
Firminy, France, CASALE	Société des Produits Chimiques de Roche-la-Molière	4725	
Choques [Marles], France, CASALE	Société des Mines de Marles		7500
Vado Ligure, Italy, CLAUDE	Società Azogeno		3000
Ougrée, Belgium, CLAUDE	Société Belge de l'Azote. Claude group with Soc. Anon. d'Ougrée-Marihaye; *also* water gas	4500	
Ostend, Belgium, CASALE	S.A. des Fours à Coke Semet-Solvay et Piette; *also* water gas	7500	
Selzaete, Belgium, CASALE	Etablisssements Kuhlmann		7500
Felguera (Asturias), Spain, CLAUDE	Sociedad Ibérica del Nitrógeno	3150	
Knurów, Poland, CLAUDE	Société Fermière des Mines Fiscales de l'Etat Polonais		3000
Electrolytic			
Syracuse, NY, USA, General Chemical	Atmospheric Nitrogen Corp. (Allied).	3800	
Niagara Falls, NY, USA (to Du Pont, 1927), CASALE	Niagara Ammonia Co.	2000	
Niagara Falls, NY, USA, FNRL	Mathieson Alkali Works	2880	
Niagara Falls, NY, USA, FNRL	Roessler-Hasslacher Chemical Co.	865	865
Pittsburg, California, USA, FNRL	Great Western Electro-Chemical	300	
Seattle, Washington, USA, FNRL (to Du Pont, 1927)	Pacific Nitrogen Corp. Subsidiary of National Ammonia Company	865	

(continued)

Table 10.6 (continued)

Source of hydrogen, location, and ammonia process	Operating company	Annual operating capacity, tons of synthetic ammonia	Capacity under construction, tons of synthetic ammonia
St Auban, France, CASALE	Cie d'Alais Froges et Camargue (Péchiney)	730	3150
Bussi, Italy, CLAUDE	Società Azogeno	1500	
Flix, Spain, CLAUDE	Sociedad Ibérica del Nitrógeno	700	
Terni, Italy, CASALE	Società Italiana Ricerche Industriali (SIRI)	1345	
Nera Móntoro, Italy, CASALE		9765	
Dugi Rat, Yugoslavia, CASALE	Società per l'Utilizzazione delle Forze Idrauliche della Dalmazia		15,000 (Completed by 1930)
Novara, Italy, FAUSER	Società Piemontese Ammonia (Montecatini)	5750	
Merano/Sinigo, Italy, FAUSER	Società Alto Adige Ammonia (Montecatini)	37,800	
Mas (Belluno), Italy, FAUSER	Montecatini	1000	
Coghinas, Sardinia, Italy, FAUSER	Società Sarda Ammonia e Prodotti Nitrici (Montecatini)	3500	
Crotone, Calabria, Italy, FAUSER	Montecatini		
Sabiñángo, Spain, CASALE	Energia e Industrias Aragonesas SA	4725	
Nobeoka, Japan, CASALE	Nippon Chisso Hiryō Kabushiki Kaisha (Nitchitsu)	19,500	
Minamata, Japan, CASALE	Nippon Chisso Hiryō Kabushiki Kaisha (Nitchitsu)	18,900	12,600
Visp (Viège), Switzerland, CASALE	Usines Electriques de la Lonza	2200	5000
Trail, British Columbia, FAUSER. Completed 1930	Consolidated Mining and Smelting Company (*and* water gas)		
Ljunga, Sweden	Stockholm Superfosfat Akt.		2500

Source: Frank A. Ernst, *Fixation of Atmospheric Nitrogen*. London: Chapman & Hall, Ltd, 1928, pp. 131–133. FNRL = Fixed Nitrogen Research Laboratory. Two factories based on coke oven hydrogen opened in Czechoslovakia in 1928. For other facilities in Japan see Chap. 15.

Piette (from 1928 merged into Union Chimique Belge). Solvay held and shared rights to the Casale process in Belgium and Holland, as well as in France. The Ostend factory, of 10,000 tons annual capacity, was the first to combine ammonia

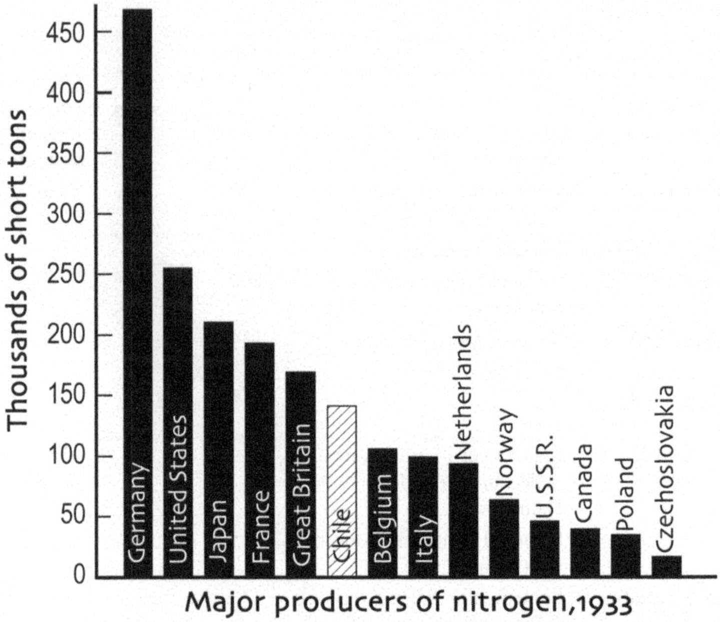

Fig. 10.11 The world's nitrogen economy. Annual production of fixed nitrogen (referred to as "chemical nitrogen") in thousands of short tons, for leading countries engaged in manufacture of nitrogen products (synthetic ammonia, calcium cyanamide, by-product ammonium sulphate), and Chilean nitrate, for the calendar year 1933, as reported in early 1934. Sources: *Chemical Nitrogen. A Survey of Processes, Organization, and International Trade, Stressing Factors Essential to Tariff Consideration, under the General Provisions of Section 332, Title III, Part II, Tariff Act of 1932. Report no. 114, second series*. Washington DC: Government Printing Office, 1937, p. 7; see also R. Norris Shreve, *The Chemical Process Industries*. New York: McGraw-Hill Book Company, 1945, p. 390 (and Table 13.1). The output for the Netherlands was around 90,000 tons, based on the agricultural years 1932–1933 and 1933–1934 (figures supplied by Ernst Homburg)

synthesis at an ultra-high pressure with the Linde system of coke oven gas fractionation for hydrogen [67]. The carbonization plants were arranged for both generation of power and utilization of gaseous by-products. The original experimental facility consisted of three 8-ton Casale reactors and three Linde units. The gas treatment process was designed by Semet-Solvay.

For the first few weeks production was low and irregular due to impurities in the coke oven gas. Following careful purification of the hydrogen, the output was 50% above that for which the plant was designed. The Ostend factory was so successful that by around 1928 four more Casale converters were installed, enabling daily production of 40–55 tons of ammonia. By this time, the circulation pumps installed in the ammonia plant at Ostend were undergoing replacement with Casale ejectors. This enabled further savings in energy. Six compressors each handled 1200 cubic metres of gas per hour. Even with the high working pressure of 750–800 atmospheres, it was found that Casale energy requirements were not much greater than those required at 200–300 atmospheres (1.2 kWh per kg of ammonia, as compared with 1.05 kWh, respectively). Since the Ostend coke oven plant also supplied gas to the town—a major seaside resort with fluctuating demand—water gas plant was also installed, though the cost of hydrogen was far greater [68]. The arrangement was described by F. A. F. Pallemaerts, research manager, division of coke ovens and ammonia at Union Chimique Belge, before the second International Conference on Bituminous Coal held at Pittsburgh in November 1928 [69]. In 1929, W. A. Dyes of Berlin estimated that coke-based processes in Germany would by 1930 account for around 100,000 tons of synthetic ammonia: Mont Cenis, 60,000 tons; Casale, 20–30,000 tons; and Claude, 12,000 tons [70]. At that time the Haber-Bosch capacity at Merseburg/Leuna was 650,000 tons.

By 1930, hydrogen produced from water gas declined to around 70%, and then in 1934 to 57%. In 1934 coke ovens were responsible for 25% of the hydrogen used in the synthesis of ammonia and would eventually become the main source of hydrogen until after World War II.

10.11.4 Natural Gas and Petroleum Gases

In 1912, Mittasch at BASF began experiments on producing hydrogen from methane (which was made from the reaction between carbon monoxide and hydrogen). After World War I there was an interest in using methane (natural gas) as a source of hydrogen. The natural gas was decomposed at elevated temperatures into hydrogen (thermal cracking).

By the late 1920s, both natural gas and petroleum gases were being investigated as sources of hydrogen by steam reforming. This provided opportunities for oil companies to enter into the manufacture of synthetic ammonia. In April 1928, the Dutch firm Bataafsche Petroleum Maatschappij, a subsidiary of Royal Dutch-Shell, acquired world rights, apart from Germany, to the Mont Cenis process based on hydrogen produced from natural or cracked gases. Natural gas had already found application in cyanamide manufacture at American Cyanamid and in Romania. Royal Dutch-Shell planned to work an ammonia pilot plant based on natural gas in Romania, but instead chose the Shell Point site of Shell Chemical, at Pittsburg, California. This was the first ammonia facility to use methane as feedstock, in 1929. Though the facility faced considerable losses and came close to being shut down, it managed to emerge out of the worst years of the depression and by 1935 was operating successfully [71]. In 1937, near Baku in the Soviet Union a new alkali and nitrogen fertilizer plant used hydrogen from local natural gas, a first for Russia [72].

From the 1940s, natural gas became increasingly important as a source of hydrogen. In the early 1950s, over 80% of ammonia produced in the United States was already made from natural gas. In Italy, the use of natural gas as a source of hydrogen grew exponentially between 1950 and 1955. By the mid-1970s, almost

70% of hydrogen was made by reforming fractions produced from distilled crude oil. Half was consumed in the synthesis of ammonia [73].

10.11.5 The Compressors

The mixed nitrogen-hydrogen gases made up in the correct proportions for the ammonia synthesis were fed from a gasometer to a series of compressors, in order to raise the pressure stepwise. Typically, in the water gas process employing Haber-Bosch conditions, three separate reciprocating compressors were used: first a low-pressure compressor, to 10 atmospheres; then an intermediate two-stage compressor, to 50 atmospheres; and finally a high-pressure compressor, to 250 atmospheres. For the Casale process, six stages of compression raised the pressure, to 3, 9, 27, 90, 270, and finally 850 atmospheres. Compressor improvements emerged with the need for high pressures, and larger equipment required in handling and processing gases for the ammonia synthesis [74].

10.12 Steels for High-Pressure Chemical Reactions

Steelmakers capable of producing high-quality steels and large forgings were critical to the success of synthetic ammonia. After 1918, they confronted problems unique to industrial reactions carried out at high pressures, particularly in the presence of hydrogen. Carl Bosch had overcome the problem of embrittlement with his double wall converter, made of ordinary steel, in which holes pierced in the outer shell enabled release of some hydrogen. The post-war steel industry, with its own research capabilities, turned to the investigation of a range of novel alloys for high-pressure processes.

For its converters, BASF relied on the Krupp steelworks at Essen. In Italy, Luigi Casale's works benefitted from the know-how of the large Terni steelmaking facility, which since 1884 had made cannons and cladding for ships of the Italian Navy. Unsurprisingly, the Casale pressure tubes bore similarities with big naval guns; early on, Ammonia Casale used ordinary steel for its ammonia converters. Another leading firm was Škoda (Skoda Steel Works), of Pilzeň (Pilsen), south-west of Prague. It manufactured ammonia converters and other high-pressure equipment for use in Czechoslovakia and for export.

The first stainless (rustless) steel was due to the work of Harry Brearly (1871–1948) who in 1913 at Brown Firth Research Laboratories, in Sheffield, England, developed an alloy containing 12.8% of chromium. In 1929, W. H. Hatfields, also at Brown Firth, described acid and corrosion-resistant (austenitic) steels based on chromium and nickel. These were soon adopted at larger works where ammonia was converted into nitric acid under pressure, especially in the United States.

Alloys of chromium, nickel, and molybdenum were developed in order to overcome the technical limitations of ordinary steels used in chemical processing equipment. Experiments undertaken by General Chemical in the United States around 1916 demonstrated the advantages of chrome-vanadium steels over a range of other steels. These were followed after World War I with research at the Midvale Steel Works, in Philadelphia, and elsewhere, including into "creep" under conditions of high pressures and high temperatures. Early experiments showed that 0.35% of molybdenum in ordinary steel produced an increase in creep strength.

In the development of novel alloys, Britain had distinct advantages, and not only for Billingham. As mentioned earlier, Claude's process was successful only after special steel converters, high in nickel, with chromium, and a small amount of tungsten, were manufactured by Vickers, and then Hadfields, both of Sheffield. No less important were the special steels developed by Société de Commentry-Fourchambault et Decazeville in France (Sect. 10.8).

Until around 1930, the composition of several steels used in the ammonia converters were often well kept trade secrets. Detailed descriptions were published following research at government, academic and industrial laboratories. The Chemical Research Laboratory of the British Department of Scientific and Industrial Research, at Teddington, Middlesex, undertook research into high-pressure materials and processes, including the synthesis of ammonia. One large autoclave in the laboratory was designed for working at 3000 atmospheres. ICI at Billingham continued in-house tests on steel alloys until well into the 1930s. Studies on the properties of steels undertaken at the British National Physical Laboratory, also at Teddington, appeared in the early 1930s.

The technological sophistication achieved reflects the extent to which the chemical industry was becoming reliant on a specialist and rapidly growing sector of the engineering industry. By the early 1930s, the weight of steel for a given production of ammonia was reduced to 15% of that in the original Haber-Bosch converters of 1913. Following the development of steels for ammonia converters, pressure vessels were introduced into several branches of the power and chemical industries.

References

1. Caro N (1927) Glossen zur Stickstoff-Industrie. Die chemische Industrie 50(6) (12 February):181–185.
2. van Rooij A (2005) Engineering contractors in the chemical industry: the development of ammonia processes, 1910–1940. Hist and Technol 21(4):345–366.
3. Waeser B (1932) Die Luftstickstoff-Industrie. Mit Berücksichtigung der chilenischen Industrie und des Kokereistickstoffs. Springer, Berlin.
4. Tongue H (1934) The design and construction of high pressure chemical plant. Chapman & Hall, London, pp 116–119, 340–353 (for Claude), 120–122, 353–359 (for Casale).
5. Travis AS (1998) High pressure industrial chemistry: the first steps, 1909–1913, and the impact. In: Travis AS, Schröter HG, Homburg E, Morris PJT (eds), Determinants in the evolution of the European chemical industry, 1900–1939: new technologies, political frameworks, markets and companies. Kluwer, Dordrecht, pp 3–21.

6. Kauffman GB (1970) Arturo Miolati (1869–1956). Isis 61(2):241–253.
7. Cariati V, Cavallone S, Maraini E, Zamagni V (eds) (2013) Storia delle società Italiane di ingegneria e impiantistica. il Mulino, Bologna, pp 55ff.
8. Crowley JF (1927) Dr Luigi Casale. J Soc Chem Ind 46(10)(11 March):220–221 (Chem Ind London).
9. Ceccarelli M (2014) Lorenzo Allievi (1856–1941). In: Ceccarelli M (ed), Distinguished figures in mechanism and machine science: their contributions and legacies. History of mechanism and machine science, vol. 26, part 3. Springer, Dordrecht, pp 1–17, on 2.
10. Casale, Italian patent, filed 18 July 1917; Casale, French patent, FP 495,725, filed 19 October 1917.
11. Casale, Italian patent, filed 21 September 1920; Casale, French patent, FP 527,818, filed 1 December 1921.
12. Miolati A (1927) Synthetic ammonia and the Casale process. Amplified edition of a lecture delivered the 27th February 1927 at the Institute of Chemistry of the Polytechnic School of Prague. Ammonia Casale SA, Rome.
13. Trinchieri G (2001) Le industrie chimiche in Italia, dalle origini al 2000. Arvan, Venice, pp 149–163, 204–213.
14. Miolati A (1927) Synthetic ammonia and the Casale process. Amplified edition of a lecture delivered the 27th February 1927 at the Institute of Chemistry of the Polytechnic School of Prague. Ammonia Casale SA, Rome, pp 9–10.
15. Miolati A (1927) Synthetic ammonia and the Casale process. Amplified edition of a lecture delivered the 27th February 1927 at the Institute of Chemistry of the Polytechnic School of Prague. Ammonia Casale SA, Rome, pp 16–17.
16. Miolati A (1927) Synthetic ammonia and the Casale process. Amplified edition of a lecture delivered the 27th February 1927 at the Institute of Chemistry of the Polytechnic School of Prague. Ammonia Casale SA, Rome, pp 20, 21.
17. Tongue H (1934) The design and construction of high pressure chemical plant. Chapman & Hall, London, p 83.
18. Fauser, Italian patent 198,374, filed 23 April 1921; 198,936 (water gas).
19. Scaglione P (2000) Il contributo di Giacomo Fauser e dei centri di recerca di Novara alla nascita ed allo sviluppo dell'Industria Chimica Italiana. De Agostani, Novara.
20. Perugini M (2014) Il farsi di una grande impresa. La Montecatini fra le due guerre mondiali. FrancoAngeli, Milan, pp 24–25.
21. Perugini M (2014) Il farsi di una grande impresa. La Montecatini fra le due guerre mondiali. FrancoAngeli, Milan, p 138.
22. Visintin M (2004) La grande industria in Alto Adige tra le due guerre mondiali. Fondazione Museo Storico Trentino, Trento.
23. Synthetic ammonia in Italy (1924) J Soc Chem Ind 43(52)(26 December):1283 (Chem Ind London).
24. van Rooij A, Homburg E (2002) Building the plant: a history of engineering contracting in the Netherlands. Walburg Pers, Eindhoven, pp 40–41.
25. Achilladelis BG (1973) Process innovation in the chemical industry. PhD thesis, University of Sussex, pp 148–153, on 151.
26. Knowles hydrogen plant at the Warfield works of the Consolidated Mining & Smelting Co. of Canada Ltd (1932) J Soc Chem Ind 51(17)(22 April):355–358 (Chem Ind London).
27. Miolati A (1927) Synthetic ammonia and the Casale process. Amplified edition of a lecture delivered the 27th February 1927 at the Institute of Chemistry of the Polytechnic School of Prague. Ammonia Casale SA, Rome, p 20.
28. Fabi L (ed) (2003) La SIRI: la fabbrica della ricerca. Luigi Casale e l'ammoniaca sintetica a Terni (catalogue of exhibition, 21 November 2003 to 21 May 2004). Centro di Documentazione sul Patrimonio Industriale. Antenna Pressa, Terni.
29. Waeser B (1932) Die Luftstickstoff-Industrie. Mit Berücksichtigung der chilenischen Industrie und des Kokereistickstoffs. Springer, Berlin, pp 140–141.

30. Crowley JF (1927) Dr Luigi Casale. J Soc Chem Ind 46(10)(11 March):220–221 (Chem Ind London).
31. Bertrams K (2013) The years of crisis (1914–1950): The making and unmaking of international alliances. In: Bertrams K, Homburg E, Coupain N, Solvay: the history of a multinational family firm. Cambridge University Press, Cambridge/New York, pp 151–329, on 204–206.
32. Baud P (1933) La grande industrie chimique française. Sa distribution et son état actuel. Annales de Geographie 42(236):127–141, on 139–141.
33. Miolati A (1927) Synthetic ammonia and the Casale process. Amplified edition of a lecture delivered the 27th February 1927 at the Institute of Chemistry of the Polytechnic School of Prague. Ammonia Casale SA, Rome.
34. The Casale ammonia process (1925) J Soc Chem Ind 44(47)(20 November):1154.
35. Isella D (ed) (1982) Carteggio dell'ing. Carlo Emilio Gadda con l' 'Ammonia Casale S.A.' 1927–1940. Stamperia Valdonega, Verona.
36. Appell MP (1918) Excerpts from RAPPORT DE M. P. Appell sur les titres et travaux de M. Georges Claude, candidat à la division des applications de la science à l'industrie, lu le 12 novembre 1918 (en comité secret).
37. Notices sur les travaux scientifiques et industriels de M. Georges Claude (1913–1924) (1924). Corbeil, Crété.
38. Thompson HL, Guillaumeron P, Updegraff NC (1952). Ammonia synthesis at 1000 atmospheres: the present-day Claude process. Chem Eng Progress 48(9)(September):468–476.
39. Sensicle LH (1927) The future of high-temperature carbonization. J Soc Chem Ind 46(2) (14 January):1T–20T, on 13T–14T.
40. West JH (1921) The Claude synthetic-ammonia process and plant. J Soc Chem Ind 40(22) (November):420R–424R.
41. Synthetic ammonia in Italy (1925) J Soc Chem Ind 44(23)(5 June):584–585 (Chem Ind London).
42. Benegiamo M (2013) Bussi e la grande chimica in Abruzzo. Un'ambizione fallita. Textus Edizioni, L'Aquila, pp 227–229.
43. London letter (1924) Ind Eng Chem 16(4)(April):427.
44. Meinzer L (1998) 'Productive collateral' or 'economic sense?' BASF under French occupation, 1919–1923. In: Travis AS, Schröter HG, Homburg E, Morris PJT (eds), Determinants in the evolution of the European chemical industry, 1900–1939: new technologies, political frameworks, markets and companies. Kluwer, Dordrecht, pp 51–63, on 54–55.
45. MacLeod R, Johnson JA (2006) The war the victors lost: the dilemmas of chemical disarmament, 1919–1926. In: MacLeod R, Johnson JA (eds), Frontline and factory: comparative perspectives on the chemical industry at war, 1914–1924. Springer, Dordrecht, pp 221–245, on 233–234.
46. Franco-German synthetic-ammonia convention (1920) J Soc Chem Ind 39(17) (15 September):305R.
47. Bertrams K (2013) The years of crisis (1914–1950): The making and unmaking of international alliances. In: Bertrams K, Homburg E, Coupain N, Solvay: the history of a multinational family firm. Cambridge University Press, Cambridge/New York, pp 151–329.
48. Curtis HA, Ernst FA (1924) United States Department of Agriculture. Supplement. Commerce Reports. Nitrogen survey. Part IV. September 29, 1924. The nitrogen situation in European countries. Trade Information Bulletin. United States Department of Agriculture, Washington DC, pp 19–21.
49. Editorial (1923) J Soc Chem Ind 42(7)(16 February):137–138 (Chem Ind London).
50. Deschiens M (1924) The position of the nitrogen industry in France. J Soc Chem Ind 43(20) (16 May):507–508 (Chem Ind London).
51. Achilladelis BG (1973) Process innovation in the chemical industry. PhD thesis, University of Sussex, pp 44–47.
52. Synthetic ammonia in France (1924) J Soc Chem Ind 43(47)(21 November):1163 (Chem Ind London).

53. Notes and news. France (1921) J Soc Chem Ind 40(10)(31 May):193R.
54. Waeser B (1926) The atmospheric nitrogen industry: with special consideration of the production of ammonia and nitric acid (trans. Fyleman E). P Blakiston's Son & Co., Philadelphia, vol. I, quoting Harker, on p xx.
55. Crowley JF, Bergstrom EM (1924) The development of national water power resources. In: The transactions of the first World Power Conference, London, June 30th to July 12th, 1924, vol. I. Power resources of the world available and unutilised. Percy Lund Humphries & Co., Ltd, London, pp 44, 349–368, on 359–363.
56. Dinner to Dr Luigi Casale (1925) J Soc Chem Ind 44(19)(8 May):477–478 (Chem Ind London).
57. Waeser B (1932) Die Luftstickstoff-Industrie. Mit Berücksichtigung der chilenischen Industrie und des Kokereistickstoffs, Band I. Springer, Berlin, pp 49–50.
58. Homburg E, Small JS, Vincken PFG (2000) Van carbo- naar petrochemie, 1910–1940. In: Schot JW, et al. (eds), Techniek in Nederland in de twintigste eeuw, II. Walburg Pers, Zutphen, pp 332–357.
59. Homburg E, van Rooij A (2004) Die Vor- und Nachteile enger Nachbarschaft. Der Transfer deutscher chemischer Technologie in die Niederlande bis 1952. In: Petri R (ed), Technologietransfer aus der deutschen Chemieindustrie (1925–1960). Duncker & Humblot, Berlin, pp 183–226.
60. van Rooij A (2004) Building plants: markets for technology and internal capabilities in DSM's fertiliser business, 1925–1970. Aksant, Amsterdam.
61. van Rooij A (2005) Why do firms acquire technology? The example of DSM's ammonia plants, 1925–1970. Res Policy 34(6)(August):836–851.
62. Homburg E (2004) Groeien door Kunstmest. DSM Agro 1929–2004. Verloren, Hilversum.
63. Forbes RJ, O'Beirne DR (1957) The technical development of the Royal Dutch/Shell, 1890–1940. EJ Brill, Leiden, pp 503–510.
64. Stokes RG, Banken R (2016) Building on air: the international industrial gases industry, 1886–2006. Cambridge University Press, New York.
65. Brown CO (1933) High pressure synthesis: basis of new chemical engineering industries. In: Kirkpatrick SD (ed), Twenty-five years of chemical engineering progress. Silver anniversary volume. American Institute of Chemical Engineers. Published by the institute and for sale by D Van Nostrand Company, Inc., New York, pp 152–168, on 167.
66. Evans RM, Newton WL (1926) Hydrogen from water gas: catalysts for its production. Ind Eng Chem 18(5)(May):513–517.
67. Tongue H (1934) The design and construction of high pressure chemical plant. Chapman & Hall, London, pp 120–122, 353–356.
68. Pallemaerts FAF (1929) Synthetic ammonia plant at Ostend. Ind Eng Chem 21(1)(January):22–29.
69. Belgian chemical industry (1929) J Soc Chem Ind 48(51)(20 December):1234 (Chem Ind London).
70. Dyes WA (1929) Chemistry, agriculture and industry in Germany. J Soc Chem Ind 48(20)(17 May):507–510, on 508 (Chem Ind London).
71. Forbes RJ, O'Beirne DR (1957) The technical development of the Royal Dutch/Shell, 1890–1940. EJ Brill, Leiden, pp 512–519.
72. Industrial notes from abroad. U.S.S.R. (1937) J Soc Chem Ind 56(39)(25 September):867 (Chem Ind London).
73. Quaderni dell'ingegnere chimico italiano (1975) 11(10)(October):155–160.
74. High pressures in the manufacture of synthetic ammonia. Recent designs in compressing plant (1929) J Soc Chem Ind 48(24)(14 June):591–598 (Chem Ind London).

Chapter 11
The United States

11.1 The Fixed Nitrogen Research Laboratory

> This laboratory has had most gifted and able directors. The fruits of its mellow years of activity have been of two kinds. First, the technical knowledge, data and designs, which have been made available are unsurpassed by any source in the world; and, furthermore, the skill and ingenuity of its staff have produced contributions to the field of pure science and of practical application, not exceeded by the research efforts of our largest corporations. Second, this laboratory has produced technical personnel and executives…who now have passed on to every fixed nitrogen plant in the United States and to some of our leading universities.
>
> Charles O. Brown, "High Pressure Synthesis—Basis of New Chemical Engineering Industries," in *Twenty-Five Years of Chemical Engineering Progress. Silver Anniversary Volume. American Institute of Chemical Engineers*, ed. Sidney D. Kirkpatrick. New York: Published by the institute and for sale by D. Van Nostrand Company, Inc., 1933, pp. 152–168, on pp. 165–166.

> An air of mystery has been thrown around the details of the [Haber-Bosch] process and we have been regaled by stories of the Krupp superlathes which turn out seamless tubes of a marvellous alloy for the new plants.
>
> "The Haber Process Pinned Down to Fact," *Industrial and Engineering Chemistry*, vol. 26, no. 11 (15 March 1922):482.

In contrast to the British government, which by late 1918 had lost interest in nitrogen research, the US government became committed to extensive support of research into nitrogen capture technologies. The aim was to stimulate further developments within private industry. In many respects this work acknowledged the uncertainties, delays, confusion and lack of progress associated with the wartime nitrogen programme, as well as the previously neglected role of applied chemical science in industry.

On 22 March 1919 a Bill for "Production of Atmospheric Nitrogen," providing for the future of government wartime nitrogen plants and ongoing research into nitrogen capture, was discussed before Congress (S3390). As a result, civilian

A.S. Travis, *Nitrogen Capture*, https://doi.org/10.1007/978-3-319-68963-0_11

consultant Arthur Graham Glasgow was appointed Fixed Nitrogen Administrator, under the Secretary of War. At Glasgow's suggestion, on 29 March 1919, the Fixed Nitrogen Research Laboratory (FNRL), was established within the War Department under the Nitrate Division of the Chief of Ordnance [1]. The FNRL was located in the McKinley Building on the campus of American University, Washington DC, which during the war had been occupied by the Chemical Warfare Service (Fig. 11.1). The apparatus and equipment delivered to the FNRL included the benchtop ammonia reactor assembled at Arlington, Virginia.

The first director of the FNRL was Colonel Arthur B. Lamb from Harvard, former assistant of both Haber and Ostwald, who previously headed the Defence Chemical Research Section of the Chemical Warfare Service (1917–1919) [2].

Like the British, the Americans were keen to gain information from Oppau. A United States Navy commission had arrived there in early March 1919, but received much the same treatment as was later accorded to British investigators; it was told to leave after three days.

One group of American visitors to Germany—Philip Drinkner, lieutenant, US Army; Samuel Iserman, of the Chemical Company of America, Inc.; and Edward C. Worden, "consulting chemist air board"—was mainly interested in coatings for aircraft wings, but during their 10-day trip in mid-June 1919 found time to survey operations at Knapsack, Ludwigshafen/Oppau, and elsewhere. At Oppau they

Fig. 11.1 The Fixed Nitrogen Research Laboratory, Washington DC, in 1926. (Travis Porter Hignett photographic collection, 2003.902.105. Courtesy of the Chemical Heritage Foundation)

appear to have been more hospitably treated, at least by the French officer Captain Hackspill, perhaps because the visit was short and less in the nature of an investigation. Worden, among his other assignments, had advised on the status of American Federal Dyestuff and Chemical Corporation in 1915. The visitors were impressed, not only with what had been achieved under wartime conditions, but by the fact that work was continuing on expanding the manufacturing facilities [3]:

> At Knapsack a large plant had been built during the war for the manufacture of calcium carbide, cyanamid, synthetic acetic acid, ammonia, etc.. . ..At Ludwigshafen we saw the nitrogen fixation portion of the Badische Chemical Co. and were conducted through by a French officer thoroughly familiar with the plant and its technical operation. Conditions are best shown by the fact that at the time of our visit they were actually constructing important additions to the plant and this was before peace had been signed. . .The nitrogen fixation plant was entirely constructed during the war and is consequently the last word on the subject. . . . It is due mainly to the efforts of this concern that the German Government was enabled to carry out their program of powder manufacture. The size and importance of this plant [Oppau] was well known to the Allies during the war and it was consequently subject to numerous air raids.

In June–July, it was the turn of a commission made up of representatives from the FNRL, headed by Glasgow, to visit Oppau. They fared little better than the navy's group. The members were forbidden to take notes. However, they could at least make use of their observations by drawing on experience already gained with the General Chemical process at Sheffield, and related wartime research [4]. In June 1921, one of the FNRL group, R. S. Tour, compared the Oppau and Sheffield facilities at the semi-annual meeting of the American Institute of Chemical Engineers, held in Detroit, Michigan [5].

On 1 July 1921, following suspension of the Nitrate Division, the FNRL was transferred to the Department of Agriculture. Shortly after, Lamb returned to Harvard. The War Department and Department of Agriculture began investigations at the FNRL into the conversion of munitions-grade ammonium nitrate into a fertilizer. Though the solid nitrate fertilizer was effective, it deliquesced and caked, and was expensive to produce. It would not enter the fertilizer market until around 1940, formulated at a lower concentration in nitrate. In Europe, surplus ammonium nitrate was mixed with gypsum and marketed as a fertilizer during the 1920s.

The FNRL was well staffed, including with up to 50 chemists. In 1922, physical chemist Frederick Gardner Cottrell (1877–1948), inventor of the electrostatic precipitator, was appointed its director. At that time, Cottrell opined, the fixed nitrogen processes were not yet able to compete with Chile saltpetre as fertilizer, unless under particularly favourable conditions [6]. Among his many duties, Cottrell was charged with investigating the future of the Muscle Shoals cyanamide plant. Despite efforts to rehabilitate the plant, even as an experimental unit, as suggested by Cottrell, it was finally written off in the early 1930s [7]. Until the mid-1920s, the Bucher cyanide process and the Serpek nitride process were investigated by the FNRL. From a commercial perspective, they were written off in 1926 [8, 9].

By 1924, the FNRL's chemists had developed a high-pressure synthetic ammonia process, working at 300 atmospheres pressure and 475 °C, capable of producing a quarter of a ton of ammonia every day. Hydrogen was obtained by electrolysis, and nitrogen by burning out oxygen from the air with hydrogen. The most significant development was in the preparation of a novel catalyst, mainly the work of Harvard educated catalyst expert Alfred T. Larson, who laboriously followed the various recipes suggested by BASF patents, mixing iron with metal oxides as promoters (see box). The FNRL catalyst, it was claimed, gave a yield of 13–14%, compared with around 7–8% for the German process [10–12]. Probably it was not much different from the BASF catalyst, of which Larson had, in common with British and other American workers, acquired samples. It was nevertheless a major contributor to the process devised by the FNRL [13].

List of Haber and Le Rossignol, and BASF patents filed in the United States during 1909–1913, in the collection of Walter Otheman Snelling (1880–1965) of Trojan Power Company. Source: Chemical Heritage Foundation, Philadelphia

The dates of filing usefully indicate when the German discoveries were made. The catalysts included osmium, molybdenum, manganese, barium, cerium, nickel, cobalt, tungsten, lithium, copper, iron, palladium, platinium, and uranium, as well as binary and ternary mixtures with these and other metals and metal oxides. This selection of patents is typical of those placed in the hands of American investigators, including Alfred T. Larson.

Fritz Haber and Robert Le Rossignol, "Production of Ammonia," 971,501. Application filed 13 August 1909, serial no. 512,679. Patented 27 September 1910. "The reaction can be carried out at ordinary pressure, but we prefer to carry it out under increased pressure, for instance from 100 to 200 atmospheres." Osmium specified as catalyst.

Fritz Haber and Robert Le Rossignol, "Production of Ammonia," 1,006,206. Application filed 1 July 1910, serial no. 569,961. Patented 17 October 1911. "Nernst carried out the reaction under pressures from 50 to 70 atmospheres." The patent specified pressures of up to 250 atmospheres.

Fritz Haber and Robert Le Rossignol, "Production of Ammonia," 1,202,995. Application filed 13 August 1909, serial no. 512,680. Patented 31 October 1916. Includes diagram of continuous method.

Carl Bosch and Alwin Mittsach, "Process of Making Ammonia," 1,053,951. Application filed 19 September 1911, serial no. 650,245. Patented 18 February 1913. Catalyst containing molybdenum.

Carl Bosch and Alwin Mittsach, "Process of Manufacturing Ammonia," 1,053,952. Application filed 15 November 1911, serial no. 662,291. Patented 18 February 1913. Catalyst containing manganese.

(continued)

Carl Bosch, Alwin Mittsach, Hans Wolff, and Georg Stern, "Production of Ammonia and Catalytic Agents for Use Therein," 1,068,966. Original application filed 24 December 1910, serial no. 599,100, claiming generic production of ammonia when making use of a catalytic agent. Subsequently divided, and this application filed 10 January 1912, serial no. 670,443. Patented 29 July 1913. "In the specification for our application, for patent Serial No. 599,101, we have described a new form of iron....catalytic power can be improved by the presence of certain bodies which we have termed "promoters." Active catalyst containing manganese and a promoter, an oxide of alkali forming metal.

Carl Bosch, Alwin Mittsach, Hans Wolff, and Georg Stern, "Production of Ammonia and Catalytic Agents for Use Therein," 1,068,967. Original application filed 24 December 1910, serial no. 599,100. Divided, and this application filed 10 January 1912, serial no. 670,444. Patented 29 July 1913. Active catalyst containing barium and a promoter, an oxide of alkali forming metal.

Carl Bosch, Alwin Mittsach, Hans Wolff, and Georg Stern, "Production of Ammonia and Catalytic Agents for Use Therein," 1,068,968. Original application filed 24 December 1910, serial no. 599,100. Divided, and this application filed 10 January 1912, serial no 670,445. Patented 29 July 1913, serial no. 670,445. Active catalyst containing cerium and promoter, an oxide of alkali forming metal.

Carl Bosch, Alwin Mittsach, Hans Wolff, and Georg Stern, "Production of Ammonia and Catalytic Agents for Use Therein," 1,068,969. Original application filed 24 December 1910, serial no. 599,100. Divided and this application filed 10 January 1912, serial number 670,446. Patented 29 July 1913. Active catalyst containing osmium and a promoter, an oxide of alkali forming metal.

Carl Bosch and Franz Lappe, "Process of Working with Hydrogen Under Pressure," 1,075,085. Application 5 June 1913, no. 771,899. Patented 7 October 1913. Describes pervious lining to iron vessel, with atmosphere of nitrogen in the space between lining and iron: a "hot pressure-sustaining wall of the reaction vessel."

Carl Bosch, "Process of Working with Hydrogen Under Pressure," 1,077,034. Application 18 April 1912, no. 691,731. Patented 28 October 1913. Maintaining temperature of pressure-sustaining wall at 100–450 °C.

Carl Bosch and Alwin Mittasch, "Catalytic Production of Ammonia," 1,083,585. Application filed 15 October 1912, serial no. 725,814. Patented 6 January 1914. Catalyst of pure iron or iron nitride prepared by treatment with ammonia.

(continued)

Carl Bosch and Alwin Mittasch, "Producing Ammonia," 1,089,185. Application filed 11 November 1912, serial no. 730,646. Patented 3 March 1914. Catalyst of nitrogen compound of molybdenum, stated as molybdenum nitride.

Carl Bosch and Wilhelm Wild, "Process of Producing Hydrogen," 1,113,096. Application filed 29 October 1913, serial no. 798,005. Patented 6 October 1914. Process for production of pure hydrogen: carbon monoxide and steam passed over nickel or cobalt catalyst. ("... over a catalytic agent in lumps containing more than 30% of nickel and a non-metallic, indifferent, refractory and porous material.") This is the shift reaction.

Carl Bosch and Wilhelm Wild, "Process of Producing Hydrogen," 1,113,097. Application filed 29 October 1913, serial no. 798,006. Patented 6 October 1914. As 1,113,096.

Carl Bosch and Wilhelm Wild, "Producing Hydrogen," 1,115,776. Application filed 30 June 1913, serial no. 776,548. Patented 3 November 1914. Attempts had been made to produce hydrogen from water gas mixed with steam over a catalytic agent so that carbon monoxide was converted into carbon dioxide and a further quantity of hydrogen was produced. Excess carbon dioxide and steam removed. Hitherto processes had not led to satisfactory results. Claims catalytic agent containing divided iron oxide.

Carl Bosch and Wilhelm Wild, "Production of Commercially Pure Nitrates," 1,115,164. Application filed 2 June 1909, serial no. 503,638. Patented 27 October 1914. Commercially pure nitrates by direct absorption of oxides of nitrogen by means of a hot dry absorbent. For pure nitrate, pass oxides over solid basic substances (oxygen compounds of the alkaline earth series). For use in the electric arc process, etc.

Carl Bosch and Alwin Mittasch, "Producing Ammonia," 1,118,628. Application filed 16 August 1912, serial no. 715,331. Patented 24 November 1914. Catalytic mixture containing a body which has a considerable power of taking up nitrogen and also a body which has a considerable power of taking up hydrogen. Catalytic mixture containing tungsten and nickel.

Carl Bosch and Wilhelm Wild, "Process of Purifying Hydrogen Gas," 1,126,371. Application filed 30 January 1912, serial no. 674,416. Patented 26 January 1915. "When, in the production of hydrogen, coal or other carbonaceous material is employed either directly or indirectly, the hydrogen obtained always contains carbon monoxide. As instances of such methods of producing hydrogen, we mention passing steam over highly heated coal or coke, also soaking coal in alkaline solution and then heating the mixture with steam at comparatively low temperatures, further heating coal with water under pressure, and also passing hydrocarbons with steam over nickel at a temperature of say 1000 °C. We have found that we can on a technical scale completely abstract carbon monoxide from hydrogen by

(continued)

heating the gas mixture with a solution of caustic alkali under pressure exceeding five atmospheres."

Carl Bosch and Alwin Mittasch, "Production of Ammonia," 1,128,843. Application filed 11 November 1912, serial no. 730,645. Patented 16 February 1915. Catalytic agent which contains a "plurality of metals," one of which belongs to the iron group. Specifies lithium and manganese.

In 1924–1925 the FNRL process, often called the "American" process, was taken up at Mathieson Alkali Works, Niagara Falls, New York, first on a pilot plant scale of one-ton daily capacity, followed by an enlarged apparatus that enabled a daily output of 10 to 12 tons. The process was used in 1924 by Pacific Nitrogen Corporation (National Ammonia Co., of St Louis and Philadelphia), at Seattle, Washington (Table 11.1). Pacific Nitrogen was acquired by Du Pont in 1927, and in 1934 its ammonia plant was sold to China. Versions were used

Table 11.1 Synthetic ammonia facilities, United States, in 1927

Name of operator	Location	Rated daily capacity, tons of ammonia	Daily operating rate, tons of ammonia	Yearly nitrogen equivalent in tons
Atmospheric Nitrogen Corp. (Allied Chemical)	Syracuse, NY	30–40	30–40	11,550
Lazote, Inc. (Du Pont)	Charleston, W. Va. (Belle), 1927	25	15	4350
Mathieson Alkali Works	Niagara Falls, NY	10	10–12	2900
Niagara Ammonia Corporation. (to Du Pont, 1927)	Niagara Falls, NY	17	8	2300
Roessler-Hasslacher Chemical Co. (to Du Pont, 1930)	Niagara Falls, NY	3	3	865
Pacific Nitrogen Corp., 1924. (to Du Pont, 1927)	Seattle, Washington (National Ammonia Company).	3	3	865
Great Western Electrochemical Co. (to Dow, 1939)	Pittsburg, California	1	1	290
Commercial Solvents Corp. Converted to methanol, 1927.	Peoria, Ill.	15	0	

Sources: Frank A. Ernst, *Fixation of Atmospheric Nitrogen*. London: Chapman & Hall, Ltd., 1928, p. 121; F. A. Ernst, F. C. Reed, and W. L. Edwards, "A Direct Synthetic Ammonia Plant," *Industrial and Engineering Chemistry*, vol. 17, no. 8 (August 1925):775–788; "Editorial," *Industrial and Engineering Chemistry*, vol. 17, no. 8 (August 1925):772; F. A. Ernst and M. S. Sherman, "The World's Inorganic Nitrogen Industry," *Industrial and Engineering Chemistry*, vol. 19, no. 2 (February 1927):196–204

elsewhere, in 1930 by Midland Ammonia Company, of Midland, Michigan, and by Roessler-Hasslacher Chemical Company at Niagara Falls, New York (daily capacity three tons). The Mathieson pilot plant was later transferred to the Great Western Electrochemical Company, at Pittsburg, California (purchased by Dow in 1939). Roessler-Hasslacher, founded in 1882, specialized in electrochemicals and other products; in 1930 it was acquired by Du Pont. Relative to coal-based processes, the operating costs for these processes that generally used by-product electrolytic hydrogen were high. They mainly fell into disuse during the depression years.

In 1927, Commercial Solvents Corporation, founded in 1919 at Terre Haute, Indiana, to work the Weizmann fermentation process for acetone and butanol, from corn, constructed at Peoria, Illinois, an ammonia plant of 15-ton capacity. For hydrogen, it used by-product gas containing 60% carbon dioxide and 40% hydrogen. Within a month of starting up, in the summer, the ammonia plant was converted to production by high-pressure synthesis of methanol from the by-product gas, following removal of part of the carbon dioxide [14]. The plant subsequently tripled the synthetic methanol capacity of the United States and contributed to the decline of the process based on the destructive distillation of wood (for "wood alcohol"). From 1932, Pennsylvania Salt Manufacturing Co. worked the Mont Cenis process, at Wyandotte, Michigan. As with the other small plants, it relied on electrolytic hydrogen, from production of caustic soda and chlorine.

The FNRL became a leading model and training ground for research carried out in the post-war US chemical industry (Figs. 11.2, 11.3 and 11.4). It also encouraged academic institutions to embark on high-pressure studies, starting, in 1925, at the MIT Research Laboratory of Applied Chemistry, followed by Yale, Illinois, and Wisconsin. On 1 July 1926, the FNRL was absorbed into the Bureau of Soils, and in 1927 its functions were assigned to the Fertilizer Research Division of the Bureau of Chemistry and Soils. Among the outstanding chemists at the FNRL was Hugh Paul Emmett (1900–1985), who received his PhD from Cal-Tech, under Arthur T. Benton. Emmett during 1926–1937 undertook work on heterogeneous catalysis, most famously the study that subsequently, in collaboration with Stephen Brunauer and Edward Teller (at George Washington University), became known as the BET theory (1938) [15]. (BET was an acronym of their names.) The theory dealt with the behaviour of gases on the surfaces of catalysts. Emmett showed that the rate determining step involved adsorption of nitrogen [16, 17]. He also worked on the nature and role of promoters. Emmett's work inspired further developments, particularly in the 1970s by Gerhard Etl (University of Munich) and Gabor Samorjia (University of California, Berkeley). In 2007, Ertl received the Nobel Prize for his studies on heterogeneous catalysis.

Chemist Travis Porter Hignett (1907–1989), specialist in fertilizer technology, undertook research at the FNRL from 1929 until 1938, when he moved to the Tennessee Valley Authority. He recalled that at the time he joined the FNRL, farmers had never heard of synthetic fertilizer. "The only fertilizer they knew about was manure, and they couldn't understand why the government would be investigating manure!" [18]. In fact until the 1940s, American farmers were far less intensive users of nitrogen fertilizer than were Europeans, Japanese, and Egyptians.

Fig. 11.2 The first stage in catalyst preparation, burning iron in an atmosphere of oxygen, with electrodes as heat sources. Fixed Nitrogen Research Laboratory, Washington DC, 1926. (Travis Porter Hignett photographic collection, 2003.902.87. Courtesy of the Chemical Heritage Foundation)

Fig. 11.3 The second stage in catalyst preparation, mixing cooled iron oxide with promoters. Fixed Nitrogen Research Laboratory, Washington DC, 1926. (Travis Porter Hignett photographic collection, 2003.902.89. Courtesy of the Chemical Heritage Foundation)

Fig. 11.4 A group of benchtop bombs for testing catalysts. Fixed Nitrogen Research Laboratory, Washington DC, 1926. (Travis Porter Hignett photographic collection, 2003.902.96. Courtesy of the Chemical Heritage Foundation)

Cottrell left the FNRL in 1930, but remained involved as a consultant. In the late 1930s he commissioned physical chemist Farrington Daniels (1889–1972), at the University of Wisconsin in Madison—who had worked on nitrogen fixation at the Bureau of Soils, and subsequently at the FNRL, where he investigated the arc process—to develop the high temperature combination of nitrogen and oxygen in a pebble-bed furnace. The furnace was based on FNRL research undertaken from the late 1920s by Percy H. Royster. Work on this nitrogen capture process and trials continued until 1954. Though offering considerable promise, it could not compete with the high-pressure ammonia synthesis [19].

11.2 The NEC Process

The Ammonia Division of the General Chemical Company—that had worked on its "modified Haber process" during the war—had by 1919 come to an agreement with the Solvay Process Company to jointly develop a synthetic ammonia process. The subsidiary Atmospheric Nitrogen Corporation was founded in December 1919, a year before the merger involving General Chemical, Solvay Process and others to create Allied Chemical. Several improvements had been introduced to the original General Chemical process, drawing on experience gained at Sheffield, Alabama, and at the Syracuse works. Atmospheric Nitrogen began producing ammonia at the daily rate of 30 tons of fixed nitrogen from August 1921 at Syracuse. Louis Jones and (Major) Charles O. Brown were the leading contributors towards development of the process. Improvements were probably made using information received from the FNRL.

No sooner had Allied Chemical been founded than its CEO, Orlando Weber, embarked on a programme of major layoffs of workers and staff and severe cutbacks in research. This in part was a response to a massive deflationary recession during 1919–1921 that was impacting on several countries. The outcome of Weber's action was the widespread dispersal of know-how concerning chemical technology, not only of the Solvay ammonia-soda process but also of synthetic ammonia. Among those departing, as already mentioned, was Louis Jones, who in 1926 with Charles Brown established the Nitrogen Engineering Corporation (NEC), which offered an ammonia converter that was far better suited to large-scale production than that of the FNRL process. The NEC process found widespread application in Europe as well as in the United States. In Europe, the corporation was represented by Hydro Nitro Soc. Anon., founded on 5 December 1928 in Geneva. In 1929, NEC was taken over by the American Cyanamid subsidiary Chemical Construction Corporation (Chemico), which offered design and construction services for high-pressure plant and equipment. The NEC process was adopted by Etablissements Kuhlmann in France, at Lille (1930), and orders were received from the Soviet Union and the Far East.

Despite the loss of staff and greatly reduced research effort, Allied's Atmospheric Nitrogen, through the Solvay and General Chemical divisions, maintained a strong interest in synthetic ammonia [20]. Success in this venture encouraged Allied during 1928–1929 to open a large ammonia facility at Hopewell, Virginia. Hydrogen was produced from coke oven gas. The principal application was as fertilizer. The only other large producer in the United States was Du Pont, whose process was based on French and Italian technologies.

11.3 Combining Casale and Claude Technologies

Niagara Ammonia Company, a subsidiary of Ammonia Corporation, licensee of Casale's process, opened a synthetic ammonia factory at Niagara Falls, New York, in 1924. It started up at 13 tons of anhydrous ammonia daily. Not long after, Du Pont obtained exclusive US rights to the Claude process [21]. On 21 May 1925 jointly with Air Liquide, Du Pont, incorporated in West Virginia the firm Lazote, Inc., to build and operate an ammonia synthesis facility at Belle, East Kanawha County, next to the Kanawha River. Claude's ammonia patent of 30 June 1925 was assigned to Lazote [22]. A catalyst testing laboratory, modelled after that of the FNRL, was installed at Du Pont's Wilmington Experimental Station where Alfred T. Larson, previously at the FNRL, conducted further investigations [23]. The ammonia factory opened on 1 April 1926, but faced several difficulties.

The promoters of the Claude process had promised lower working costs than were found to be the case—and the advantage of requiring smaller equipment—but, as Du Pont found to its great cost, this was not always realizable. Serious explosions, with loss of life, hindered progress, as did problems encountered with compressors. This led Du Pont in 1927 to acquire Niagara Ammonia Corporation, and with it access to the Casale process. The "best features" of the French and Italian processes "were adapted to American conditions" [24]. In the same year, National Ammonia was acquired, and the Belle plant was expanded. An editorial in *Industrial and Engineering Chemistry* noted that in the United States, "A belated start in the fixation of atmospheric nitrogen seems to have been overcome by the development of our own processes—so satisfactory as to more than hold their own in world competition" [25]. The growth of synthetic ammonia led to the introduction of special railway tank cars for the shipment of anhydrous ammonia, at first of 30,000 lb. capacity, and by 1928 of 50,000 lb. capacity. Before 1925, ammonia was transported in cylinders of up to 150 lb. capacity.

The Du Pont Ammonia Corporation was established in 1929, at a time when ICI and Du Pont were intent on drawing up an agreement for the exchange of know-how. Since ICI had an interest in Allied Chemical through Solvay Process, the agreement was strongly opposed by Orlando Weber of Allied Chemical. The outcome was that ICI severed its ties with Allied in order to collaborate with Du Pont. Weber had already cut out exchanges of technical information between the American Solvay Process Company and European Solvay interests, in addition to dismissing or releasing many technical staff, including Louis Jones, who was busy peddling abroad the Nitrogen Engineering Corporation ammonia process. After the 1929 agreement with ICI was signed Du Pont sent over a delegation to England to inspect the Billingham factory in order to gain access to Haber-Bosch type technology. Du Pont later bought out Air Liquide's interest in Lazote. Subsequently, Du Pont continued to develop its own ammonia technology. The Claude process, through innovations adopted from the Casale process and know-how acquired from ICI, became the modified process of Du Pont.

The fall in the price of ammonia after the mid-1920s encouraged American firms to take up its oxidation for the manufacture of nitric acid, thereby displacing the method based on imported sodium nitrate. Du Pont and American Cyanamid's Chemical Construction Corporation (Chemico) developed processes suited to both anhydrous ammonia and ammonia liquor. Among the advantages of using anhydrous ammonia was the lower cost, per unit of ammonia, of transportation in high-pressure railway tank cars. Brunner, Mond's Herbert Humphrey had been struck by the extensive use of aluminium paint in American factories. It was especially useful for ammonia storage tanks that were required to be painted in a light colour in order to radiate heat.

In the United States during the early years of the Great Depression, as elsewhere, farmers were often unable to purchase synthetic nitrogen fertilizer especially since without suitable lines of credit the cost could not be recovered until after the harvest. Many were mired in debt. Also, as a seasonal trade, manufacturers of fertilizers required extensive storage facilities, which became overloaded as business fell off. By 1930, the price of fertilizer had halved from that achieved in 1928. In September 1931, the Du Pont Ammonia Corporation became the Ammonia Department, effectively a downgrade similar to the situation at Billingham. From 1932, however, ammonia for fertilizer use was in demand again and thereafter there was steady growth. Starting in September 1932, Du Pont manufactured urea-ammonia liquor, and from early 1935 solid urea (Du Pont Crystal Urea), a soluble white crystalline compound, thereby releasing the United States from its reliance on urea imported from Germany. Anhydrous ammonia and urea-ammonia were converted into non-acid forming fertilizers that, combined with phosphorus and potash, found extensive use in the eastern half of the United States, where slightly acidic soils gave the best yields of crops. Acidity levels of fertilizers were adjusted through the controlled addition of limestone. The relatively high nitrogen-content of urea-ammonia was a particular advantage. Another important use of Du Pont ammonia was in the production of urea-based resins. In 1927, Du Pont worked on a pilot plant scale the reaction between the carbon monoxide removed from ammonia synthesis gas and hydrogen to produce, via a high-pressure reaction, methanol, and then the ammonia. This single synthesis gas stream process was later introduced in Europe, though dropped by Du Pont [26].

In the mid-1930s, Du Pont and Allied together produced over 80% of synthetic ammonia made in the United States, and they would remain the nation's leaders, with Allied, by the end of the decade, producing over 50%. For the fiscal year 1937–1938, annual consumption of "chemical nitrogen," including imports, was 543,000 tons, of which over 350,000 tons were produced in the United States. The American Cyanamid facility at Niagara Falls had a capacity of 355,000 tons of calcium cynamide (75,000 tons of nitrogen).

Table 11.2 Synthetic ammonia producers, and annual capacities, in tons, United States, 1940

Company	Year manufacture began	Process	Source of hydrogen	Capacity, tons
Allied Chemical	1928	Atmospheric Nitrogen; General Chemical/Solvay	Coke oven	250,000
Du Pont	1926/1928	Claude-Casale	Water gas	146,000
Shell	1931	Mont Cenis	Natural gas	27,000
Midland	1930	American	By-product	9000
Mathieson	1925	NEC	By-product	5800
Pennsylvania Salt	1932	Mont Cenis	By-product	4900
Niagara (Du Pont)	1926	Casale	By-product	4500
Great Western (Dow)	1927	NEC	By-product	1100
Hercules	1940	Claude	Natural gas	10,900

Notes: NEC = Nitrogen Engineering Corporation. American = based on the Fixed Nitrogen Research Laboratory method. By-product = from electrolysis of salt. Source: Willard C. Cope, "Ammonia. Part I. US Production facilities," *Chemical Industries*, vol. 64, no. 6 (June 1949):920–925

In 1939, Du Pont's "Claude" process was licensed to Hercules Powder Co. for use at Pinole, California. This was the first plant to use steam reforming of methane in the production of hydrogen. Around this time the Du Pont process was modified. The unreacted gas stream was recirculated following passage through two converters made of ordinary carbon steel, as found suitable in practice. The series arrangement of converters required a lower capacity circulation pump than used with parallel converters. The usual iron oxide catalyst was employed [27]. Nine synthetic ammonia plants in the United States were producing 459,000 tons of ammonia in 1940 (Table 11.2). Prior to entry into and during World War II, the US government ordered the construction of ten new strategic ammonia plants, one at the old Muscle Shoals cyanamide site [28]. Ammonia production grew exponentially. Five of the ten government plants were sold shortly after 1945. The booming post-war economy propelled the United States to first place among global ammonia producers at least until 1980.

References

1. Lamb AB (1920) The Fixed Nitrogen Research Laboratory. Chem Met Eng 22(21) (26 May):977–979.
2. Emmett PH (1983) The Fixed Nitrogen Research Laboratory. In: Davis BH, Hettinger WP, Jr (eds), Heterogeneous catalysis. American Chemical Society, Washington DC, pp 195–221.
3. Drinkner P (1919) Dyestuffs: Hearings before the Committee on Ways and Means, House of Representatives, on H.R. 2706 and H.R. 6495, June 18–20 and July 14–18, 1919 [66th Cong., 1st Sess., 1919]. Committee on Ways and Means. Government Printing Office, Washington DC, pp 513–534, on 534.
4. Clarke MJ (1976) The federal government and the fixed nitrogen industry 1915–1926. PhD dissertation, Oregon State University, pp 125, 130.
5. Tour RS (1922) The German and American synthetic-ammonia plants. I–V. Chem Met Eng 26(6)(8 February):245–248; 26(7)(15 February):307–311; 26(8)(22 February):359–362; 26(9) (1 March):411–415; 26(10)(8 March): 463–465.
6. News and notes. The Fixed Nitrogen Research Laboratory (1923) J Soc Chem Ind 42(4) (26 January):80 (Chem Ind London).
7. Hubbard PJ (1961) Origins of the TVA: The Muscle Shoals controversy, 1920–1932. Vanderbilt University Press, Nashville.
8. Guernsey EW, Lee JY, Braham JM, Sherman MS (1926) Some factors affecting the fixation of nitrogen as sodium cyanide. Ind Eng Chem 18(3)(March):243–248.
9. Krase HJ, Thompson JG, Lee JY (1926) The fixation of nitrogen as aluminum nitride.Ind Eng Chem 18(12)(December):1287–1290.
10. Larson AT (1924) Increasing ammonia production with improved catalysts. Ind Eng Chem 16 (10)(October):1002–1004.
11. Larson AT, Richardson CN (1925) Preparation of fused iron oxide for use as a catalyst. Ind Eng Chem 17(9)(September):971–972.
12. Larson AT, Brooks AP (1926) Ammonia catalyst. Ind Eng Chem 18(12) (December):1305–1307.
13. Ernst FA, Reed FC, Edwards WL (1925) A direct synthetic ammonia plant. Ind Eng Chem 17 (8)(August):775–788.
14. Kelly FC (1936) One thing leads to another: the growth of an industry. Houghton Mifflin Company/Riverside Press, Cambridge MA.

15. Emmett PH (1983) The Fixed Nitrogen Research Laboratory. In: Davis BH, Hettinger WP, Jr (eds), Heterogeneous catalysis. American Chemical Society, Washington DC, pp 195–216.
16. Emmett PH (1930) Studies on the mechanism of ammonia synthesis over iron catalysts. J Chem Ed 7(1930):2571–2582.
17. Garten RL (1983) Paul H. Emmett: six decades of contributions to catalysis. In: Davis BH, Hettinger WP, Jr (eds), Heterogeneous catalysis. American Chemical Society, Washington DC, pp 45–60.
18. New career at 65 brings Hignett honors (1980) Times Daily (Florence AL) 25 August. Travis Porter Hignett photographic collection, FNRL. Chemical Heritage Foundation archives.
19. Stranges AN (1992) Farrington Daniels and the Wisconsin process for nitrogen fixation. Social Studies of Science 22(2)(May):317–337.
20. Annual report of the director of the Fixed Nitrogen Research Laboratory, Washington, 6 August 1924. Year ended June 30, 1924, director FG Cottrell. Solvay Process Company archives, microfilm roll 1. Sidney M. Edelstein Library, Israel National Library.
21. Claude's first ammonia patent filed in the United States was no. 1,332,460 of 2 March 1920.
22. Claude US patent, 1,544,373, filed 8 September 1920.
23. Killeffer DH (1925) Catalyst testing: the basis of new industries. Ind Eng Chem 17(8) (August):789–792.
24. Jasper E. Crane, of Du Pont, to George W. Norris, senator, Nebraska, 8 May 1933. In: Tongue H (1934) The design and construction of high pressure chemical plant. Chapman & Hall, London, pp 376–378.
25. Significant trends in nitrogen fixation (1927) Ind Eng Chem 19(12)(December):1307.
26. Cope WC (1949) Ammonia. Part I. US production facilities. Chemical Industries 64(6) (June):920–925.
27. Thompson HL, Guillaumeron P, Updegraff NC (1952) Ammonia synthesis at 1000 atmospheres: the present-day Claude process. Chem Eng Progress 48(9)(September):468–476.
28 Pressure-synthesis operations of the Du Pont Ammonia Corporation (1930) Ind Eng Chem 22(5)(May):433–437.

Chapter 12
New Ideologies and National Security in the 1920s

12.1 Italy, and Central and Eastern Europe

Here we consider the way in which the nitrogen industry emerged during the 1920s in five reshaped or new European countries—Italy, Czechoslovakia, Romania, Hungary, and Poland—whose land and mineral resources and geographical bearings dictated industrial and security policies far more so than in the western, generally more industrialized, maritime nations that bordered the Atlantic Ocean. This was particularly the case for the Austrian succession states after 1918. Created by the unification of two kingdoms in 1867, the Austro-Hungarian Empire at the end of the nineteenth century encompassed much of central Europe and the Balkans: in the north Bohemia, Moravia, Silesia and Galicia; in the south Croatia, Dalmatia and Bosnia-Herzegovina; in the west Austria and the Trieste region; and in the east Transylvania and Bukovina.

The breakup of the Austro-Hungarian Empire gave Italy, as one of the victorious allies, the Tyrol south of the Brenner Pass, the Veneto region, Trieste, the Isonzo Valley up to the Julian Alps, and the Istrian peninsula with the ports of Pola and Fiume; Italy thereby gained control of the northern Adriatic, while Austria and Hungary both lost their outlets to the sea. The Republic of Czechoslovakia arose out of the fusion of contiguous fragments of the empire that included in the west Bohemia, home to Kladno, the industrial heart of old Austria, and Moravia, and in the east the agricultural lands of Slovakia and Ruthenia (Trans-Carpathian Ruthenia), formerly in Hungary. There was also, north of Moravia, the former Duchy of Upper and Lower Silesia (Austrian Silesia). The political borders were mainly in accord with the Treaty of Saint-Germain signed by Austria on 10 September 1919. By the terms of the same treaty, Romania, in south-eastern Europe, received Bukovina and the eastern part of former Hungary from the Banat to the Carpathians. Yugoslavia (Kingdom of Serbs, Croats and Slovenes) was created by the merging of Serbia with Bosnia, Herzegovina, and Dalmatia. The new Austria and Hungary, the latter shaped by the Treaty of Trianon of 4 June 1920,

A.S. Travis, *Nitrogen Capture*, https://doi.org/10.1007/978-3-319-68963-0_12

were truncated fragments of the defunct empire. As for Poland, in north-central Europe, it received from Germany, according to the Treaty of Versailles of 28 June 1919, the narrow "Polish Corridor," providing access to the Baltic Sea, most of the Province of Posen and the lower Vistula Valley, formerly part of Prussia. The great prize from 1922 was coal-rich Upper Silesia, second only to the Ruhr among the former German Empire's coal-based industrial districts. The port city of Danzig came under international supervision, a free city, under the League of Nations.

New and existing neighbours, as well as mineral and energy resources, dictated strategic policies. Italy, with a population of 40.5 million (1925), shared four land frontiers, totalling 1851 km, with France, Austria, Yugoslavia, and Switzerland. The Italian seaboard was approximately 8000 km. Czechoslovakia, with a population of over 14 million, was totally landlocked. To the west and north-west its land frontier was shared with Germany, which created political tensions and concerns over the defence of its western reaches. There were also questions over economic protection. The reduced post-Habsburg Hungary, with a population of just over 8 million, and Romania, with a population of almost 17 million (1923), were also landlocked. Poland, with a population of over 29 million, had a border that extended 5011 km, of which 1662 km were shared with Germany and 1412 km with Russia. The Peace of Riga signed on 18 March 1921 divided disputed territories between Poland and Soviet Russia. The Polish sea coast was just 130 km, but included the new port of Gdynia, which became the export gateway for the newly acquired Upper Silesian industrial belt, via a French-financed direct railway through the corridor.

From the perspective of comparative approaches, a review of the commercial applications of nitrogen technologies in Italy is particularly apt. It contains all the elements that enable an insight into the pursuit of agendas based on national needs and resources, and of how and why rival Italian ammonia processes, the only viable competitors to the Haber-Bosch process before the mid-1920s, were adopted elsewhere. There was also the growing political element. Italy under a fascist government pursued a policy of self-sufficiency, with particular reference to increasing the yield of grain.

12.2 The Fauser Process and Political Developments in Italy

Benito Mussolini came to power following the fascist March on Rome in October 1922. It was the beginning of a single-party dictatorship, established in 1925. This would provide unprecedented opportunities for industrial concerns such as Montecatini, where Donegani, with ready access to lines of credit and the media, and with political influence, hastened and consolidated control of the greater part of the Italian nitrogen industry. As far as was possible, he used these powers to

Table 12.1 Calcium cyanamide production and imports, Italy, 000's of metric tons

Year	Production	Imported
1921	21.4	–
1922	30.8	0.9
1923	44.3	5.0
1924	52.0	14.3
1925	47.0	32.8

Source: *League of Nations. Armaments Year-Book, Third Year. 1926–1927*. Geneva: League of Nations, 1927, p. 574

crush, or at least contain, rivals. Thus while he lost interest in the Casale process, this was not the case for his interest in control of Ammonia Casale output by the Terni company. Montecatini, through its control of the market in nitrogen fertilizers, was enabled to dictate terms of business when it came to Terni. In 1925, an agreement was reached between Montecatini and Terni: it created Montecatini's (near) monopoly in synthetic ammonia. According to this agreement, Terni was allowed to produce 3500 metric tons of anhydrous ammonia yearly until Montecatini had produced at least 17,000 metric tons. Beyond this, Terni was permitted to produce up to 7000 tons. The agreement was due to last until 1935 [1, 2]. However it would appear that Montecatini was unable to place restrictions on the Terni company's output of cyanamide from its Carburo division.

Mussolini's government was installed at a time when the foreign trade balance and high imports had brought the Italian economy to a state of severe recession, with a weakening lira increasing the price of imports, especially of wheat, coal and industrial products. Autarky and the development of national infrastructure became priorities. Among major projects was electricity generation, notably with construction of the Marlengo hydro-electric facility that served the Merano ammonia factory in South Tyrol. In July 1925, with the inauguration of Il Duce's "Battle of Grain" (Battaglia del Grano) and soon after protective tariffs on imported grain, the government supported growth of the agricultural sector, which included the increased application of artificial fertilizer. This had an immediate impact on the production of ammonium sulphate from synthetic ammonia. In addition, the manufacture and import of calcium cyanamide grew considerably (Tables 12.1 and 12.2).

Montecatini's holdings in cyanamide producing and marketing concerns from the early 1920s, and the adoption of Fauser ammonia, generated substantial profits [3].[1] By the late 1920s, the powerful conglomerate, profiting from Mussolini's economic policies and government contracts, was producing 70% of all nitrogen fertilizer manufactured in Italy, with almost complete control over

[1]By 1930, Montecatini's near domination of the Italian nitrogen industry was through ownership or holdings in four companies: Società Ammmonia e Derivati; Società Meridionale Ammonia; ASED (50% holding); and SIPA (62% holding).

Table 12.2 Calcium cyanamide capacities, and production, expressed as fixed nitrogen, at European facilities in 1927

Location	Annual capacity, short tons	Production, short tons
Italy		
Terni-Papigno (Terni company)	10,000	5500
Domodossola (Montecatini)	3500	1000
Ascoli Piceno	2500	1000
San Marcel	1000	500
Germany		
Piesteritz	35,000	30,000
Trostberg	35,000	35,000
Waldshut	12,000	8000
France		
Lannemezan	15,000	10,500
Bellegarde	5000	3250
Marginac	6000	2500
Brignoud	6000	3000
Modane	4000	1000
Switzerland		
Martigny	2500	2500
Norway		
Odda	15,000	4500
Sweden		
Ljunga	3000	2500
Alby	3000	2500
Yugoslavia		
Sebenico	10,000	14,000 to 15,500
Dugi Rat	10,000	
Maria Rast	7000	
Poland		
Chorzów	30,000	17,000
Czechoslovakia		
Falkenau	6000	4000
Romania		
Dicsö St. Marton (Dicsöszentmárton, Ungar-Altenberg)	5000	3000

Source: Frank A. Ernst, *Fixation of Atmospheric Nitrogen*. London: Chapman & Hall, Ltd, 1928, pp. 129–130. Considerably higher figures, probably inflated, are given for Germany and France in H. W. Bates, "Economic Relationships Between Nitrogen and Fertilizers," *Chemical and Engineering News*, vol. 20, no. 11 (November 1928):1133–1139, on 1133 (see Table 12.3). For Poland, see *League of Nations, Armaments Year-Book. Third Year. 1926–1927*. Geneva: League of Nations, 1927, p. 685, which gives cyanamide production for 1925 at 86,000 metric tons. Published statistics for Yugoslavia vary, and are probably unreliable. The factories in Dalmatia, at Sebenico and Dugi Rat, were controlled by the Terni company, until 1929, when they were acquired by the Société Française des Forces Hydro-électriques de la Dalmatie, an affiliate of the Soc. des Phosphates Tunisiens et des Engrais et Produits Chimiques (Société Pierrefitte).

distribution, from all sources [4, 5]. From 1925, Montecatini's involvement with hydro-electricity and electrochemistry led to the production of aluminium [6].

In 1928, six synthetic ammonia factories in Italy used the Fauser process, two the Claude process, and (apart from the small unit at Terni) one the Casale process. The estimated annual Italian synthetic ammonia capacity, expressed as fixed nitrogen, was around 64,000 tons [7].[2]

From 1934 the chemical autarky programme, in the hands of Montecatini, which effectively designed the programme, cemented the interests of the fascist state and its leading chemical corporation when it came to nitrogen fertilizers, as well as other products. The oligarchy that owned Montecatini allowed the dictatorial Donegani to take all decisions, including those that affected affiliate companies [8]. Montecatini's power stretched much further. Thus when the industrial firm A. Brambilla of Milan in the early 1930s planned to construct an ammonia facility using Nitrogen Engineering Corporation technology at Verrès in the Aosta Valley a principal concern was the possibility of a clash with Montecatini.

For Montecatini and similar large corporations that co-operated to satisfy the needs of the state's self-sufficiency programmes the benefits were enormous. Italian industrialists relied on a system of awarding special favours to purchasers from, and middlemen connected with, manufacturers who won lucrative contracts. In addition, subsidies were provided to Italian farmers for the purchase of fertilizer, seeds, and machinery. Through unusual corporate ownership structures—particularly extended families and byzantine share structures—complex webs of companies were controlled by a few individuals. This enabled price and market fixing, which was encouraged by protectionism introduced in 1925, around which time the Committee for the Preparation of the National Mobilization was established. Fauser's hydrogen and hydrogenation patents became important bargaining chips when Montecatini invested in petroleum refining.

Lack of competition, in part the result of Montecatini's monopoly, encouraged complacency, with the result that imports of grain remained significant. Nevertheless, cereal production doubled from 1922 to 1939, and wheat imports were reduced by 75% in the decade after 1925 [9]. Italian nitrogen and related technologies, at least Montecatini's, contributed to the supply of food, and a seemingly satisfied population, which allowed the corporation in its advertising and promotional material around 1940 to draw on Il Duce's pronouncement that harmony was constituted of three "principles"—capital, technology and labour—that underscored the fortunes of Italy ("....solo dall'armonia costituita dai tre principî: capitale, tecnica, lavoro, vengono le sorgenti della fortuna").

There was another significant feature of fascist ideology associated with the location of the ammonia facility at Merano: the first major population shift

---

[2]The figure of 44,000 metric tons given in this source was a misprint. Five of the factories converted ammonia into ammonium sulphate. Cyanamide production in Italy contributed 20,000 metric tons of nitrogen; output remained at consistently high levels until the late 1930s. The import of ammonium sulphate declined considerably, as was the case in other European countries that took up industrial nitrogen capture.

connected directly with industrial expansion. When in 1919, Italy annexed South Tyrol (Südtirol), its population consisted mainly of ethnic Germans. Under Mussolini's government, Italian culture and language were imposed on the entire population. By the end of 1925, Italian law had effectively excluded all manifestations of the German language and culture. Italianization was also pursued through developments in industry. The model factory township of Sinigo near Merano was created during the mid-1920s to provide homes for Italian workers, as part of a larger scheme of fascist-inspired internal migration, generally based on land reclamation (such as Maremma, the Pontine Plain, Sardinia, Sicily, the Ferrera district, and Puglia). The Sinigo community served as a satellite of the large ammonia facility and power station. Through the merging of hydro-electric power, synthetic ammonia, and agricultural organization—all backed with state subsidies—and the transfer of Italian workers from the south, a political function was served: increase of the Italian population of South Tyrol [10]. In sum, Montecatini's extended empire comprised a network of influence that stretched deeply into corporate and political life, and was closely aligned with the interests of the state. By the end of the 1930s, following diversification, Montecatini controlled over 80% of the Italian chemical industry, comparable perhaps only to the position of IG Farben in Germany. Montecatini's market dominance in nitrogen fertilizers did however cause Mussolini to intervene and promote the entry of other firms into different and innovative sectors of chemical manufacture, including through international collaborations. As a result Montecatini confronted competition from state-owned firms, but not in nitrogen products, though some of its dominance was diminished. The two Italian nitrogen fertilizer cartels were reorganized in June 1936, in the synthetic ammonia sector with Montecatini controlling just over 71% of Consorzia Italiano Azoto, and the Terni company the balance; and in calcium cyanamide with Montecatini controlling 22%, Società Industriali Carburo 15.6%, and Terni 61.5% (Società Anonima Calciocianamide), though Montecatini controlled marketing. By then, Terni, with its interests in ammonia, cyanamide, steel, coal, and electricity, was controlled by the fascist state.

The failure of sanctions against Italy in the mid-1930s following Mussolini's adventure in East Africa, the 1935–1936 war with Ethiopia (following the Abyssinia crisis)—encouraged by the failure of the League of Nations to take action against Japan's takeover of Manchuria and Chinese provinces—stimulated further moves towards autarky, particularly increased production of machinery and chemicals.

12.3 Czechoslovakia

The Austro-Hungarian Empire's leading chemical firm was the Austrian Association for Chemical and Metallurgical Production (Oesterreichischer Verein für chemische und metallurgische Produktion; Aussiger Verein; present-day Spolchemie), founded in February 1856 by the German entrepreneur Christian Gustav Clemm, at Aussig on the River Elbe (Labe), in northern Bohemia (Aussig is present-day Ústí nad Labem, Czech Republic) (Fig. 12.1). From early 1918, as a contribution to the Austrian war

Fig. 12.1 Czechoslovakia and surrounding countries, according to political boundaries in the inter-war period. The names shown and designations are those commonly used in English language publications. Moravia was also called Silesia-Moravia as a result of Czechoslovakia incorporating a part of the former Duchy of Upper and Lower Silesia

effort, Aussiger Verein undertook cyanamide manufacture at Falkenau (Falknov nad Ohří; since 1948, Sokolov) in western Bohemia, drawing on lignite mined at nearby Lastner. Work on developing the mines began in 1916, carried out by Aussiger Verein in collaboration with the Buschtĕrader Eisenbahn, backed by Allgemeine Osterreichische Bodenkreditanstalt. After the war, the cyanamide factory manufactured carbide alone, until 1920, when it recommenced production of cyanamide in the hands of Czechoslovakian entrepreneurs [11].

Demand for nitrogen products was so great that imports of Chilean nitrate remained high, mainly to support the beet sugar industry. The Falkenau site was close to the frontier with Germany, with implications for national security. In 1925 a ministerial commission considered a proposal that new nitrogen factories should be constructed in the interior, in addition to expansion of cyanamide production at Falkenau [12]. Cyanamide output was increased from 13,662 tons in 1926 to 30,974 tons in 1929, in line with the commission's suggestion. However, its policy was not, at first, adopted for ammonia factories. Czechoslovakia, the only true inter-war democracy in central-eastern Europe, though concerned about German ambitions, feared the Soviet Union even more and was engaged in border disputes with Hungary. As a result, the two new nitrogen factories that were built before the 1930s lay in the north and west, close to the frontier with Germany and access to coal.

In 1928, Spolchemie introduced the Nitrogen Engineering Corporation ammonia process at its original Aussig factory. Spolchemie was also associated with introduction in February 1928 of the Claude process by Société Tchécoslovaque des Produits Azotés (Tschechoslowakische Stickstoff-

Werke AG) in Czech Silesia, near Marienberg (Mariánské Hory). Daily production was 30 tons, increased to 40 tons by 1930. Hydrogen came from the Moravia-Ostrava (Mährisch-Ostrau) coke oven works; it was later made by the water gas reaction [13, 14]. The factory's main product was ammonium sulphate, for agriculture. Some ammonia was converted into nitric acid, for production of ammonium nitrate, which by reaction with limestone gave "Ostrava Saltpêtre." Other products included superphosphate, phosphoric acid and ammonium phosphate. Škoda manufactured both ammonia converters and other high-pressure equipment for use in the chemical industry. By the mid-1930s, just as the economy began to improve, Czechoslovakia was producing 18,000 tons of fixed nitrogen each year.

After 1933, revised national security considerations dictated the sites of two important nitrogen factories. In 1936, ICI drew up plans for an ammonia plant at Bratislava (capital of present-day Slovakia), in the south. By 1937, the Bratislava facility and another ammonia factory at Pardubice (east of Prague) were producing intermediates for explosives for Dynamit AG (owned by IG Farben) and "Pardubice Chemical and Dynamite Works" at Semtin (founded in 1920; present-day Explosia SA), respectively.

In 1936 Hitler, in violation of the Versailles and Locarno treaties, seized the Rhineland's demilitarized zone. Škoda turned to the large-scale production of arms. The future was uncertain for Spolchemie, Europe's fourth-largest chemical corporation, in which the Belgian Solvay & Cie held an interest; there were also substantial investments from Britain and elsewhere. During 1937 Spolchemie's headquarters were relocated to Prague; but its manufacturing centres remained in the north. In October 1938, Germany annexed the Sudetenland, which included the city of Aussig. In December 1938, Spolchemie came, directly and indirectly, under the ownership or control of IG Farben [15].

12.4 Romania

Late in World War I, a substantial cyanamide works, based on the Polzenius process, using the tunnel furnace, was constructed at Dicsöszentmárton (Diczö St. Marton, central Transylvania; Ungar-Altenburg) in the Kingdom of Hungary by the Ungarische Stickstoffdünger Industrie AG, founded in 1916. It never served the needs of the Austro-Hungarian War Ministry. After the Treaty of Trianon, the factory was in Romanian territory (at what is today Tărnâveni) and in a bad way. From March 1921, it was put back to work and managed by the engineer and town planner Imre Forbáth (1875–1944), citizen of the former Hungary, who had to ward off attempts by the Romanian War Ministry to take over the facility. Claims and counter claims lasted until the mid-1930s. By then, the main source of power was local natural gas, delivered to the factory by an 11-km pipeline. Nitrogen was obtained by burning air in natural gas. The factory produced ammonia, which was sent to the Blumen powder works for conversion into nitric acid [16, 17].

12.5 Hungary

The new Hungary possessed a single cyanamide works, the Ungarnwerke, at Magyaróvár, close to the border with Czechoslovakia (south of Bratislava). However, output was not great. Hungary was an important customer for Romanian cyanamide until the early 1930s, when the government placed restrictions on imports of nitrogen fertilizer in order to protect the new Hungarian Ammonia Factory (Magyar Ammóniágayár Rt.) at Pétfürdő, authorized in 1928 by the minister of defence and that worked the Nitrogen Engineering Corporation process. It opened in 1932 and relied on lignite from mines at Várpalota, and local dolomite. The state, to maintain sufficient nitrogen supplies as a matter of national security, was involved through a second, associated facility, the Hungarian Nitrogen Fertilizer Factory (Magyar Műtrágyagyár Rt.). Hungarian production of nitrogen in 1934 was 3760 tons. In 1935, the Bubiag-Didier water-gas type apparatus, manufactured by Didier-Werke AG, was installed for production of hydrogen from lignite [18]. The Pétfürdő facility produced nitric acid for conversion into ammonium nitrate, used in explosives, in addition to ammonium sulphate fertilizer. As elsewhere, railway communication was critical for ease of distribution, in this case the Székesfehérvár-Veszprém line. Farmers in the Nile Delta, Egypt, were major consumers of Hungarian fertilizer. The two Hungarian firms were later merged. Today the facility is the Hungarian Nitrogen Works Co., Ltd (Hungarian Nitrogénművek Zrt.).

12.6 Poland

During 1916–1917, Ignacy Mościcki supervised construction of a nitrogen factory based on his electric arc process at Jaworzno in the industrial region west of Krakov, Austrian Galicia. The nitric acid was reacted with coke oven ammonia to produce ammonium nitrate. Annual production was 3400 tons. From the mid-1920s the factory produced cyanide. The wartime experience was put to good use when Mościcki from 1918 was placed in charge of reconstruction of the former German cyanamide factory at Chorzów.

Though Chorzów, located in Upper Silesia, had been assigned to Poland, under Article 256 of the Treaty of Versailles, the German government claimed ownership of the factory. Chorzów and Piesteritz in 1920 were transferred to the Stickstofftreuhand GmbH, a company created by the Bayerische and Deutsche Petroleum AG, as trustee on behalf of the German government. The Chorzów plant was then registered in the name of a new company, Oberschlesische Stickstoffwerke AG, registered at Königshütte (Chorzów) and then in Berlin in 1920. The convention regarding Upper Silesia was signed between Germany and Poland on 15 May 1922. On July 1, the local Polish court declared the German registrations null and void, and assigned ownership of the factory to the Polish

Treasury. Two days later, under the direction of Mościcki, work began on putting the cyanamide factory back into operation as the Polish State Factory of Nitrogen Compounds (Państwowa Fabryka Zwiazkow Azotowych). Mościcki designed the carbide ovens. In 1925, the restored Chorzów factory produced 86,000 metric tons of cyanamide [19–21]; in 1928, its annual capacity (and production) was stated to be 150,000 tons of cyanamide [22].

Germany, however, had claimed compensation for loss of the facility. The resulting inter-state dispute dragged on for several years at various tribunals, eventually reaching the Permanent Court of International Justice in 1928, after which a compromise was reached [23].

For Poland, security considerations remained uppermost, following the Polish-Soviet War (1919–1921) and conflict with Czechoslovakia in 1919. There were also concerns over the long borders shared with Germany and the Soviet Union. Following a military coup in 1926, Mościcki was appointed president of the Second Republic of Poland, which became increasingly authoritarian. A programme of economic centralization was inaugurated, involving considerable state involvement in the country's industries.

The first Polish synthetic ammonia factory, a private venture at Knurów, north-west of Krakov, operated the Claude process and opened in 1928 (Société Fermière des Mines Fiscales de l'Etat Polonais). It was located at the site of a former state owned Prussian coal mine. Hydrogen was obtained from coke oven gas [24]. The factory produced 25 tons of ammonia daily until closure not long after 1 September 1931 as a result of the depression and technical problems in operation.

More successful was the second, state owned, ammonia factory, which employed the Fauser process. It was based at Tarnów, east of Krakov, at a site named Mościce. Planning was undertaken at the Chemical Research Institute, Warsaw, founded by Mościcki. Construction began on 5 May 1927, and the factory, claimed to be the largest of its kind in the world, was completed on 2 October 1929. The main source of funding became available after October 1927, in the form of loans for national infrastructure development made available from a consortium of American banks, with a smaller contribution from European banks.

Eight converters produced 60 tons of ammonia daily. Hydrogen was derived from water gas. As a (relatively) low pressure process, the product ammonia, as in the Haber-Bosch process, had to be absorbed in water to enable its separation. The nitric acid required for preparing ammonium nitrate was made by the usual oxidation of ammonia over a platinum catalyst. Ammonium sulphate and ammonium nitrate were the principal products; the latter was mixed with phosphate rock to give the compound fertilizer "Nitrophos." Engineer Mieczyslaw Kalous worked out a process for producing salts from ammonia, while the Warsaw research institute developed a method for making ammonium sulphate using local gypsum. At the official opening on 18 January 1930, the facility was named the Mościcki State Nitrogen Products Factory (Państwowa Fabryka Związków Azotowych w Mościcach), in honour of President Mościcki, who presided over the event [25]. Mościcki personally designed the nitric acid absorption towers, filled with granite [26]. On the same day the London *Times Trade and Engineering Supplement*, published a special edition devoted to “Poland and Free City of Danzig.” Aimed at attracting foreign investment capital, at a time when it had dried up, the articles presented a rosy picture of industrial developments. In fact, far too rosy when it came to nitrogen products [27].

During the first two years of the depression, Polish consumption of Chilean nitrate dropped from 79,200 to 29,500 tons. Partly this was due to the fact that in 1930 Poland was the fifth largest global producer of nitrogen products (Table 12.3). A decree was passed that prohibited imports of nitrogen products into Poland. The depression, however, had a major impact on two new ammonia factories, both at Wyry, south of Chorzów.

In February 1929, the Fauser process was introduced at Wyry-Łaziska, using coke oven hydrogen. The capacity was 20 tons of ammonia daily. The main investor was Ammonium AG, of Shaffhausen, Switzerland. The investing companies soon after faced financial difficulties. The second factory at Wyry operated the Nitrogen Engineering Corporation (NEC) process, with two converters producing 25 tons of ammonia daily; hydrogen was obtained electrolytically. As a result of falling prices and demand the workers were dismissed in 1930, and the factory was closed down [28].

Around 1930, the Chorzów facility also took up the NEC ammonia process and was merged with the Mościcki factory. Soon after they were joined by the nitrogen facility at Jaworzno, that had originally worked the Mościcki electric arc process. In 1934, Polish fixed nitrogen production reached 35,165 short tons, placing the country 13th among the global producers.

The Mościcki factory was subsequently expanded, and a research laboratory and pilot plant were installed. A license was obtained from ICI in Britain for the production of crystalline ammonium sulphate. At the end of the decade hydrogen was produced from both coke oven gas and natural gas, for which the Hercules Powder Company’s catalytic methane decomposition process was adopted [29]. The successor to the Polish facilities (Zakłady Azotowe w Tarnowie-Mościcach SA) is Grupa Azoty SA, Poland’s largest chemical corporation, in which the state treasury holds a substantial interest.

12.7 Austria and Yugoslavia

Austria lost its wartime cyanamide works in Styria, at Maria-Rast, near Marburg (Maribor), opened in July 1918, which went to Slovenia (Yugoslavia). The Pauling works at Patsch, Innsbruck (Luftverwertungs-Gesellschaft), continued to manufacture nitric acid and sodium nitrate, until it was closed in 1928, as a result of restrictions on the electrical power supply, which was required by the growing city [30].

Table 12.3 Estimates of annual nitrogen capacities, in short tons, late 1920s

	Synthetic ammonia	Cyanamide	By-product ammonium sulphate	Electric arc
Germany	450,000	114,000	100,000	
France	55,000	53,500	5000	
Great Britain	55,000		100,000	
United States	25,000	40,000	147,000	
Italy	63,700	[17,000–] 20,000	3500	
Japan	43,500	20,000	5000	
Belgium	22,000		10,000	
Spain	7200			
Poland	3000	30,000		
Czechoslovakia	4500	6000		
Norway		15,000		46,500
Canada		60,000	5000	
Sweden	2500	6000		
Switzerland	2200			
Yugoslavia		14,000		
Romania		5000		
USSR	7300 [completed 1928]			
Total capacity	740,900	383,500	375,000	46,500
Actual production	700,000	200,000	370,000	30,000

Source: H. W. Bates, "Economic Relationships Between Nitrogen and Fertilizers," *Chemical and Engineering News*, vol. 20, no. 11 (November 1928):1133–1139, on 1133. French synthetic ammonia capacity, expressed as nitrogen, was around half the figure given in this source (116,700 tons). French cyanamide capacity for the year ending 30 April 1925 was approximately 55,000 tons. *League of Nations, Armaments Year-Book. Third year. 1926–1927*. Geneva: League of Nations, 1927, p. 470. For Poland, the cyanamide capacity in 1928 (stated to be the output) was 150,000 tons, according to: "The Polish Nitrogen Fixation Industry," *The Polish Economist*, vol. 5, no. 2 (February 1930):40–43; and Tadeusz Zamoyski, "Chemical Products," *The Times Trade and Engineering Supplement, Poland and Free City of Danzig Industrial Number*, 18 January 1930, pp. 22–23. Canadian production of calcium cyanamide was from the Niagara Falls works of the American Cyanamid Company

Finally, Yugoslavia remained committed to cyanamide, produced at two sites in Dalmatia, at Sebenico, and Dugi Rat (both Frank-Caro process), and the one in Slovenia, at Maria Rast (Polzenius process). As a result of the earlier Italian holdings in Sebenico and Dugi Rat, these two factories passed into the hands of the Terni company, and then in 1929 into French ownership. Maria Rast was originally owned by a consortium made up of Bosnische Elektrizitäts-AG, Dynamit Nobel, and Prager Eisenindustrie. All three factories received hydro-electricity, generated, respectively, at power stations on the Kr[e]Ka, Cetina, and Drava rivers. Cyanamide output in 1928 was 30,000 metric tons, of which 27,000 tons were exported [31].

By 1927, plans had been drawn up for the introduction of the Casale process at Dugi Rat, which was achieved in 1930.

The factories in Dalmatia, engaged in production of calcium carbide, cyanamide and synthetic ammonia, also commenced manufacture of ammonium phosphates as well as other nitrogen products. Through its control of the French operating company (Société Française des Forces Hydro-électriques de la Dalmatie), the Soc. des Phosphates Tunisiens intended, apparently, to exceed the output of ammonium phosphates produced in the French Pyrenees, there using the ammonia from its Soulom factory (Casale process), at the Pierrefitte-Nestelas works [32]. This was in keeping with the strong emphasis in France on superphosphates, to some extent based on phosphate deposits in Algeria and Morocco, as well as in Tunisia. Guido Donegani, head of Montecatini, as the administrative director of Soc. Phosphates Tunisiens, had managed the changeover of the factories in Dalmatia from Italian to French ownership. The Montecatini connection was significant. It had long relied on North Africa for the supply of phosphate rock; Italy imported more than 850,000 tons per year for superphosphate in the 1930s. No doubt much cyanamide made in Yugoslavia was imported into Italy through Montecatini. The synthetic ammonia made at Dugi Rat was converted into nitric acid by the Fauser-Montecatini process.

The Wall Street crash and entry into cartel arrangements frustrated the plans for export-led expansion in Yugoslavia, where home demand for cyanamide was almost insignificant. In September 1930, Maria Rast cut back production of cyanamide. Recovery was slow, at least until just before the mid-1930s.

In 1934, total nitrogen production in Yugoslavia was 20,025 short tons. Yugoslavia was the world's second largest exporter of cyanamide after Canada (from the American Cyanamid works at Niagara Falls).

References

1. Trinchieri G (2001) Le industrie chimiche in Italia, dalle origini al 2000. Venezia, pp 204–213.
2. Trinchieri G (2011) Giacomo Fauser (1892–1971) e Luigi Casale (1882–1927): persone della chimica, Anno internazionale della Chimica, http://chimica-ieri-e-domani.blogspot.co.il/2010/11/giacomo-fauser-1892-1971-e-luigi-casale.html (accessed 10 January 2017).
3. Perugini M (2014) Il farsi di una grande impresa. La Montecatini fra le due guerre mondiali. FrancoAngeli, Milan, pp 39–40.
4. Cerruti L (2013) L'industria dei composti azotati (Il contributo italiano alla storia del pensiero: Tecnica), www.treccani.it/.../l-industria-dei-composti-azotati (accessed 27 September 2015).
5. Perugini M (2014) Il farsi di una grande impresa. La Montecatini fra le due guerre mondiali. FrancoAngeli, Milan.
6. Amatori F (2002) The fascist regime and big business: The Fiat and Montecatini cases. In: James H, Turner J (eds), Enterprise in the period of fascism in Europe. Ashgate, Aldershot, pp 62–67.
7. Nitrogen fixation in Italy (1929) J Soc Chem Ind 48(49)(6 December):1181 (Chem Ind London).
8. Petri R (1998) Technical change in the Italian chemical industry: markets, firms and state intervention. In: Travis AS, Schröter HG, Homburg E, Morris PJT (eds), Determinants in the

evolution of the European chemical industry, 1900–1939: new technologies, political frameworks, markets and companies. Kluwer, Dordrecht, pp 275–300, on 283–286.
9. Schmidt CT (1936) The Italian 'battle of wheat'. Journal of Farm Economics 18(4) (November):645–659
10. Steininger R (2003) South Tyrol: A minority conflict of the twentieth century. Transaction Publishers, New Brunswick.
11. Waeser B (1932) Die Luftstickstoff-Industrie. Mit Berücksichtigung der chilenischen Industrie und des Kokereistickstoffs. Springer, Berlin, pp 148–149.
12. Nitrogen fixation in Czechoslovakia (1925) J Soc Chem Ind 44(17)(24 April):429–430 (Chem Ind London).
13. Industrial notes from abroad. Czechoslovakia (1937) J Soc Chem Ind 56(12)(20 March):284 (Chem Ind London).
14. Nitrogen fixation in Czech Silesia (1937) J Soc Chem Ind 56(38)(18 September):848–849.
15. Hayes P (2001) Industry and ideology. I.G. Farben in the Nazi era. 2nd edition. Cambridge University Press, Cambridge, pp 233–245.
16. Manufacture of cyanamide in Rumania (1922) J Soc Chem Ind 41(8)(29 April):178R–179R.
17. Waeser B (1932) Die Luftstickstoff-Industrie. Mit Berücksichtigung der chilenischen Industrie und des Kokereistickstoffs. Springer, Berlin, pp 152–153.
18. Brownlie D (1938) Bulk production of hydrogen. Ind Eng Chem 30(10)(October):1139–1146, on 1141–1143.
19. Stanik W (2007), [O]n the 140th anniversary of the birth of Ignacy Mościcki. Ignacy Mościcki, engineer-inventor, holder of honorary doctorates, president of Poland. Chemistry & Chemical Technology 1(3):xi–xiii.
20. Mościcki I (1993) Ignacy Mościcki, prezydent RP. Auotobiographia. Wydaw, Belloma (in Polish).
21. League of Nations (1927) Armaments year-book: general and statistical information. Third year. 1926–1927. League of Nations, Geneva, p 685.
22. Zamoyski T (1930) Chemical Products. The Times Trade and Engineering Supplement, Poland and Free City of Danzig Industrial Number, 18 January 1930, pp 22–23.
23. Case concerning the factory at Chorzów (Claim for indemnity) (Merits) (1929) A/17, 1 W.C. R.2119.
24. The Polish nitrogen fixation industry (1930) The Polish Economist 5(2)(February):40–43.
25. Lichwała R (2012) Technical and technological history of Zakłady Azotowe w Tarnowie-Mościcach S.A. Chemik 66(10):1039–1056.
26. Polish notes (1930) J Soc Chem Ind 49(20)(16 May):423 (Chem Ind London).
27. Zamoyski T (1930) Chemical Products. The Times Trade and Engineering Supplement, Poland and Free City of Danzig Industrial Number, 18 January 1930, pp 22–23.
28. Waeser B (1932) Die Luftstickstoff-Industrie. Mit Berücksichtigung der chilenischen Industrie und des Kokereistickstoffs. Springer, Berlin, pp 67, 145–148.
29. Lichwała R (2012) Technical and technological history of Zakłady Azotowe w Tarnowie-Mościcach S.A. Chemik 66(10):1039–1056, on 1050.
30. Nitrogen plant closes (1928) J Soc Chem Ind 47(16)(20 April):419 (Chem Ind London).
31. Haber LF (1971) The chemical industry 1900–1930: international growth and technological change. Clarendon Press, Oxford, pp 328–330
32. Waeser B (1932) Die Luftstickstoff-Industrie. Mit Berücksichtigung der chilenischen Industrie und des Kokereistickstoffs. Springer, Berlin, p 153.

Chapter 13
International Conferences, and an Adriatic Cruise

13.1 The Nitrogen Cartels

The expansion of the post-war nitrogen industry had been fostered mainly by strategic concerns: economic nationalism and security in case of war. However, the accelerating trend towards large-scale production posed certain challenges that were less based on these concerns than on the markets and, in particular, the often slow uptake of novel fertilizers in agricultural sectors. Notwithstanding their rivalries, the European manufacturers established networks of communication for those engaged in what had become the globalized nitrogen business. The "First International Nitrogen Conference," to discuss issues arising took place at Biarritz, France, in April 1926. Overproduction of nitrogen products and regulation of competition were high on the agenda. However, there was little agreement. Thus in 1927, IG Farben attempted to get ICI to hold back on expansion at Billingham, but Pollitt, passionately engaged with the new high-pressure technology, refused [1].

More significant was the second international conference, a ten-day Adriatic cruise held on board North German Lloyd's SS *Lützow*, which departed from Venice on 30 April 1928 [2]. Among those on board were Alfred Mond (ennobled in May 1928) chairman of Britain's ICI and Professor Carl Bosch of IG Farben, who "added lustre and importance to the meeting" [3]. Sponsors of the event included—in addition to IG Farben (through the Stickstoff-Syndikat) and ICI (through Nitram)—Montecatini, Norsk Hydro, and the Comptoir français de l'azote. The latter organization was founded in 1907 as a consortium of mainly coke and gas works firms whose prices for ammonium sulphate were fixed by the state. Later the cyanamide and synthetic ammonia producers were included. France, though the third leading producer of synthetic nitrogen in the mid-1920s, was, unlike Germany and Britain, not an exporter of ammonium sulphate. It was a leading importer, which for a time included reparations shipments.

A.S. Travis, *Nitrogen Capture*, https://doi.org/10.1007/978-3-319-68963-0_13

The big issue in 1928 was the looming threat of overcapacity resulting from over-zealous adoption of a new technology, as well as expansion in cyanamide. Demand growth was already slackening. The decade-long expansion, encouraged by the appeal of repeat orders and continued growth in home agriculture and exports, in addition to security issues, had come to a sudden halt. Nitrogen production outstripped demand. Prices had fallen, and cutbacks had been made. Plants that had been run hard could at least undergo maintenance, that is if they had not already been mothballed.

Though the Adriatic cruise was very much a European affair, its implications were strongly felt across the Atlantic. American Cyanamid remained a major player in the nitrogen business, and the US output of ammonium sulphate from by-product coke ovens had increased considerably. The latter represented a major export source, in direct competition with both Britain and Germany. As for synthetic ammonia, of the two main facilities, those of Du Pont and the new Hopewell works of Allied Chemical, only the former had gone on stream. The situation was reviewed, with echoes of both optimism and concern, at an "Economic Symposium on Nitrogen," before the Division of Industrial and Engineering Chemistry at the 76th meeting of the American Chemical Society, held at Swampscott, Massachusetts during 10–14 September 1928 [4–8]. Reflecting interest in the newest source of hydrogen, C. J. Ramsburg, of the Koppers Company, Pittsburgh, spoke on the "Economic Status of the By-Product Coking Industry with Reference to the Nitrogen Situation," while Walter S. Landis, vice president and director of American Cyanamid, addressed "The International Nitrogen Problem." Jasper E. Crane, chairman of the board of Lazote, Inc., discussed "New Economic Aspects of Nitrogen," and E. M. Allen, president of The Mathieson Alkali Works, Inc., dealt with "Synthetic Nitrogen."

The industry-wide slump had become a global problem. Given the size and prominence of the German and British nitrogen industries, and their reliance on exports, how could their conflicting interests be resolved? There was but one answer to the entangled web of nitrogen politics, one that also included the industry in Chile: a cartel arrangement, starting with an agreement between the German and British nitrogen producers: the Stickstoff-Syndikat, effectively controlled by IG Farben, which acted as the marketing organization for ammonium sulphate produced by most German cyanamide, coal gas and coke oven works, in addition to IG Farben's two factories; and the British Sulphate of Ammonia Federation, controlled by ICI through Nitram Ltd. Since around 1926, they had worked closely as a European duopoly [9]. During 1929, ICI and IG Farben discussed arrangements for market sharing in nitrogen products, including from the Norsk Hydro facilities that had come under the control of IG Farben. A pact with the Chilean nitrate industry was agreed in the summer. The main agreement between the Germans and the British, creating what was known as the DEN-Gruppe (Germany-England-Norway), was signed on 25 February 1930. This British-German cartel (in which Norwegian interests were subsumed under IG Farben) was intended to last for ten years.

The Wall Street stock market crash of 29 October 1929 and the start of the Great Depression had greatly aggravated the situation. The United States and Canada, followed by Germany, suffered from lack of demand at home due to great dependence on their agricultural sectors, which were badly hit. Britain's ICI, as with IG Farben, suffered from a collapse in the export market. It was a failure of excess.

A sales agreement among European manufacturers of cyanamide was agreed on 3 May 1930, involving mainly Germany, France, Italy, Poland, Norway, Sweden, Yugoslavia, and Czechoslovakia. They represented over 90% of world production, and their outputs demonstrated that despite the rapid emergence of synthetic ammonia, the cyanamide industry had continued to flourish. Thus in France, cyanamide manufacture grew from 14,000 metric tons in 1921 to 55,000 metric tons in 1925; imports of cyanamide grew from 7400 to 29,500 metric tons during the same period [10].

These arrangements were followed with meetings held in Ostend and Paris during the summer of 1930 involving producers of all nitrogen products. On 1 August 1930, the terms of what became known as the Convention de l'Industrie de l'Azote (Nitrogen Industry Convention, CIA) were agreed in Berlin. Its original members included the important cyanamide and/or synthetic ammonia producers Germany, Britain, France, Belgium, Holland, Poland, Italy, and Czechoslovakia, as well as Chile. (Norway was included through IG Farben interests.) The main purpose of the CIA was to cut back production among, and in coordination with, all these producers of nitrogen products. Through the DEN group, ICI and IG Farben dominated the CIA; they were in fact the controlling members in what was effectively a two-tiered cartel system. The DEN group acted as a powerful kernel around which other firms were gathered in.

The negotiations had been complex, and involved both corporations and state institutions. Montecatini, representing most Italian nitrogen production, had to be encouraged by the DEN group to join the CIA. Belgian and French producers, confronting deteriorating economic conditions at home (the main impact of the recession was felt in France during 1931), were represented by organizations closely affiliated with the state.

World nitrogen capacity in 1930 was 2.8 million metric tons, but only around 1.7 million metric tons were produced. The fall in demand was considerable. Total fixed nitrogen used in fertilizers for the fiscal year 1931–1932 was 182,000 tons (compared with 316,000 tons in 1929–1930) [11]. Worldwide cyanamide production during 1930–1931 was around 200,000 metric tons, as fixed nitrogen.

Farmers had no cash to purchase fertilizer, and exports declined, along with prices. Nitrogen output was now dependent on a network of international agreements involving almost every European nation. The DEN group drew up separate contracts with all other parties that had been brought into the CIA, thus controlling over 70% of world production. The group represented, as sales agent, Belgian, Czechoslovakian, Polish, and Swiss members. Prices in all cases were agreed. Further arrangements were reached with the producers in Chile. A common fund was set up, and exports to countries outside the cartel were strictly regulated through a quota system. Goals for how much production capacity had to be eliminated were set. The fund provided compensation to manufacturers whose output was kept below the agreed level. As an example of the impact of the 1930 agreement, the firms in Czechoslovakia (represented by an association, or Spolek) that had been drawn in by the Stickstoff-Syndikat were obliged to agree on reduced levels of exports to Hungary, Romania, Yugoslavia, and Austria. However, production to satisfy domestic demand was allowed [12]. Not only were sales and prices for members of the cartel fixed, but steps were taken to prevent other countries from entering the industry.

The United States remained nominally outside the nitrogen cartel as a result of the 1890s Sherman Anti-trust Act, which, as *Chemistry and Industry* pointed out in July 1929, prohibited concerns from "entering a group whose endeavour would be to control prices in that country" [13]. No less relevant was the fact that the United States was not yet a significant producer of synthetic ammonia. Cartels involving American and European firms did exist, notably a potash syndicate, but arrangements tended to involve American subsidiaries in other countries or discrete agreements [14]. More acceptable were pooling arrangements for sharing technologies. There was another factor that made American chemical firms less likely to openly join European cartels—apart from becoming targets of regulators—the nation's turn to isolationism, including restrictions on immigration (1924), effective from 1929.

In 1928, *Industrial and Engineering Chemistry* warned against collusion with foreign monopolies [15]. The American chemical industry was exhorted to create a solid front to counter the threat posed by European firms. Bankers were criticized for making substantial investments in Europe, which could only harm American enterprises. At home, the chemical industry was equally concerned over proposals for a state-owned fertilizer facility at the Muscle Shoals complex. The situation in Europe was used to argue against the opening up of Muscle Schoals. Open competition was encouraged, apart from when firms combined in the export trade, an exemption allowed under the Webb-Pomerene Act of 1918.

In early 1929, *Industrial and Engineering Chemistry* warned that the European nitrogen and other cartels were focused on taking "the field against the United States in world markets" [16]. However, one year later, and despite the fact that members of the European cartel were apparently allowed a free hand in exporting to the United States, there was a note of optimism, since the moves by Allied Chemical and Du Pont into large-scale production of synthetic ammonia had introduced "a new factor which should strengthen fixed nitrogen production in the United States against probable competition with the new international cartels" [17]. In early 1931, the question was raised regarding the US nitrogen industry: "either to sit in or stay out" of the European cartel [18]. As it happened, the Europeans did fear the impact of the arrival of Allied and Du Pont on the nitrogen scene, which led to an uneasy compromise whereby markets on both sides of the Atlantic were not invaded. Du Pont had in 1929 formalized its technology exchange arrangement with ICI, which offered one way of controlling markets without violating or breaching anti-trust regulations. In addition, the difficult economic situation dictated a more liberalized regulatory attitude; American firms began to collaborate on their home territory. By the end of the 1930s, the Barrett Division of Allied Chemical, by far the largest US producer of synthetic ammonia, almost completely controlled the distribution of nitrogen fertilizers in the United States [19].

13.2 Survival of the Nitrogen Industry Convention

The CIA agreement was for one year, until 30 June 1931. On maturity, the CIA collapsed, at a meeting held in mid-1931 at Lucerne. Belgium was unhappy with the terms, particularly the restrictions on exports (as were Italy, Czechoslovakia and Poland). France, Holland and Denmark were unhappy with any restrictions on the import of Chilean nitrate; France demanded that its home production of nitrogen should be determined by the needs of its military and not the market in fertilizer. The DEN group, however, remained intact. ICI and IG Farben further controlled markets by continuing to act as main selling agents for the by-product firms, mainly those engaged in making ammonium sulphate from gas and coke oven works. In the case of IG Farben, sales arrangements also included cyanamide (through the Stickstoff-Syndikat).

In April 1932, Britain introduced a 20% duty on imports of nitrogen fertilizer, to prevent dumping from abroad. Elsewhere tariffs were introduced and quotas were reviewed. By July the CIA was resurrected. The Stickstoff-Syndikat brought French and Belgian interests back into the fold, partly through diplomatic channels. Both countries agreed to renegotiate agreements over customs duties on nitrogen imports: the duties became void. In exchange their manufacturers were compensated. Germany raised customs duties on imports of nitrogen, as fertilizer, and Chilean nitrate, to 120 marks per ton, again mainly to prevent dumping [20, 21].

The affairs of the CIA were controlled by a "committee," the Compagnie Internationale de l'Industrie de l'Azote SA, that had been set up on 15 August 1930 in Basel. Its responsibilities covered the allocation and gathering in of funds. Effectively, most of the world's export markets were split between ICI and IG Farben, and they shared the substantial markets in China, Egypt and Japan: 40% to ICI, and 60% to IG Farben. ICI retained the British Empire market, along with the Dutch East Indies (Indonesia), Spain, Portugal, and the Canary Islands (the latter three through former Nobel connections), while IG Farben sold to the rest of Europe, South and Central America, Asiatic Russia, and the Philippines. By prior agreement, ICI and IG Farben retained their existing sales outlets.

The CIA made separate arrangements with each of Belgium, Czechoslovakia, France, Holland, Italy, and Poland. They covered one fertilizer year (beginning in the spring) and had to be continually renegotiated. This included the quota arrangements. The French, however, under government pressure, soon pulled out. Poland, which rejoined in 1932, was permitted the export of 32,000 tons/year of nitrogen fertilizer.

Apart from IG Farben and ICI, whose nitrogen outputs by the mid-1930s were around 20% of operating capacity, the greatest percentage drop was in Holland, to around 60% of capacity by 1935, except at the large Sluiskil operation of Coppée-Montecatini, which negotiated separately and produced almost 80% of its capacity (Table 13.1). As in Czechoslovakia, producers that catered for home markets were less exposed to the vagaries brought on by the depression. However, Belgian firms, and the Consolidated Mining and Smelting Company at Trail, British Columbia, were hard hit. In general, smaller firms in countries not dependent on exports were working at 60–80% capacity. They were, relatively, certainly more fortunate than members of the DEN group, whose investments and over-capacities

Table 13.1 Global production of nitrogen, including from by-products, in short tons, for the calendar year 1933

Country	Production, short tons
Germany	462,500
United States	256,700
Japan	208,000
France	187,555 (47% capacity)
Great Britain	175,000
Chile	141,755
Belgium	109,835 (44% capacity)
Italy	98,620 (92,780)(74% capacity)
Netherlands	90,405 (synthetic ammonia: 62% capacity, except 78% at Sluiskil)
Norway	65,505
Soviet Union	45,000
Canada	41,080
Poland	35,165 (38% capacity)
Yugoslavia	20,025
Switzerland	9465 (64% capacity)
South Africa	8085
Spain	8000
Sweden	8000
Hungary	3760
Australia	3000
India (British)	2500
Austria	1500
Total	2,000,000 (approximate)

For a number of members of the CIA cartel, total capacities for 1934–1935 are shown in parentheses

Sources: *Chemical Nitrogen. A Survey of Processes, Organization, and International Trade, Stressing Factors Essential to Tariff Consideration, under the General Provisions of Section 332, Title III, Part II, Tariff Act of 1932. Report no. 114, Second Series.* Washington DC: Government Printing Office, 1937, p. 74; A. Cressey Morrison, *Man in a Chemical World: The Service of the Chemical Industry*. London: Scientific Book Club, 1937, p. 74; W. J. Reader, *Imperial Chemical Industries: A History, vol. II. The First Quarter-Century, 1926–1952*. London: Oxford University Press, 1975, p. 138. Canadian production includes the American Cyanamid Company's Niagara Falls works. The figure for the Netherlands was provided by Ernst Homburg and is for the agricultural year, 1 July 1932 to 30 June 1933

were tremendous. Governments supported their nitrogen industries, including through state-backed trade groups, and did not oppose the CIA's measures. This second cartel lasted until the outbreak of war in 1939 [22].

Japan, a major producer from 1930, showed no interest in joining the CIA. European and American manufacturers of ammonium sulphate competed aggressively for the Japanese market, which was less constrained by the cartel arrangement. Demand remained robust despite the depression. Attempts were made to bring Japan into line through its curbing production and controlling the price of ammonium sulphate. The Fujiwara-Bosch draft agreement, drawn up on the German side by a brother of Carl Bosch, proposed that Japan refrain from exports. Also that foreign firms cease exports to Japan. However, it failed due to opposition in Japan [23]. A second draft ammonium sulphate agreement was drawn up in April 1931, but could not be ratified. The Japanese were not willing to negotiate terms. In any case the international cartel was about to collapse. Three years later, in March 1934, an agreement with Japanese manufacturers was finally reached, mainly through the efforts of IG Farben. It became effective in 1935. Imports from Britain ceased, at a time when Japan was almost self-sufficient. The agreement was renewed in February 1935, and again in November 1935. Exports to Japan (and China) from Germany were restricted by the agreement, but continued until 1938, with lesser amounts after that time, at least until 1940. The reduced export market for Germany, in part resulting also from the Japanese expansion in home production, contributed towards IG Farben's change of policy in the 1930s whereby it engaged in the sale of Haber-Bosch know-how to Japanese firms.

The impact of the international cartel and the need for an orderly supply of fertilizer to importers served the DEN's efforts to suppress the emergence of the nitrogen industry in countries where the agricultural sector was critical to the economy and industrial developments were lacking. Thus in Denmark around 1933 the government, under pressure from the DEN, refused to support a modified and considerably improved Birkeland-Eyde process invented by the engineer Johan Ernst Nyrop (1892–1959), fearing that the CIA members might restrict supplies [24]. This strategy did not work with Hungary.

ICI in Britain, in order to enhance awareness of the benefits of its nitrogen fertilizer and promote sales, had made great efforts to educate farmers in the use of ammonium sulphate. This called for scientific investigations into agricultural improvements. In 1927, ICI began to purchase land near Maidenhead, west of London, for an agricultural experiment station, Jealott's Hill, opened in 1929. Rotational grazing of pastures and grassland, successively dressed with nitrogen fertilizers, was studied. The results were used to encourage farmers to spend more on products that helped enlarge their yields. But to little avail (at least in the 1930s, though the station's work was significant in later decades). The depression hit hard, impacting greatly on export of nitrogen products, the mainstay of Billingham. The cost of introducing a revolutionary technology had turned into massive losses on the balance sheet, and perhaps contributed to the death of Alfred Mond in December 1930 [25]. At the time of his death Mond was president of the Society of Chemical Industry, which mourned the passing "of the great outstanding figure in chemical industry; he was our great leader... He was an Englishman and a Jew; he was not ashamed to be a Jew; and he had no reason to be" [26].

Jealott's Hill suffered severe budget cutbacks, causing its director to resign. Plans for future expansion at Billingham drawn up in 1926–1927 were not to be fulfilled, even after the economic situation improved during 1932. In 1934, ICI's George Pollitt, who had played a major role in the creation of Billingham, was dropped as an executive director; he retired and took up farming at his estate in the New Forest. No doubt his overoptimistic production projections for ammonia, which turned out to be an expensive burden for ICI, was a factor in his departure. Between 1929 and 1931, Billingham lost 6000 staff, almost one-third of the total laid off by ICI. Also in 1934, Roland Slade moved to London, where he reported to McGowan and later became a research manager. In October 1935, the new board of the ICI subsidiary Fertilizers and Synthetic Products Ltd. (Fertilizers and Synthetic Products Group) emphasized the role of Billingham as a chemical complex within which was contained fertilizer capacity.

In 1933, just after recovery began, Charles O. Brown, of American Cyanamid's Chemical Construction Corporation, summarized the situation thus [27]:

> As one could easily predict, this epoch making development unchecked through 20 years, led to the usual reaction—over production. This is today manifest, both in the birth of too many 'processes,' as well as in too many individual plants. The year 1932 completed a program of more than $137,000,000 of invested capital, whose annual products are valued at over $56,000,000. During this period there has been a break in the prices of synthetic ammonia and of methanol to a steady level of one-sixth to one-quarter of their former quotations, and all prices, formerly subject to wide and ruinous fluctuations, are now quite steady from season to season and year to year. But the bottom has probably not been reached. These changes in basic commodities have also rearranged the importance of their compounds and changed the financial stability of nations, particularly of Chile. Germany was the first to demonstrate that, through high pressure synthesis, it was possible to feed a nation and carry on a major war, without outside help. Today we find a relatively large number of countries independent of nitrate supplies from without their boundaries—having developed atmospheric nitrogen fixation through high pressure synthesis.

Synthetic nitrogen fertilizer had fallen ignominiously from grace, it seemed. However, for ICI there were two useful contracts for the construction of ammonia facilities from among the countries of the Dominions, one in South Africa, at Modderfontein, near Johannesberg, for serving the mining industry (1932), and another in Australia, completed in 1939. In addition, designs were prepared for an ammonia factory in Czechoslovakia, and assistance was given to the Sociedad Ibérica del Nitrógeno in connection with its facility at Valladolid. Major production improvements at ICI from 1937 were driven by the growing threat of war.

In the spring of 1940, concerns that the Billingham factory might become a target of the Luftwaffe led to the opening of two synthetic ammonia factories in Canada, both based on NEC technology, one at Calgary, Alberta, and the other at Welland, Ontario. For hydrogen, the Calgary facility relied on natural gas, using an ICI process, while Welland relied on coke oven gas. Both plants began operating in 1941. Reflecting changes in the status of synthetic ammonia and high pressure processes, Billingham in 1944 became known as the Billingham Division and from

1963 as the Agricultural Division. By this time Billingham and the nearby ICI oil-based Wilton facility, on the opposite side of the River Tees, was one of the world's largest chemical complexes. ICI had recently redesigned its outdated ammonia technology. Typically, yields from the high-pressure process were around 15%.

As for ammonia production at Oppau, the facility was working at 30% capacity during 1931–1932. IG Farben's Ammonia Laboratory was scaled back. New directions had been sought to exploit the extensive facilities for production of hydrogen that were available as an outcome of the growth of synthetic ammonia. The reactive gas was seen as a valuable source of future production, particularly through high-pressure hydrogenation of coal to oil. The main research and development had taken place at IG Farben, based on the process of Friedrich Bergius (Chap. 16).

During 1927, the decision was made to engage in the hydrogenation of coal at Merseburg (Leuna). In the summer of 1929, as sales of nitrogen and other products began to fall, Carl Bosch began a process of rationalization at IG Farben; it was intensified following the Wall Street crash. This included the creation of three product divisions (Sparten). Division I covered the high pressure processes—nitrogen, methanol, coal to oil—and mining. Coal-to-oil would have been a massive financial failure had not Hitler's autarky programmes favoured synthetic gasoline.

Following his retirement Colonel Pollitt remained on the board of ICI and increasingly promoted imperial preference and protectionism at the cost of trade with nations outside the British Empire [28]. During World War II, he served as a member of ICI's secret War Committee.

13.3 Revival of Chilean Nitrate

The considerable involvement of Chilean interests in the CIA arose from revival of the country's nitrate industry. From the mid-1920s, the Chilean nitrate business was controlled by the Guggenheims, who invested at least $17 million in new production facilities, based on a labour-saving and yield-enhancing process, and a purer product, as developed by Danish-American metallurgist Elias Anton Cappelen Smith (1873–1949). The new procedures, introduced in 1926, did not involve chemical processing. As with the Shanks process, a series of mechanical steps, now aided by electrical operations, removed rock and overburden. Only at the final crystallization stage did any enhanced removal of minor components occur. The Guggenheims dominated the nitrate cartel known as the Compania Salitre de Chile; members were encouraged to introduce the new process [29]. The batch processes would remain unchanged—including centrifuges, nitrate crystallizers, etc.—until the end of the twentieth century.

The United States continued to rely on imported Chilean nitrate, which until well into the 1920s was the most economical starting material for producing nitric acid. The nitrate remained a strong and effective threat to synthetic ammonia and ammonium sulphate—peaking at over half a million metric tons of fixed nitrogen in 1928—which is why Chile was brought into the international cartel arrangements during the summers of 1929 and 1930. Following a period of political and financial turmoil, Chilean nitrate production declined until the mid-1930s. It was exported to several countries, including members of the CIA.

13.4 Norway: Hafslund, Norsk Hydro, and IG Farben

Norway was an early producer of electrothermal and electrochemical products, based on abundant hydro-electric power. Large-scale production began with calcium carbide (1899) followed by Norsk Hydro's working of the Birkeland-Eyde process for the electric arc capture of nitrogen (1905), aluminium production (1906–1909), and the manufacture of calcium cyanamide. As in Italy and Japan, foreign equipment and machinery were required—in the case of Norway, German and Swiss turbines and generators, installed at both Notodden and Rjukan for the electric arc process. Capital came from Sweden, France, Britain and Germany [30]. Odda Smelteverk, part of the Hafslund group, which in 1924 had taken over the works of Nitrogen Products and Carbide, continued to manufacture cyanamide, mostly for the export market.

There were also new innovations. In late 1927, IG Farben introduced a highly concentrated three-component nitrogen-phosphorus-potassium fertilizer called Nitrophoska (Fig. 13.1). On learning of this product, Odda Smelteverk's chief chemist Erling Johnson (1893–1968), previously at North Western Carbide/Nitrogen Products and Carbide (1919–1921), invented an improved process by the end of 1928. Failure to commercialize the product by the Hafslund group resulted from the depressed economic situation, management changes (with accompanying difficulties) and failure to set up a joint venture with American Cyanamid [31].

At Norsk Hydro, the emerging crisis in the nitrogen industry caused the Birkeland-Eyde electric arc process, which required vast amounts of energy, to come under intense competition. Through a 1927 arrangement with IG Farben, which was just beginning to loosen its exchange of know-how policy on synthetic ammonia, Norsk Hydro changed over almost exclusively from the Birkeland-Eyde process to synthetic ammonia. However, the first ammonia facility to open, in February 1928, at Notodden, worked the Nitrogen Engineering Corporation process. By 1929, Haber-Bosch converters were in operation at Rjukan, and at Herøya, near Porsgrunn, Lower Telemark. The latter site was chosen due to the nearby source of limestone required in the manufacture of calcium nitrate, rather than proximity to hydro-electric power which by then could be transmitted over long distances. Hydrogen was generated by electrolysis. (A few electric arc furnaces continued in operation—until 1934 at Notodden and 1940 at Rjukan.)

Commercially, there was a heavy price to pay. Norsk Hydro's adoption of Haber-Bosch technology was conditional on IG Farben having control over output. IG Farben acquired a quarter of the shareholding in Norsk Hydro, reducing holdings

Fig. 13.1 BASF/IG Farben Limburger Hof research station, with a range of nitrogen products, including mixed fertilizers. The three-component Nitrophoska, introduced in 1927, is second from the right in the lower row. Ammonium sulphate is in the upper row at left. (*Erzeugnisse unserer Arbeit*. Frankfurt am Main: I.G. Farbenindustrie, 1938, opposite p. 156)

Fig. 13.2 Railway tank wagons for the transport of ammonia from Norsk Hydro's Haber-Bosch Rjukan facility, opened in 1929 and closed, along with the railway and ferry connections, in the late 1980s. Here seen at Tinnoset in 2009. (Photo by Nick Lera)

from elsewhere. Similar conditions had been specified by IG Farben when other firms attempted to gain access to its ammonia process. Norsk Hydro copied Odda's three-component fertilizer process in the early 1930s. An important market for Norwegian nitrogen fertilizer was Egypt.

Odda Smelteverk, subsidiary of Hafslund, was sold to the British Oxygen Company in 1938. The Haber-Bosch process was worked at Rjukan until the late 1980s (Fig. 13.2).

13.5 Discussion

In late 1918, if not before, Carl Bosch, then grappling with the uncertainties arising from Germany's collapse, laid out his vision for the future of BASF. It was to become the global leader in the supply of nitrogen fertilizers, based on the belief that its high-pressure chemical technology could not readily be imitated. That confidence was born of personal experience over several years struggling to bring together and expand the Haber-Bosch process. It echoed another famous episode in industrial history, the skill, vigour and ambition that enabled BASF to bring synthetic indigo to the market-place in 1897.

For a few years the power and organizational might of BASF, notwithstanding the post-war recession and the turbulent political times, seemed to bear out Bosch's hopes and aspirations. The greatest obstacle to any competitor was design and manufacture of the converter, though there were many other technical challenges. As a result, licensing payments and royalties on future sales were deliberately set high, and included clauses that gave IG Farben a say in output. These terms were considered extortionate in every case.

After 1918, the Haber-Bosch process stimulated inventions based on wartime research that had taken place in Britain, France, Italy, Japan and the United States. Each of these countries had the scientific and technical skills, and the engineering, energy and financial resources—as well as access to markets, at home and often abroad—required to motivate research and development programmes. By the late 1920s the existence of seven viable synthetic ammonia processes encouraged massive expansion in synthetic nitrogen. The outcome, in the West at least, was overcapacity, that had to be reigned in.

In the inter-war period, large cartels of various sorts, often backed by state interests, were common in national and international trade [32]. Nitrogen cartels were prominent among them. The main European nitrogen cartel arrangements started out as an exclusive club with just two members, the most powerful, ICI and IG Farben. They already controlled their nationwide nitrogen industries, in Britain through ICI's subsidiary, Nitram Ltd, which took over sales of coke and gas works ammonium sulphate, and in Germany through the Stickstoff-Syndikat, which served a similar function and included cyanamide.

In general it has been argued that policy and not leaders create cartels [33]. For nitrogen, however, the DEN group was the outcome of negotiations initiated by Carl Bosch and Alfred Mond. The arrangements established from 1928 were focused on a system of regulating production and prices. They received state support, including through raising, lowering and removing tariffs. Smaller countries received concessions in return for the concessions they made, sometimes in separate deals. This avoided messy situations at a time of crisis. Only in the case of Japan—and to a lesser extent Norway—were the arrangements related to the direct transfer of high-pressure nitrogen technology.

By 1930, outside the world of nitrogen products, upheavals were taking place in Western capitalist society accentuated by: the Wall Street crash and the suspension of American loans to Europe; the withdrawal of French and Belgian troops from the Rhineland in June 1930; protectionism in America following passage of the 1930 Smoot-Hawley Tariff Act, which led to global trade war; collapse of the German economy in 1931; and formation of a national government in Britain and abandonment of the gold standard (September 1931). Britain strengthened "imperial preference" to support its empire. Like the Billingham ammonia works, certain of these events resonate with Huxley's *Brave New World*. His skepticism underscored the opposing visions the world's industrial nations had about future security and new political ideologies. The world-wide economic depression had far-reaching consequences. Famine and discontent bred nationalism. In 1932, Manchuria became a puppet state of Imperial Japan. On 30 January 1933, Adolf Hitler became chancellor of Germany. Nitrogen featured in autarky programmes, including in fascist-leaning Spain, where in 1940 Sociedad Ibérica del Nitrógeno established, with the assistance of ICI, Nitratos de Castilla at Valladolid [34].

References

1. Bertrams K (2013) The years of crisis (1914–1950): The making and unmaking of international alliances. In: Bertrams K, Homburg E, Coupain N, Solvay: the history of a multinational family firm. Cambridge University Press, Cambridge/New York, pp 151–329, on 195–196, 201, etc.
2. Reader WJ (1975) Imperial Chemical Industries: a history, vol. II. The first quarter-century, 1926–1952. Oxford University Press, London, p 138.
3. Bear FE (1928) The second international nitrogen conference. Ind Eng Chem 20(10) (October):1112–1116.
4. Crane JE (1928) The new economic aspects of nitrogen. Ind Eng Chem 20(11) (November):1128–1130.
5. Allen EM (1928) Synthetic ammonia. Ind Eng Chem 20(11)(November):1131–1133.
6. Bates HR (1928) Economic relationships between nitrogen and fertilizers. Ind Eng Chem 20 (11)(November):1133–1139.
7. Ramsburg CJ (1928) Economic status of the by-product coking industry with reference to the nitrogen situation. Ind Eng Chem 20(11)(November):1139–1144.
8. Landis WS (1928) The international nitrogen problem. Ind Eng Chem 20(11) (November):1144–1147.
9. Reader WJ (1975) Imperial Chemical Industries: a history, vol. II. The first quarter-century, 1926–1952. Oxford University Press, London, p 145.
10. League of Nations (1927) Armaments year-book: general and statistical information. Third year. 1926–1927. League of Nations, Geneva, p 470.
11. Norris Shreve R (1945) The chemical process industries. McGraw-Hill Book Company, New York, p 390.
12. Teichova A (1974) An economic background to Munich: international business and Czechoslovakia, 1918–1938. Cambridge University Press, Cambridge, pp 315–320.
13. Editorial: nitrates and nitrogen (1929) J Soc Chem Ind 48(27)(5 July):672 (Chem Ind London).
14. Travis AS (2004) Dyes made in America. the Calco Chemical Company, American Cyanamid and the Raritan River. Edelstein Center, Jerusalem, pp 94–95, 477–478, n. 62.
15. Foreign cartels and American industries (1928) Ind Eng Chem 20(3)(March):231–232.
16. Some important events of 1928 (1929) Ind Eng Chem 21(1)(January):4–8, on 6.
17. The chemical kaleidoscope for 1929 (1930) Ind Eng Chem 22(1)(January):4–10, on 5.
18. The chemical kaleidoscope for 1930 (1931) Ind Eng Chem 23(1)(January):4–9, on 5.
19. Lamer M (1957) The world fertilizer economy, issue 10. The Food Research Institute of Stanford University, Stanford, pp 175–176.
20. Schröter V (1984) Die deutsche Industrie auf dem Weltmarkt, 1929–1933. P Lang, Frankfurt am Main, pp 313–331.
21. Schröter HG (1997) Fields of competition and cooperation: cartel structures in the international chemical industry. In: Pohl H (ed), Competition and cooperation of enterprises on national and international markets (19th–20th centuries). Franz Steiner Verlag, Stuttgart, pp 35–56.
22. Schröter HG (1991) Privatwirtschaftliche, Marktreguliering und staatliche Interessenpolitik. Das internationale Stickstoffkartell, 1929–1939. In: Schröter HG, Wurm CA (eds), Politik,

Wirtschaft und internationale Beziehungen. Studien zu ihrem Verhältnis in der Zeit zwischen den beiden Weltkriegen. Philipp von Zabern, Mainz, pp 117–137.
23. Kudo A (2000) Dominance through cooperation: I.G. Farben's Japan strategy. In: Lesch JE (ed) The German chemical industry in the twentieth century. Kluwer, Dordrecht, pp 243–283, on 272–273.
24. Kragh H (1998) The take-off phase of Danish chemical industry, ca. 1910–1940. In: Travis AS, Schröter HG, Homburg E, Morris PJT (eds), Determinants in the evolution of the European chemical industry, 1900–1939: new technologies, political frameworks, markets and companies. Kluwer, Dordrecht, pp 321–339, on 333–334.
25. Reader WJ (1976) Personality, strategy and structure: some consequences of strong minds. In: Hannah L (ed), Management strategy and business development: an historical and comparative study. Macmillan Press, London, pp 108–129, on 114–112.
26. Editorial. Lord Melchett (1931) J Soc Chem Ind 50(1)(2 January):1(Chem Ind London).
27. Brown CO (1933) High pressure synthesis—basis of new chemical engineering industries. In: Kirkpatrick SD (ed), Twenty-five years of chemical engineering progress. Silver anniversary volume. American Institute of Chemical Engineers. Published by the institute and for sale by D Van Nostrand Company, Inc, New York, pp 152–168, on p 153.
28. Reader WJ (1975) Imperial Chemical Industries: a history, vol. II. The first quarter-century, 1926–1952. Oxford University Press, London, pp 229–230.
29. Glaser-Schmidt E (1995) The Guggenheims and the coming of the Great Depression in Chile, 1923–1934. Business and Economic History 24 (1995):176–185.
30. Nerhem G (1992) Patterns of technological development in Norway. In: Hult J, Nyström B (eds), Technology & industry: a Nordic heritage. Science History Publications, Canton, MA, pp 58–72, on 62–66.
31. Sogner K (1998) Norwegian capitalists and the fertiliser business: the case of Hafslund and the Odda process. In: Travis AS, Schröter HG, Homburg E, Morris PJT (eds), Determinants in the evolution of the European chemical industry, 1900–1939: new technologies, political frameworks, markets and companies. Kluwer, Dordrecht, pp 239–256, esp 248–256.
32. Nussbaum H (1986) International cartels and multinational enterprises. In: Teichova A, Lévy-Leboyer M, Nussbaum H (eds), Multinational enterprises in historical perspective. Cambridge University Press, Cambridge, pp 131–145.
33. Schröter HG (2013) Cartels revisited: an overview on fresh questions, new methods, and surprising results. Revue économique 64(66)(November):989–1010, on 1003.
34. Puig N (1998). The frustrated rise of Spanish chemical industry between the wars. In: Travis AS, Schröter HG, Homburg E, Morris PJT (eds), Determinants in the evolution of the European chemical industry, 1900–1939: new technologies, political frameworks, markets and companies. Kluwer, Dordrecht, pp 301–320, on 316.

Chapter 14
Synthetic Nitrogen in the Soviet Union

. . .. I became interested in Russia professionally as it seemed the only country where any large scale chemical development was to be expected during the depression.

Alcan Hirsch, *Industrialized Russia*. New York: Chemical Catalog Company, 1934, p. 21.

Yet at the very time when nitrogen companies in western Europe were confronting problems of over-production, and chemists and chemical engineers were being laid off, a large new market for nitrogen technologies appeared, the Soviet Union. This new entrant, prepared to work with proponents of an opposing ideology, provided openings for foreign chemical and engineering firms with know-how in nitrogen and coke oven processes around 1930 when the nitrogen industry in the West was in the doldrums.

The inter-war transfer of technology from democratic to totalitarian states has attracted considerable scholarship and aroused considerable controversy. This includes polemics over scientific and technological developments in authoritarian systems, including the sort of failures that in democratic societies are seen to be unproblematic [1–4]. Within the various discussions, chemical technologies have received short shrift. Writing their balanced histories remains challenging. One major exception is Peter Morris's exhaustive study of the successful development of acetylene chemistry and synthetic rubber in National Socialist Germany [5]. There is, however, an account of the first successful production of synthetic rubber (polybutadiene) in the Soviet Union, achieved in 1932 by a process that involved high-pressure reactions, but, unusually in the area of industrial chemistry, at least of the type discussed here, with hardly any contributions from the West [6].

What Morris does clearly demonstrate is the ongoing leading role of chemistry in science-based industry, far more so than physics, within the sociopolitical contexts of both authoritarian and democratic societies, at least until 1940. However, and in contrast to studies on physics and its application, there are few similar studies on chemistry and its application within the Soviet system. Since we are here concerned with the transfer of technology from the West, we do have two useful sources to draw upon, though both are politically skewed. The first is Alcan

A.S. Travis, *Nitrogen Capture*, https://doi.org/10.1007/978-3-319-68963-0_14

Hirsch's 1934 book *Industrialized Russia*. It included an upbeat account of progress made in the chemical and other industries between 1917 and 1933, based on the author's experience as an American chemical engineer working on various projects in the Soviet Union during 1931–1933 [7].[1]

A similar viewpoint was that of Louis Segal in 1934 when describing economic progress; he drew attention to the fact that the Soviet Union occupied third place in the world production of fertilizers, as compared with twelth place in 1929 [8].

A quite different approach, and with a completely different agenda, was taken by Antony C. Sutton in the first and second volumes of his trilogy *Western Technology and Soviet Economic Development*, which cover the periods 1917–1930, and 1930–1945 [9, 10]. Sutton here and elsewhere argued that the military-industrial complex in the United States, in supplying the Soviet Union (and other countries) with machinery and manufacturing plant, was responsible in the long run for creating America's greatest enemy after 1945. (In Sutton's opinion the state-military conglomerate of the United States handed out hugely lucrative contracts to its arms industries for the supply of equipment to friend and foe alike.) While Sutton's conclusions, which might have been welcomed in the context of the Cold War era, have since been challenged, including the argument that had the United States not assisted Russia, then Stalin would never have been able to defeat Germany, his scouring of archival material, including on the nitrogen industry, was comprehensive.

14.1 Towards the First Five-Year Plan

In the summer of 1921, schoolchildren in London donated pocket money to aid victims of famine in Bolshevik Russia, a situation brought on by severe drought [11].[2] Nationwide insurrections had led Vladimir Lenin to introduce the New Economic Policy (NEP), or "state capitalism," in March. It replaced the austere programme known as War Communism, and was a partial reversal of the policy of total collectivization in order to improve the food supply. But for many people liberalization was too late to help. The American government was soon sending surplus grain to the Bolsheviks [12]. The outcome was a novel economic relationship between the Bolsheviks and the West, including direct foreign investment from and provision of concessions to Western businesses. America was looked upon as the model to emulate in large-scale mechanization and electrification.

[1]The noted Russian-American writer Maurice Gershon Hindus (1891–1961) provided the introduction to *Industrialized Russia*.

[2]The author's father, then living in East London, remembered the day at school assembly when the head teacher raised the issue of collecting funds for Russia. Following the ensuing debate, it was agreed to support the relief effort.

In March 1922, Josef Stalin was appointed Secretary General of the Central Committee of the Communist Party. In December the Union of Soviet Socialist Republics (USSR) was formed. On 21 January 1924, Lenin died. Two days later Leon Trotsky was defeated by Stalin and others in the executive committee of the Communist Party (the Politburo). Stalin now consolidated his power as an autocratic ruler. Centralized control of industrialization was to become a principal feature of Soviet policy with, in the chemical field, a special focus on fertilizers.

Also in 1924, the Amtorg Trading Corporation was founded in New York to enable trade between the United States and the Soviet Union. A similar firm was established in London. In January 1925, Trotsky was dismissed as war commissar, and in May appointed head of both the Soviet electrotechnical board and the scientific-industrial board of industry, in addition to chairman of the chief concessions committee. Trotsky, though strongly opposed to the NEP, particularly its emphasis on the granting of concessions, was nevertheless forced to hand them out, which somewhat compromised his own ideological position. No doubt that was Stalin's intention. Apart from that the NEP was deemed to be a failure.

Trotsky, as far as was possible, set about introducing his alternative strategy. Concessions were no longer given to Western firms. In their place, contracts were drawn up for know-how, licensing and material supplies, and the installation of manufacturing plant. American finance houses provided arrangements for long-term credit. Soviet scientists visited capitalist countries to learn about scientific and technological advances relevant to industrialization.

The leading chemist engaged in the acquisition of ammonia and nitrogen technologies from the West during the mid-1920s was the expert in high-pressure technology Vladimir N. Ipatieff, formerly a professor at the Mikhaylovskaya Artillery Academy (1899–1917), who under the tsarist regime had been appointed in 1910 to the rank of major general, and in 1914 to lieutenant general (Fig. 14.1). His role in promoting the Soviet chemical industry was particularly prominent, at least as long as a specialist from the time of tsarist rule was needed.

Within the new geopolitical context, and the growing need for know-how rather than concessions, in 1925 Ipatieff visited German, French, Italian, and Swedish facilities to investigate possibilities for licensing nitrogen processes for use in the Soviet Union. Ipatieff was not impressed with the Claude process. Probably he had been informed of its earlier problems and was aware of the rivalry in France with the Casale process. He had no success with BASF. He did succeed with Bayerische Stickstoff-Werke, whose cyanamide expert Nikodem Caro usefully spoke Russian, as did the laboratory researcher Carl Freitag [13].

Ipatieff and Caro had met once before, in 1909, at the Seventh International Congress of Applied Chemistry, held in London. Cyanamide was not new to Ipatieff. During the war, Russian chemists had developed a manufacturing process at Kharkov, and also oxidized ammonia into nitric acid there, using Kaiser's process.

As a result of negotiations with Caro, from 1927 Ipatieff acted as consultant to the Bayerische, visiting three times a year to work with Caro, Albert Frank and Freitag [14]. It was agreed that the Bayerische would file patents for Ipatieff's

Fig. 14.1 Vladimir Nikolayevich Ipatieff. (Edelstein Collection)

inventions jointly with the Soviet government. The projects included the catalytic oxidation of phosphorus under pressure and various high-pressure processes. One method investigated would lead to a route to cyanamide that started—instead of with calcium carbide—with producer gas (for nitrogen) and water gas (for hydrogen). Under Haber-Bosch conditions they gave ammonia, which was then reacted with calcium carbonate in the presence of carbon monoxide. The process was found to be favoured by catalysts based on heavy metals. This ammonia route to cyanamide was followed up over the following years, though never applied commercially [15].

It was probably with the help of Ipatieff that Caro and Frank worked on the manufacture of concentrated nitric acid from nitric oxides, including liquid nitrogen tetroxide, under pressure at 50–200 atmospheres [16]. Based on Ipatieff's studies at the Bayerische, the Piesteritz cyanamide factory (Mitteldeutsche Stickstoffwerke AG), which suffered from loss of demand, turned to the manufacture of phosphoric acid and mixed fertilizers, including nitrogen products, mainly the ammonium phosphates, and in particular superphosphate, mixed with ammonium sulphate [17].

Ipatieff also visited Fauser at Novara and Casale at Terni. Montecatini was not yet licensing the Fauser process. This left the Casale process, through which a programme for the construction of nitrogen factories in Russia was inaugurated in

1925 [18]. Ipatieff first had to convince Trotsky that large-scale removal of nitrogen from the atmosphere would not pose a threat to its composition. There was also the greater problem of dealing with a fascist state with its opposing ideology. However, Ipatieff overcame the latter difficulty by pointing out that the Casale process was all that was available, and reliable, apart from the Haber-Bosch process. In Italy, Casale Ammoniaca Società Ammonia Italiana (CASAI) was established for drawing up licensing and equipment supply arrangements with Chimstroi, the Soviet Union's chemical construction agency. During 1926, ammonia converters and related equipment were purchased and a Russian technician worked at Terni. The first major Soviet synthetic ammonia plant, located in the Central District (the then administrative region which in part corresponds with the later Volga Region), 400 km east of Moscow, was for the Severnii Khimicheskii Trest (Severny Chimichesky Trest; Northern Chemical Trust) and built under license from Ammonia Casale at Dzerzhinsk, just east of Nizhny Novgorod (Nijni, Nizhiny), birthplace of Maxim Gorky (and called Gorky between 1932 and 1990). This facility, part of the Chernorechenskii (Black River) plant, opened in 1928, with three converters each capable of producing 8 tons of ammonia daily; later, capacity was doubled. Hydrogen was obtained from water gas by the process designed earlier in the century by Julius Pintsch of Berlin.

In 1926, following negotiations between Ipatieff and Soviet officials and Stockholms Superfosfat Fabriks Aktiebolag, the latter's cyanamide process was acquired for the Dzerzhinsk facility, with a planned annual capacity of 30,000 tons. The Swedish manufacturer was chosen in preference to the Bayerische because the Russians wished to start on a small scale, for which the German firm's license fee was too expensive [19]. In 1927, Soviet trade with foreign firms was encouraged when the United States allowed long-term private credits to the USSR. In December, the collectivization of agriculture was decreed. Also in 1927, the State Institute of High Pressures, in Leningrad (formerly Petrograd), was established by Ipatieff. There, studies into synthetic ammonia and cyanamide were conducted. Ipatieff left Russia in 1930, apparently in the belief that he was about to be arrested.

14.2 Western Technologies

In the spring of 1928 there was a massive shortfall in supplies of grain. Under the First Five-Year Plan (October 1928–December 1932), Stalin now increasingly forced through the programme of collectivization of agriculture, seized grain from the Kulak farmers, and pushed for greater production of fertilizers. The plan, introduced on October 1, just four weeks before the Wall Street crash, had brought an end to the NEP and called for far more rapid industrialization than was envisaged by Trotsky. On the shop floor, superhero or shock workers encouraged greater output through leading by example. From the autumn of 1929, the expansion of state and collective farms brought about the rapid mechanization of agriculture, for which many thousands of Fordson tractors were purchased. Nitrogen products as fertilizers were especially important after the Communist Party formulated its “chemisation plan” for the state economy [20]. The main industrial belt in which the modern chemical industry developed stretched from coal-rich southern Ukraine to Moscow and Leningrad. During the Third Five-Year Plan (1938–1941), which placed great emphasis on the chemical industry, the new centres were in the east, along the Volga, and in the Urals and Siberia (Fig. 14.2). This reduced the distances covered by rail transportation of minerals found in these locations to sites of manufacture, and dispersed factories, which, as elsewhere, was increasingly a matter of strategic planning. Industries east of the Urals benefited from the double-tracking of the Trans-Siberian Railway by the early 1930s.

New Western-designed nitrogen facilities were added until 1933 as part of a very determined reform effort in agriculture intended to serve the home market, though much grain was sent to the West in exchange for industrial equipment (however the sharp fall in the price of grain from the end of the 1920s was hardly offset by the more limited decline in the cost of machinery). During 1928–1932, this programme involved bringing in leading American and European chemical corporations and contractors with gaps in their order books; skilled Western chemical engineers and technicians arrived in Moscow to provide expertise suited to the needs of the Soviet Union.

American synthetic ammonia technology was supplied by Chemical Construction Corporation (Chemico), the subsidiary of the American Cyanamid Company that owned the Nitrogen Engineering Corporation (NEC) process. Technical staff headed by Colonel Frederick Pope, aided by NEC founders Charles O. Brown and

Fig. 14.2 Western Russia, showing the location of synthetic nitrogen factories in the Soviet Union, around 1933

Louis Jones, were heavily involved at two new plants, Berezniki, in the Ural District, and Bobroki (later Stalinogorsk), in the then Central District, 200 km south of Moscow. The first contract, with Chimstroi, was signed on 11 November 1928 and extended in June 1931 [21]. The considerable involvement of Chemico, whose NEC process operated at 200–350 atmospheres, probably arose from centralization of the chemical industry under a programme of the Supreme Economic Council. The Berezniki plant, opened in stages during 1931–1933, was part of one of the largest chemical combines in the world, the Berezniki-Solikamsk, located on the Kama River, and consisting of ten plants [22, 23]. It included a large nitric acid plant, with stainless steel absorption towers designed by Uhde and manufactured by Bamag (Fig. 14.3). Just as was the case with electrical technologies, German, British and American firms contributed towards both supply and erection of chemical plant. Charles Brown organized building of the ammonia facility at Bobroki. It opened on 24 December 1933, on which date Bobroki was renamed Stalinogorsk (present-day Novomoskovsk).

South-east of Berezniki was the Magnitogorsk steel factory, also in the Ural District, incorporating the largest by-product coke plant in Europe, planned and built at the end of the 1920s with assistance of the Koppers Engineering Company of Pittsburgh, though some chemical manufacturing equipment was designed and constructed locally [24]. This facility produced ammonia, nitric acid, and ammonium sulphate. Coke oven gas came from the Zaporozhye metallurgical plant.

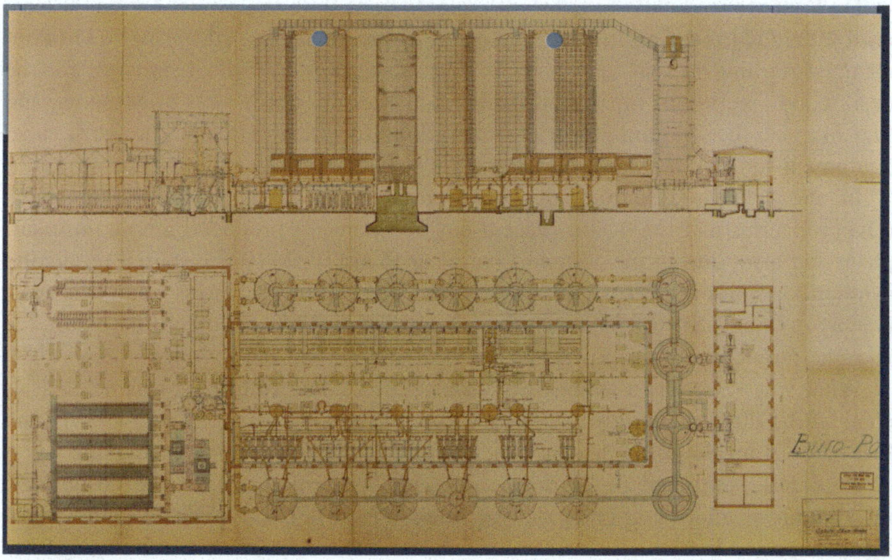

Fig. 14.3 Plan dated 12 March 1932 for the nitric acid facility at Berezniki, Soviet Union, designed by Friedrich Uhde. (ThyssenKrupp, Corporate Archives, Duisburg)

Fig. 14.4 Gorlovka ammonia factory, Soviet Union, c. 1940. Today, Horlivka, eastern Ukraine. (Ostchem/Group DF International)

Du Pont resisted calls to enter into negotiations with the Soviet Union, at least until 1929 [25]. Fearing loss of markets at a time of depressed business, Du Pont changed its tune in that year and drew up an agreement with Chimstroi for the design and erection of nitric acid facilities at Kalinin, north-west of Moscow, and Shoshka, in the Ukraine District. According to Alcan Hirsch, the nitric acid equipment for Kalinin was made in the United States [26] (Table 14.1).

In the Ukraine District, the coke and chemical works at Kamenskoye (Dneprodzerhinsk from 1936; present-day Kamianske) produced ammonium sulphate and also, during the Second Five-Year Plan (1933–1937), started producing synthetic ammonia. Electrical power for local industry came from the vast Dneproges dam, on the Dnieper River, opened in 1932. The fertile soil of the Ukraine provided both summer and, in the south-central part, winter wheat.

What was referred to as the "Soviet Chemical State Industry" was headed, as chair of the directorate during 1929–1930, by Mikhail Tomsky (1880–1936), who had fallen from Stalin's grace as a member of the Soviet Union's leadership.

Contracts continued to be drawn up with Western firms for the design and installation of additional chemical facilities [27, 28]. The A. Borsig company of Berlin supplied the Soviet Union with compressors for use in the ammonia synthesis. The third ammonia facility constructed under the First Five-Year Plan used Montecatini's Fauser process. This time there was no problem in dealing with fascist Italy, since, as with the United States, the main interest was not ideological but transactional. The location was Gorlovka in the coal-rich industrial heartland of

Table 14.1 Western contributions to the Soviet Union's nitrogen industry during the First Five-Year Plan, 1928–1932

Place	Company/process	Date of contract	Year work started	Completion	Notes
Berezniki	Chemico, Nitrogen Engineering Corporation/synthetic ammonia. Uhde, Bamag/ nitric acid from ammonia	11 November 1928 (with Chimstroi)		1931	Ural Mountains. Full operation by mid-1930s. 25,000 employees in 1937.
Bobroki Chemical Combinat. Known as Stalinogorsk (Stalinogorak) from 1933.	Chemico, Nitrogen Engineering Corporation/synthetic ammonia. Nitric acid from ammonia	29 June 1931 (with Veschimprom, All-Union Trust for Chemical Industry)		1933	200 km south of Moscow. Town of Bobriki founded 1930. Present-day Novomoskovsk
Shoshka (Ukraine), first of five plants. Second at Kalinin, used German equipment	DuPont/nitric acid from ammonia	1929	1929	Kalinin, 1930; Shoshka, 1932	
Gorlovka	Montecatini-Fauser synthetic ammonia. Koppers, USA/Coke ovens		1929	1933-1934	Donbass (eastern Ukraine). Today Horlivka
Karakliss	Stockholms Superfosfat Fabriks AB/cyanamide			October 1931	Armenia; 10,000 tons/ year, later increased to 20,000 tons
	Koppers, USA/Coke ovens				For hydrogen
	Borsig				Compressors

the Donetsk District (Donetsk Oblast, or province; Donbass, eastern Ukraine), where major coking operations were in operation (Figs. 14.4, 14.5 and 14.6). It opened in April 1933 and produced 25–40 tons of ammonia daily [29–31]. By 1933, at Chirchikskaya (Chirchick), north-east of Tashkent in the Central Asia and Kazakstan District, work had started on a synthetic ammonia factory employing electrolytic hydrogen. However it and an ammonia facility at Kemerovo probably opened during the Third Five-Year Plan.

To gain an insight into the level of Western involvement we draw from the perspective of the chemical engineer Alcan Hirsch as an American participant

Fig. 14.5 Assembled frames for providing access to unhoused vertical processing equipment, "Gorlovka Nitrogen Fertilizer Works," Soviet Union, 1930s. (Ostchem/Group DF International)

engaged in technology transfer during the years when the Soviet nitrogen industry underwent its first major programme of expansion. Hirsch, moreover, had unrivalled access to information, as clearly demonstrated in his *Industrialized Russia*. It is worth understanding how he became involved and his motive for writing this tract, particularly since in part it is self-serving, and in general hardly impartial. Was he promoting Kremlin propaganda, and if so what was in it for him? And does this place a limitation on our reliance on Hirsch? For a partial answer, we need to look at his background, in both chemistry and politics.

Fig. 14.6 Administration building under construction, "Gorlovka Nitrogen Fertilizer Works," Soviet Union, early 1930s. (Ostchem/Group DF International)

14.3 Reporting on the Soviet Industrial Revolution

Alcan Hirsch (1885–1938) studied chemistry at the University of Texas at Austin, where he received his undergraduate degree in 1907, in which year he joined the American Electrochemical Society. He undertook research on pyrophoric alloys at the University of Wisconsin, for which he was awarded the PhD in 1911; in the same year he described the preparation and properties of metallic cerium. Hirsch then set up a consultancy service in New York. With the aid of a Carnegie Fellowship from the Iron and Steel Institute of Great Britain he worked on improvements in electric furnaces for use in the steel and metallurgical industries. The results were given before the institute on 7 May 1914. During 1916–1918, he took a special interest in chemical developments in Japan. In New York, with Takamine Jōkichi, owner of Takamine Laboratories, Inc., he planned the development of fermentation processes for use in Japan. He may well have worked with Tamaru Setsuro, Haber's former assistant in Berlin, who was then in New York. During World War I, Hirsch undertook research into the constitution of Novocaine (the anaesthetic effect of which was discovered by Alfred Einhorn in 1905). In 1917, Hirsch and his brother, mining expert Marx Hirsch (1898–1964), filed a patent for a method of producing castings of rare earths and their alloys. In 1920, Alcan Hirsch established Hirsch Laboratories, Inc., with offices in New York and two laboratories in Brooklyn. Two years later there were five chemists offering consultancy services in organic chemistry and rare earth metals—notably cerium, thorium and molybdenum, and their alloys. Alcan Hirsch's partner was Marx. The business was sold in 1923. The brothers participated in the formation of the smelter and refiner Molybdenum Corporation of America, founded in 1919, and later known as Molycorp (afterwards a part of Union Oil of California), which was managed by Marx. Alcan continued his connection with the firm, and with work on a number of inventions, including a novel electric battery.

Outside of his chemical interests, Alcan Hirsch was politically left leaning, and closely involved with the Farmer-Labor Party, for which in 1920 he was a candidate for New Rochelle, Westchester, New York, where he resided. As an ethnic Jew, Hirsch no doubt had welcomed the Russian Revolution of 1917 that brought to an end the pogroms and anti-semitic policies tolerated under the tsarist regime. This must have aroused strong sympathies for the communist state. Moreover, in the early 1930s capitalism was facing a crisis, no more so than in the United States. In cities, soup lines were lengthening, in the Midwest farms were turning to dust—an erie prelude to "the Dust Bowl" years—and in squatter camps could be found a growing number of homeless and unemployed (totalling 12 million, one-quarter of the workforce, in 1932). For many, including intellectuals who had lost their faith in the American system, communism offered some semblance of hope. In Germany W. A. Dyes suggested that "sympathy with Russian ideas" arose from the negative impact of "American methods of rationalisation" that were leading to job losses, including among chemists at firms where previously staff were not dismissed during periods of recession. This phenomena was spreading elsewhere [32].

Alcan Hirsch stated in his book that it was the impact of the Great Depression that contributed towards his becoming a chemical consultant in the Soviet Union during the last stage of the First Five-Year Plan, in which the steel and metallurgical industries received top priority. Closely connected to steel was the production of coke, by-products, including ammonia, and the use of coke in the production of hydrogen.

For Hirsch the first step was undoubtedly as much for profit as it was for political reasons. It followed business transactions probably from early 1931 in which Marx Hirsch arranged for the Molybdenum Corporation to begin delivering ferro-tungsten to the Soviet Union's steel industry via Amtorg. (This enabled restarting of the Nevada-Massachusetts Company's ore-producing operations that had been closed down because of the depression) [33]. The Soviet Union's great need for ferro-alloys (including molybdenum, titanium, tungsten, and vanadium alloys) was such that during the First Five-Year Plan imports exceeded 33,000 tons each year.

As a result of the Molybdenum Corporation's orders, from the spring of 1931 until 1933, Alcan Hirsch made several visits to Russia, the longest lasting six months.

In 1932, on one of his return trips to the United States he lobbied for recognition of the Soviet Union, as reported in the *New York Times* of July 24, under the heading: "Roosevelt to get Soviet trade plea; Dr. Alcan Hirsch returning here to ask his support for recognition movement. He scores our policy. Chemical engineer says United States is ignoring a great and safe market."[3] Franklin D. Roosevelt was

[3]In June 1932 the Russians met with Frederick Pope in an effort to enlist his support for recognition of the Soviet Union.

then Governor of New York. He defeated Herbert Hoover in the November 1932 presidential election. Following difficult negotiations that began in October 1933, the United States officially recognized the Soviet government on 16 November 1933, in an effort to both improve trade links and counter Japanese expansion in Asia, following the takeover of Manchuria. However, almost from the start the relationship was far from smooth.

As for Hirsch's status in the Soviet Union, there is some confusion, of his own making. He presented himself as so influential that he was appointed to act as chief engineer of chemical trusts, including the chemical engineering group Giprochim and the nitrogen group Giproazot (Engineering Company for Design of Nitrogen Based Plants, GIPROazot), which in 1943 merged with the State Research Institute of the Nitrogen Industry). He also trumpeted his purported role as chief consultant to the chemical industry, a claim which on the evidence available is highly suspect. He was undoubtedly in touch with all these bodies, through an engineering office in the Giprochim Building, Moscow, but his main activity, by his own account, was under contract, assisting in the design of machinery and equipment for use in chemical plants. Hirsch's contribution to the nitrogen industry was probably through assisting Chemico to introduce Nitrogen Engineering Corporation (NEC) ammonia technology at Bobroki, and in applying his expertise in production of alloys required for converters and other equipment.

To what extent was Hirsch engaged in objective reporting? Was his book an exercise in propaganda, or a form of political advertising, perhaps an attempt to mend damage caused by events such as the Metropolitan-Vickers case (see below), and make the United States aware of the vast commercial opportunities still to be had through technology transfer? That is hard to say. Certainly his knowledge of chemistry and chemical industry was used to advantage, making his account in many ways superior to those of later Western historians and critics with little or no knowledge of chemical industry. Thus the assumption that after importing Western chemical technologies in the 1930s new Soviet advances were held back until the late 1950s has little foundation, since, as we shall see, much the same applied elsewhere. This was particularly the case for nitrogen fixation technologies which had reached a stage of maturity by the early 1930s and would undergo few changes until around 1960.

The conclusion is that while Hirsch may have exaggerated his roles, and received uncommon favours, there is no doubt that he had access to a tremendous amount of information concerning industrial programmes and plans for expansion. Enthused by his experience, Hirsch repeatedly drew attention to progress made since the collapse of tsarist rule in 1917. The Soviet Union appealed to Hirsch politically, scientifically and technologically, though he drew attention to shortcomings and failures, and shortages of food as a result of the payment of foreign debts, some of which included for imported machinery. Hirsch had no qualms about the fact that "[t]he Soviet Government is frankly a dictatorship, and makes no pretense of being democratic" [34]. More than anything else, however, Hirsch managed to capture the wide sweep of Stalin's massive and seemingly unrelenting industrialization programme aimed at mimicking and then overtaking the United States.

In Moscow, Hirsch's discussions with Frederick Pope, Chemical Construction Corporation's administrator, suggested, according to Hirsch, that there were no reservations in doing business with the Soviet Union. That was shortly after Hirsch's arrival, in 1931. American dealings with the Soviet Union and other Western countries were at their peak. The end of the First Five-Year Plan was followed with a tense period in Soviet dealings with foreign firms, including claims of espionage and wrecking, notably in 1933 against British employees of the Metropolitan-Vickers Electrical Export Company in the Soviet Union. In 1934, Pope's complete frustration while dealing with the demands of the authorities led to a parting of the ways with Soviet central planners. Du Pont's experience was not altogether different [35]. By 1936, Du Pont executives echoed Pope's frustrations [36]. Alcan Hirsch, it is worth noting, contributed to science fiction, including the article "Within Sight of Hell," published in *Amazing Stories*, perhaps appropriately, in December 1933.

14.4 Problems of Rapid Industrialization

The growing anti-capitalist ideology favoured elevation of the status of young scientists and engineers in order to reduce dependence on Western firms for the more innovative technologies. The outcomes included improvements in catalysts and catalytic processes made at Russian research institutes that increasingly focused on the application of science. The State Research Institute of the Nitrogen Industry in Moscow opened in 1931, again in an effort to bring about independence from foreign specialists and imported equipment.

Still, the introduction of foreign technology during the period of inception of the Soviet nitrogen industry was not without its problems.

Marrying untested pieces of equipment and machinery from different foreign suppliers and plant designers proved highly challenging. The training and skills of Russian technicians were in the early years not always up to the management of complex and expensive processes. Some facilities were blighted by mechanical failures and accidents. In March 1930, the exiled Trotsky, then living in Turkey (he had been expelled from the Communist Party in 1927 and deported in 1929), condemned the pace of what he called "super-industrialization." Hastened industrialization and collectivization had claimed many victims, apart from the Kulaks. Trotsky called into question the wisdom and motivations for such rapid expansion, for which the Soviet Union, in his opinion, was ill prepared [37]. Similar sentiments appeared elsewhere. In 1934, *Chemistry and Industry* commented: "Considering the difficulties Russia has had to face it is possibly a great achievement that she has already gone so far and done so much. On the other hand it is impossible to count the cost, the loss, and the waste of this gigantic and sudden experiment" [38].

Delays and difficulties in the construction and operation of chemical plant to accompany American-built coke ovens at the Gorlovka works (completed in 1933-1934), contributed towards the arrest of, or were the pretext for the arrest of, chemical engineer Leonid E. Tamm (1902-[?]), deputy chief engineer of the Soviet Nitrogen Industry—and brother of later Nobel laureate Igor E. Tamm (1895–1971; physics, 1956). This was late in 1936, the first year of the mass purges of scientists and engineers (1936–1938). Tamm, who had previously been in charge of production at the Gorlovka works (mid-1932 until February 1936), was hauled before the Supreme Court of the Soviet Union on the evening of 27 January 1937 and accused, along with others, of anti-Soviet Trotskyite acts of sabotage—namely, explosions that had occurred in the hydrogen department and coke oven section at Gorlovka in 1934 and 1935. He was indicted on the basis of trumped up charges, and disappeared soon after. The court also considered the cause of the "breakdown of the Berezniki Combined Chemical Works" [39].

Meantime, from 1935, manufacturing production in the Soviet Union was dictated by a new programme that rewarded those who exceeded production norms, called Stakhanovism [40]. This sometimes resulted in damaged tools and machines, and accidents, arising from the increasingly challenging production targets.

The several developments had not been sufficient to prevent the major famine of 1932 in the Ukraine and other regions. It was brought on by a repressive policy aimed at those who refused to adopt collectivization, by poor planning and apparently poor weather that devastated harvests; production fell far short of expectations. Famine would continue until 1934.

By the end of 1937, total ammonia output in the Soviet Union was estimated at around 200,000–250,000 tons [41]. Ammonia increasingly contributed to large-scale cultivation of the land, along with tractors and the new combine harvesters. Russia had become a major factor in mechanized agriculture. Its industries benefited from new research and development institutes [42]. Natural gas was, for the first time in the Soviet Union, used in the production of hydrogen at a new ammonia facility near Baku [43, 44]. With the aid of "huge electro-chemical-metallurgical combinats," Soviet fertilizer production grew from 234.1 thousand tons in 1928, to 922.8 thousand tons in 1932, and 3216.3 thousand tons in 1938 [45].

From 1939, new national borders emerged in Europe, though in the form of occupations and invasions. In March 1939, Germany, following takeover of the Sudentenland, occupied the rest of Czechoslovakia, which was divided into the protectorate of Bohemia and Moravia and the Slovak Republic, a client state. In a secret protocol to the German-Soviet nonagression pact of August 1939 (the Molotov-Ribbentrop Pact), Stalin and Hitler divided Poland between Germany and Soviet Russia, and gave Russia the three Baltic states—Estonia, Latvia and Lithuania—that had gained their independence after World War I. Germany invaded Poland on 1 September 1939; Britain and France declared war on Germany. Germany invaded Russia in June 1941; by August, the Wehrmacht had occupied the Baltic countries. In December, Imperial Japan attacked Pearl Habour; the United States declared war on Japan.

Synthetic nitrogen was a commodity much needed in time of war, as much in World War II as in 1914. Nitrogen factories in the western part of the Soviet Union were hurriedly emptied of equipment, which was taken east by train, and put to work in remote locations. Total Soviet synthetic ammonia production by 1941, expressed as nitrogen, was estimated at 350,000 metric tons, around 110,000 tons less than the United States.

Rebuilding of the Soviet Union's ammonia industry after World War II, and expansion in the 1970s, enabled the communist nation to surpass the output of the United States during the 1980s.

References

1. Siegel KAS (1993) Technology and trade: Russia's pursuit of American investment, 1917–1929. Dipl Hist 17(3)(July):375–398.
2. Siegel KAS (1996) Loans and legitimacy: the evolution of Soviet-American relations, 1919–1933. University Press of Kentucky, Lexington.
3. Melnikova-Raich S (2010) The Soviet problem with two 'unknowns': how an American architect and a Soviet negotiator jump-started the industrialization of Russia, Part I: Albert Khan. J Soc Industrial Archeology 36(2):57–80.
4. Melnikova-Raich S (2011) The Soviet problem with two 'unknowns': how an American architect and a Soviet negotiator jump-started the industrialization of Russia, Part II: Saul Bron. J Soc Industrial Archeology 37(1–2):5–28.
5. Morris PJT (1982) The development of acetylene chemistry and synthetic rubber by I.G. Farbenindustrie Aktiengesellschaft: 1926–1945. DPhil thesis, University of Oxford.
6. Lewis RA (1979) Innovation in the USSR: The case of synthetic rubber. Slavic Review 38 (1):48–59.
7. Hirsch A (1934) Industrialized Russia. The Chemical Catalog Company, New York.
8. Segal L (1934) Economic progress in the U.S.S.R. (1934) J Soc Chem Ind 53(21) (1 June):486–488 (Chem Ind London).
9. Sutton AC (1968) Western technology and Soviet economic development, 1917–1930. Hoover Institution, Stanford.
10. Sutton AC (1971) Western technology and Soviet economic development, 1930–1945. Hoover Institution, Stanford.
11. Tooze A (2015) The deluge: The great war and the remaking of global order. Penguin Books, London [Allen Lane 2014], pp 424–425.
12. See for example 1921 newsreel, International News-442, US National Archives and Records Administration, ARC identifier 88967.
13. Ipatieff VN (1946) In: Ipatieff J, et al. (eds), The life of a chemist: the memoirs of Vladimir N. Ipatieff (trans. Haensel V, Lusher RH). Stanford University Press, Stanford, p 409.
14. Ipatieff VN (1946) In: Ipatieff J, et al. (eds), The life of a chemist: the memoirs of Vladimir N. Ipatieff (trans. Haensel V, Lusher RH). Stanford University Press, Stanford, pp 327–328, 339–340.
15. Caro N, Frank AR, Franck H[ans] H[einrich]. Process for preparing alkaline earth cyanamides. United States patent, US 1,999,289, filed 26 February 1935. First German patent filed 8 February 1929.
16. Caro and Frank, British patent, BP 342,068, December 1929.
17. Chemical industry in Germany (1929) J Soc Chem Ind 48(26)(28 June):650–652 (Chem Ind London).
18. Zaitseva EA, Homburg E (2005) Catalytic chemistry under Stalin: science and scientists in times of repression. Ambix 52(1):45–65.
19. Ipatieff VN (1946) In: Ipatieff J, et al. (eds), The life of a chemist: the memoirs of Vladimir N. Ipatieff (trans. Haensel V, Lusher RH). Stanford University Press, Stanford, pp 425, 428.
20. Brooks NM (2005) Growing links between chemistry and industry in Russia and the Soviet Union, 1900–1953. Ambix 52(1)(2005):27–43.

21. Sutton AC (1971) Western technology and Soviet economic development, 1930–1945. Hoover Institution, Stanford, pp 102–103.
22. Hirsch A (1934) Industrialized Russia. The Chemical Catalog Company, New York, pp 73–79.
23. Sutton AC (1971) Western technology and Soviet economic development, 1930–1945. Hoover Institution, Stanford, pp 100–101.
24. Hirsch A (1934) Industrialized Russia. The Chemical Catalog Company, New York, pp 81–82.
25. Siegel KAS (1993) Technology and trade: Russia's pursuit of American investment, 1917–1929. Dipl Hist 17(3)(July):375–398, on 394–396.
26. Hirsch A (1934) Industrialized Russia. The Chemical Catalog Company, New York, p 93.
27. The synthetic ammonia industry in the USSR (1951) Central Intelligence Agency, Office of Research and Reports, Provisional report No. 5 (CIA/RR PR-5), 8 October, p 2. http://www.foia.cia.gov/sites/default/files/document_conversions/89801/DOC_0000232458.pdf (accessed 20 September 2015).
28. Hirsch A (1934) Industrialized Russia. The Chemical Catalog Company, New York, p 86.
29. Lamer M (1957) The world fertilizer economy, issue 10. The Food Research Institute of Stanford University, Stanford, pp 398–400.
30. Sagers MJ, Shabad T (1990) The chemical industry in the USSR: an economic geography. ACS Professional Reference Book. Westview Press, Boulder, esp. pp 115–117.
31. For Gorlovka, see the Ostchem 80 site, http://en.stirol.net/history/ (accessed 27 September 2015).
32. Dyes WA (1931) Official reports of the German chemical industries. J Soc Chem Ind 50(4) (23 January):59–61.
33. Limbaugh R (2010) Tungsten in peace and war, 1918–1946. University of Nevada Press, Reno, pp 78–82.
34. Hirsch A (1934) Industrialized Russia. The Chemical Catalog Company, New York, p 229.
35. Sutton AC (1971) Western technology and Soviet economic development 1930–1945. Hoover Institution, Stanford, pp 99–100.
36. Travis AS (2017) Globalising synthetic nitrogen: the interwar inauguration of a new industry. Ambix 64(1):1–28, on 15–18.
37. For Trotsky's criticism of Stalin's industrialization programme, see http://www.marxists.org/archice/trotsky/1930/03/newcourse.htm (accessed 14 January 2017).
38. Editorial: American and British problems (1934) J Soc Chem Ind 53(15)(13 April):319–320 (Chem Ind London).
39. Report of court proceedings in the case of the anti-Soviet Trotskyite centre (1937) Peoples Commissariat of Justice of the U.S.S.R, Moscow, pp 414, 439–442.
40. Cohen Y (1996) Politics and passion for production: France and the USSR in the 1930s. In: Fox R (ed), Technological change: methods and themes in the history of technology. Routledge, London, pp 215–227.
41. The synthetic ammonia industry in the USSR (1951) Central Intelligence Agency, Office of Research and Reports, Provisional report No. 5 (CIA/RR PR-5), 8 October, p 3. http://www.foia.cia.gov/sites/default/files/document_conversions/89801/DOC_0000232458.pdf (accessed 20 September 2015).
42. Lewis RA (1972) Some aspects of the research and development effort of the Soviet Union, 1924–35. Science Studies 2:153–179.
43. Industrial notes from abroad. U.S.S.R. (1937) J Soc Chem Ind 56(39)(25 September):867 (Chem Ind London).
44. Hemy G (1971) The Soviet chemical industry. Barnes & Noble, Inc., New York, pp 239–241.
45. Baykov A (1950 [1946]) The development of the Soviet economic system: an essay on the experience of planning in the U.S.S.R. Cambridge University Press, Cambridge, pp 155, 331.

Chapter 15
Imperial Japan: From Cyanamide to Synthetic Ammonia

15.1 High Pressures

Physical chemistry was introduced into Japan in the late nineteenth century by Sakurai Joji (1858–1939) at Tokyo Imperial University, and by the early 1900s had a strong following among Western-trained scholars [1, 2]. The leader among Japanese academics in nitrogen capture and the ammonia equilibrium was Tamaru Setsuro, who in 1913 introduced chemist Suzuki Tatsuji—head of Yokohama Technical College—then in Germany to Fritz Haber. This was enough to convince Suzuki of the superiority of Haber's method over other methods of nitrogen capture then in use—that is, cyanamide and electric arc processes.

World War I cut off the import of essential supplies of fertilizers, particularly ammonium sulphate, as well as other strategic products, encouraging home production in Japan. Noguchi's cyanamide enterprise expanded considerably and was enabled to increase capital by 22 million yen. The war enabled tremendous expansion of the Japanese chemical industry, as well as encouraging research and development in state, private and industrial laboratories, at first mainly on nitrogen fixation.

Much interest had been shown in the high-pressure ammonia synthesis from around 1912, following the BASF presentation at the Eighth International Congress of Applied Chemistry in New York. Ammonia research was undertaken during the war, including by Suzuki Tatsuji at the Yokohama institute's Chemical Research Laboratory (from 1915). Though the Japanese government had sequestered the Haber-Bosch patents, the information therein, as other Allied scientists found, was not sufficient to develop industrial manufacture [3]. This was a spur to academic studies. In 1917, work began on construction of the Institute of Physical and Chemical Research (RIKEN), in Tokyo. It was backed by the entrepreneur Shibusawa Eiichi (1840–1931), who had lobbied for such an institute since 1913, and funded by industry, the government and the imperial family; its purpose was similar to that discussed earlier by Tamaru and Takamine in the United States. Tamaru headed the division of physical chemistry, set up in 1917, where the main

A.S. Travis, *Nitrogen Capture*, https://doi.org/10.1007/978-3-319-68963-0_15

focus was on nitrogen capture. In 1918–1920 he was in Europe acquiring equipment for the new institute. This included, especially, apparatus for the synthesis of ammonia. The year 1918 saw the opening of the Special Nitrogen Research Laboratory, of the Department of Commerce and Agriculture (later Tokyo Industrial Experimental Laboratory, TIEL, and then the Chemical Engineering Research Institute), headed by Kodera Fusajirō, who had worked with Fritz Haber from 1908. In 1920, Kodera's colleague and high-pressure expert Yokoyama Buichi, while in Europe, attempted to gain access to Oppau, but was turned away. In Paris, he investigated the Claude process, but was not impressed.

In December 1920, Suzuki Masaya, president of Sumitomo, arrived at Oppau in an attempt to negotiate a license for the Haber-Bosch process, but Carl Bosch demanded far more than the visitor was prepared to pay. Earlier, at the suggestion of Takamine, Suzuki had considered the General Chemical process, but the Americans were not prepared to collaborate, since, it was said, German input was required [4]. Technical experts from four Japanese companies inspected the BASF Oppau facility in 1919, while it was still occupied by the French [5].

According to a later report of the US Department of Commerce, the Japanese government was unable, "for fiscal reasons"—namely, the large sum and conditions over licensing demanded by BASF in return for technical information on the Haber-Bosch process—to come to an agreement with BASF. As a result, in 1921 the Oriental, or Eastern, Nitrogeneous Fertilizer Company was established (Toyo Nitrogen Company; Oriental Nitrogen Association, ONA), with collaboration from Mitsui, Mitsubishi, Sumitomo, Furukawa, Sankyo, and three other firms. Several others were interested. The sequestered BASF nitrogen patents were then made available to private companies, without charge, provided that they were able to construct operating Haber-Bosch type plants within ten years. Following lack of success, a Fertilizer Investigation Committee claimed that the Japanese firms did not intend to apply the Haber-Bosch patents but instead use their arrangement "to control the import of nitrogenous fertilisers from Germany [6]." In effect, from 1923, the ONA became an importing cartel [7].

Despite the failure to come to an arrangement with BASF, Japanese industrialists had no problems in negotiating with other European innovators in and producers of synthetic ammonia and nitrogen products. This was facilitated by the accumulated wartime profits and devaluation of European currencies. Licenses could be acquired at acceptable prices, though sometimes the bargaining was tough. Noguchi Shitagau was in Europe during early 1921, in part to renew his cyanamide license with the Rome-based holder of patents for the Frank-Caro process, but also to investigate other nitrogen capture processes. He was accompanied by two managers. They were especially interested in Casale's process. Noguchi and one of his managers spent some time at the Terni workshop, though exercising caution, and probably even a little suspicion. After all, the workshop was quite small, the general appearance was sparse and certainly not organized in the style expected of an engineering laboratory; nothing like the great palaces of research that existed in Germany.

Funds from the sale of licenses, as we have seen, were critical to the growth and expansion of Casale's business. In his dealings with Noguchi, Casale demonstrated that he was an aggressive and capable businessman. Before the arrival of Noguchi, Casale had already been approached by competitors Suzuki Shōten of Kobe and Kuhara Fusanosuke, whose technical people inspected the Terni apparatus [8]. Noguchi's early reluctance, and procrastination, over taking on the Casale process changed after being made aware of the apparent interest from his rivals. Following difficult negotiations, and payment of the very large sum of 10 million lira demanded by Casale, exclusive rights to Japan were sold to Noguchi in December 1921. It was a tremendous boost for Casale, in fact his first major licensing arrangement (Fig. 15.1). Funding for the purchase was raised at the new branch of Mitsubishi Bank in London. Casale himself went to Japan in September 1923 to open the Nitchitsu Nobeoka plant on the east coast of Kyūshū. He arrived on the day a massive earthquake devastated both Tokyo and Yokohama. The main industrial manufacturing areas, including Osaka, and the nitrogen factories to the west, were spared. The Nobeoka ammonia plant started operating at the end of the month with a daily capacity of 60 tons of synthetic ammonia (Fig. 15.2).

Despite the high price paid for the Casale license, Noguchi had a free hand in working the ammonia process. In addition, the liquid ammonia from the Casale process was ideal for use as a refrigerant. (Some 96,000 lb. of anhydrous ammonia were imported into Japan during 1925, mainly for this purpose.) Interestingly,

Fig. 15.1 Luigi Casale (centre) and representatives of Ammonia Casale with their guest Noguchi Shitagau (second from left), probably at Terni, 1922. (Casale SA)

Fig. 15.2 Nobeoka ammonia factory (Casale process), Japan, around 1924. (Casale SA)

cultural differences played little or no role in the dealings between Japanese and Italian industrialists over cyanamide and ammonia. Both sides had a common purpose, that of maintaining strong positions in fast-growing markets. The success of the Ammonia Casale process rested on its very significant advantages over the cyanamide process. It was a clear-cut case of Noguchi's faith in Casale's process. Noguchi made further visits to Terni and built up an excellent relationship with Casale prior to the latter's death early in 1927.

In February 1925, Noguchi opened a second Casale synthetic ammonia facility at the Minamata (cyanamide) site, also with a capacity of 60 tons daily. Minamata was the main producer of synthetic ammonia for Nitchitsu; in 1927 the facility was expanded to enable production of 100 tons daily (Fig. 15.3). Cyanamide production at Minamata had been transferred to a subsidiary [9]. Growing demand for synthetic ammonia in part resulted from the decline in imports of bean cake from China, including Manchuria. The cake was increasingly used for animal feed [10]. In 1925, its import to Japan was half that of the 1920 level, as a result of government support of the home nitrogen industry. Preferences were shown for domestic production, for which prices were high even at a time of free imports [11]. Noguchi relied on electrolytic hydrogen. Westinghouse and Siemens contributed to the development of the Japanese electrical industry on which nitrogen production relied.

Fig. 15.3 Newly installed ammonia converters, Casale process, at Minamata, prepared for an inspection by the Japanese Emperor, probably 1927. The total output of converters shown here was 40 tons per day, raising the factory's output from 60 to 100 tons per day. (Casale SA)

15.2 Korea

Noguchi Shitagau, through introducing the Casale process into Korea, played a leading role in the somewhat forced industrialization of the peninsula. It had been annexed by Imperial Japan in 1910 and placed under administrative, and, by the 1920s, growing military control. Noguchi's plans for a Korean enterprise as an overseas division of Nitchitsu first met with strong resistance from the Mitsubishi zaibatsu, a conglomerate that until 1921 had backed him. Mitsubishi, by acquiring water rights in Korea that were essential to Noguchi, schemed to bankrupt the upstart. Noguchi responded by convincing the Japanese military authorities in Korea, who were no friends of old established zaibatsu such as Mitsubishi, to amend, in part, the apportioning of water rights. The wildly ambitious Noguchi gained access to a tributary of the Yalu River though he had no information about its water flow, which turned out to be problematic for a time. It made no difference to his master plan. In 1927, he began work on a 320-acre chemical production complex (Nippon Chisso Hûngnam; Chōsen Chisso Hiryō) at Hûngnam, a small fishing village in the north-east, with a planned ammonia capacity of 120 tons per day. Noguchi's technical people came from the Minamata factory. With help from

Fig. 15.4 Hûngnam ammonia factory, northern Korea, 1930s. (Edelstein Collection)

the army, and support from the Japan Oriental Development Company and the state, Noguchi created the Far East's largest hydro-electric power complex, mainly to serve his factory. Ammonia and ammonium sulphate were produced towards the close of 1929. Other chemicals, including phosphoric acid, were manufactured at the site. Hûngnam became the world's third-largest ammonia factory, with 45,000 workers (Fig. 15.4) [12]. Hydrogen, as on mainland Japan, was produced by electrolysis.

Chemical manufacture, power generation, and extensive railway and port facilities were brought together to create a distinctly modern infrastructure. Hûngnam became a centre for heavy industrial development. Noguchi also diversified in Japan, including in 1929 into rayon manufacture with production of cuprammonium (Bemberg) artifical fibre. This was an important step in enabling Japan to become the leading producer of synthetic fibres. By 1936 a second Korean factory, at Pon'gung, close to Hûngnam, was producing both synthetic ammonia and cyanamide. The Hûngnam factory became the basis of modern industry in present-day North Korea. Noguchi's enterprise also brought about the electrification of Korea, where Nitchitsu was the main Japanese industrial concern. Noguchi's new zaibatsu later branched out into other chemical products and massive hydro-electric schemes in Manchuria.

15.3 Suzuki Shōten

After the Suzuki Shōten zaibatsu lost interest in Ammonia Casale's process in 1921, attention was drawn to Claude ammonia. In 1922, Suzuki became Japanese licensee of the French process, which was worked at Suzuki's Claude Method Nitrogen Industries, on Hikoshima Island, Yamaguchi Prefecture. An associated plant, also on Hikoshima, was Hikoshima First Nitrogen Industries. As at other early Claude factories there were numerous technical difficulties. Suzuki Shōten also faced difficulties of a financial nature, in the form of a substantial unsecured debt owing to the Bank of Taiwan, whose collapse precipitated the third and main

1927 Japanese banking crisis. Suzuki folded, along with its ammonia business. Another factor that had harmed Suzuki Shōten was in the arena of national politics and rivalry. The Mitsui zaibatsu was a powerful opponent of Suzuki. During the 1918 rice riots, Mitsui used the media to, wrongly, discredit Suzuki for profiteering. Mitsui subsequently continued to undermine Suzuki's credibility.

Mitsui, an early supporter of Denka, had earlier expressed an interest in the General Chemical ammonia process, but found it wanting, and later joined with Oriental Nitrogen (ONA) in its attempts to master Haber-Bosch technology. Mitsui Mining Company (Toyo Koatsu Industries) in 1928–1929 began trials at the Suzuki Shōten ammonia factories, which it purchased in 1929. Mitsui's attempts to use the Claude process did not meet with success [13]. In 1933, with government support, Mitsui started up ammonia production at Miike, Fukuoka Prefecture, using Du Pont technology, based on both Claude and Casale processes [14].

Mitsui Mining had joined the trend in which during the 1930s mining companies diversified into the manufacture of chemicals. The Suzuki Shōten concern managed to sort out its financial affairs, but did not return to synthetic ammonia [15].

15.4 Fauser, NEC, TIEL, and Haber-Bosch Processes in Japan

In 1926, Dai Nippon Artificial Fertilizer, successor to Tokyo Artificial Fertilizer, the first Japanese superphosphate manufacturer, negotiated with Montecatini for rights to the Fauser process. It was one of the first firms to gain access to Fauser technology outside Italy. The acquisition price was cheaper than that for the Haber-Bosch process, and royalties were no more than around one-tenth of what was demanded by IG Farben. The Fauser factory was erected at Toyoma, Toyoma Prefecture, a site that offered good railway connections and access to a reliable source of hydro-electricity from Japan Electric Power Co. Synthetic ammonia was produced at Toyoma from April 1928. The source of hydrogen was electrolytic. Following investigations of European processes, from 1938 hydrogen was also obtained from coke ovens [16]. (In 1927, Nitchitsu's cyanamide facility at Kagami had been transferred to Dai Nippon.)

Early in 1928, Sumitomo acquired a license to work the Nitrogen Engineering Corporation's ammonia process. In March a team from Sumitomo visited the NEC demonstration facility. Under the supervision of NEC staff, a pilot plant was put together by November 1928. A 15-ton-per-day ammonia plant was installed by the end of 1929 at Niihama, Ehime Prefecture. Manufacture of ammonia commenced in December 1930, and of ammonium sulphate shortly after [17]. Expansion took place in 1933, including the introduction of finished fertilizer and liquid ammonia. Coke oven gas and the water gas shift reaction were used to produce hydrogen, at far less expense than the electrolytic process. These coal-based processes were advantageous for Sumitomo and Mitsui, which, unlike Noguchi, controlled coal mines. Denka, also with access to coal mines, was producing cyanamide at close to full capacity in the second half of the 1920s. Another newcomer was Shin-Etsu Fertilizer, founded in 1926 at Nagono City, Nagono Prefecture, to manufacture cyanamide, based on local calcium carbonate deposits and water power.

At the TIEL nitrogen laboratory, catalyst expert Shibatu Katsutarō, collaborating with Yokoyama Buichi, developed a synthetic ammonia process for which a patent was filed in 1922. This first Japanese home-designed process was investigated by Mitsui, but it just missed acquiring a license, which went to Shōwa Fertilizer Company (Shōwa Hiryō; later Shōwa Denkō), established by Mori Nobuteru (1884–1941) and Suzuki Saburosuke (1867–1931) in 1928 [18]. Though Mori had limited financial resources, he managed as a result of the depression to acquire equipment on very favourable terms. By demonstrating the viability of the TIEL process, as introduced in 1931, Mori's enterprise emerged as a new zaibatsu. By 1936, it was the third-largest manufacturer of synthetic ammonia, after Nitchitsu and Mitsui (Table 15.1).

Another newcomer to synthetic ammonia was Ube Nitrogen Industry, Ltd, established in 1933. This was an offshoot of a mid-1890s mining company based at Okinoyama, Yamaguchi Prefecture, that exploited coal in the Ube area. Ube Nitrogen was originally assisted by the South Manchuria Railway, but then turned to Dai Nippon, whose technical input was based on its experience with Fauser technology. The manufacture of ammonia, using a process developed in collaboration with coke specialist Heinrich Koppers, of Essen, and incorporating a low temperature carbonization plant, began in July 1934; nitric acid production followed in 1936. Ube Nitrogen eventually became the fourth-largest manufacturer of ammonia in Japan, until its plant was destroyed during a United States Army Air Forces strategic air raid in July 1945.

In 1919, the cost of BASF know-how created a formidable barrier to entry by Japanese firms. A decade later, in 1929, Julius Bueb, former German wartime commissioner for nitrogen products, now representing the German nitrogen industry controlled by IG Farben, was on a second visit to Japan and China in an attempt to encourage both countries to accept increased imports of nitrogen from Germany [19]. The Japanese were not particularly interested and certainly rejected requests to join the international nitrogen cartel. At this time, overproduction, a matter of considerable concern at IG Farben, contributed to a review, as elsewhere, of its sale of know-how policy. IG Farben's shift in strategy was determined by market conditions arising from the several high-pressure ammonia factories operating in Japan, and the fact that there were Japanese firms, particularly those that had not already taken up ammonia processes, eager to deal with the German conglomerate. The several arrangements between IG Farben and Japanese firms in the mid-1930s over synthetic ammonia were without parallel elsewhere. They were in part favoured by the moves of Germany and Japan in similar political directions, and Japanese agreement to join the cartel in the mid-1930s [20, 21]. Still, IG Farben laid down its usual stringent conditions in the use of the Haber-Bosch process when it was adopted in Japan. Mitsubishi Mining's Japan Tar began operating the Haber-Bosch process in 1937, with hydrogen from a coke oven plant. Japan Tar was

Table 15.1 The Japanese synthetic ammonia industry, 1921–1940

Place	Company/Process	Date of license/ contract	Completion	Notes
Nobeoka, Miyazaki Pref.	Nitchitsu/Casale	December 1921	1923	Nippon Chisso Hiryō
Minamata, Kumamoto Pref.	Nitchitsu/Casale		1926	Nippon Chisso Hiryō
Hûngnam/Korea	Nitchitsu/Casale		1930–1931	Nippon Chisso Hûngnam; Chōsen Chisso Hiryō
Pon'gung/Korea	Nitchitsu/Casale		1936	Nippon Chisso Hûngnam; Chōsen Chisso Hiryō.
Hikoshima Island, Yamaguchi Pref. (Honshu)	Claude Method Nitrogen Industries, incorporated April 1922 (Suzuki Shōten; Kurodo-shiki Chisso Kōgyō KK)/Claude	1921		With Hikoshima First Nitrogen Industries (Daiichi Industries). Closed 1927, plant sold to Mitsui Mining in 1929 *(see below)*
Toyoma, Toyoma Pref.	Dai Nippon Fertilizer/ Fauser	1926	1928	
Kawasaki, Kanagawa Pref.	Showa Fertilizer, incorporated 1928 (Mori Nobuteru)/ TIEL	1928	1931	Operated cyanamide factory at Kanose, Niigata Pref. Showa Denko, from 1939
Niihama, Ehime Pref.	Sumitomo/Nitrogen Engineering Corporation	1928	1929–1930	Ammonia, December 1930; nitric acid, 1934. Sumitomo Chemical Industries from 1934
Hikoshima Island, Yamaguchi Pref.	Mitsui Mining/Miike Mining Industries. Claude, then Du Pont. To Miike, Fukuoka Pref.		1933	Miike Nitrogen Industries Inc., founded August 1931
Befu, Hyōgo Pref.	Taki Fertilizer Works (Taki Seihisho)/IG Farben	1935	1938	Later to Sumitomo Chemical Industries
Nagoya, Aichi Pref.	Yahagi Kōgyō/IG Farben	1935	1938	Later Toa Gosei Kagaku Kōgyō
Kurosaki, Fukuoka Pref.	Mitsubishi Mining/ Japan Tar/IG Farben	1936	1939	Later Japan Chemical
Yokohama, Kanagawa Pref.	Dai Nippon Tokkyo Hiryō/IG Farben	1937	1939	Later Nitto Kagaku Kōgyō
Hachinone, Aomori Pref.	Dai Nippon Seito/IG Farben	1937	1940	Later Nitto Kagaku Kōgyō
Ube, Yamaguchi Pref.	Ube Development Company (Ube Yuka Kōgyō)/Heinrich Koppers, Essen		July 1934	Ube Nitrogen Industry/Ube Ammonia

(continued)

Table 15.1 (continued)

Place	Company/Process	Date of license/ contract	Completion	Notes
Dalian (Darien)/ Manchukuo (Manchuria)	Manchuria Chemical Industry Co./Mont Cenis	1933	1935	

Sources include: http://www.sumitomo.gr.jp/english/history/origin/chemical02.html (accessed 24 January 2017); https://www.mitsui-kinzoku.co.jp (accessed 24 January 2017); "Denka Milestones," www.denka.co.jp/eng/corporate/about/history/ (accessed 8 September 2015); Akira Kudo, "Dominance through Cooperation: I.G. Farben's Japan Strategy," in *The German Chemical Industry in the Twentieth Century*, ed. John E. Lesch. Dordrecht: Kluwer, 2000, pp. 243–283; Barbara Molony, *Technology and Investment: The Prewar Japanese Chemical Industry*. Council on East Asian Studies, Harvard University. Distributed by Harvard University Press, 1990

renamed Japan Chemical (Nihon Kasei). Compared with 1928, Japanese production of calcium carbide, mainly for cyanamide, by 1936 had increased 130% and was second only to Germany, which also experienced consistent growth in carbide and cyanamide.

15.5 Japan's "East India Company": The South Manchuria Railway

During the first four decades of the twentieth century, railways were the preferred, and often only, means of efficient land transport for the extractive and heavy chemical industries, including coal and coke, as well as nitrogen products. They were as much part of the infrastructure of manufacture as were reliable supplies of water, for cooling and power generation, electricity grids, navigable canals and deep sea harbours. Special lines were laid down to serve the industries described here, some extensive, and often reaching remote locations, such as those systems serving the Chilean nitrate industry and the Norwegian Rjukan factories of Norsk Hydro. Construction work was often difficult, including the challenges of mountainous and rocky terrain and absence of water. In 1898, Sir William Crookes had emphasized the critical role of railways, existing or new, in vastly increasing agricultural productivity, particularly if new areas for the cultivation of wheat were to be exploited. In Russia the absence of an adequate railway system before the mid-1930s had severely hindered industrial growth and the distribution of grain at times of famine. In contrast, Belgium had a higher concentration of railways than any other country. By 1927, its system of light and narrow gauge railways, exceeding 3000 miles in length, and still growing, carried annually around four million tons of fertilizer, coal, farm produce and bulk products.

In sparsely populated Korea, Noguchi laid down railways to the otherwise isolated Hûngnam site in the late 1920s, and in the 1930s built railways that served

Fig. 15.5 Steam packet and express train of the South Manchuria Railway, around 1920. When Fritz Haber and his wife visited Manchuria in 1924 they would have travelled by a similar ship and train. (Author's collection)

the world's second largest dam, on the Yalu River. Few railways however played the unique role in innovation and industrialization as did the Manchurian railway system while under Japanese control.

The origins of the Manchurian railways lay in the Chinese Eastern Railway, constructed by Russian engineers to the track gauge of 5 feet, and completed in 1901. Following the Russo-Japanese War (1904–1905), in which Japan was the victor, and the Treaty of Portsmouth, negotiated in the United States in September 1905, Japan controlled Manchuria, mainly through a government company, the South Manchuria Railway (SMR), established early in 1906 (Fig. 15.5). This was the southern section of the Chinese Eastern Railway [22, 23].[1]

The SMR, as an instrument of the Japanese government, was the single most important corporation engaged in the industrial and economic development of Manchuria. Its first governor, physician-turned-administrator Gotō Shinpei, compared its functions with those of the British East India Company and other colonial enterprises, and encouraged introduction of the latest scientific and technological

[1]During the conflict, the Japanese had converted much of the railway to the Japanese gauge, of 3 ft 6 in, and constructed a light railway from Mukden (Shenyang) to Antung of 2 ft 6 in gauge. From 1906, these lines were converted to the international standard gauge, 4 ft 8.5 in. Since Japanese manufacturers were not equipped to construct equipment for the standard gauge, the first orders for new railway equipment were placed with American firms.

Fig. 15.6 Cover of a folder containing three postcards presented to tourists and other visitors at the Fushun mine complex, probably around 1930. (Author's collection)

developments into the Manchurian economy [24]. To protect the railway and Japanese enterprises, Tokyo dispatched the Kwantung Army to Manchuria.

Known as Mantetsu, the SMR was to become Japan's largest corporation, operating one of the most modern transportation systems in the world. It set up shipping services, hotels, scientific laboratories, agricultural research stations, experimental farms, and scientific and technical programmes in many disciplines. Other activities included espionage and collaboration with the Imperial Japanese Army.

New railway routes, notably the Hailung line, served and enabled expansion of agricultural districts. Other lines enabled the exploitation of mineral resources, particularly those under the control of the SMR, none more so than the vast open cast Fushun coal complex, the largest in Asia. At Fushun the railway introduced beehive coke ovens in 1914, and the first of at least three Mond gas units, which were fed with waste coal, in April 1916 (Figs. 15.6, 15.7, 15.8 and 15.9). Production of ammonium sulphate fertilizer from the first Mond gas unit at Fushun was well over 5000 tons annually. The second Mond gas unit was installed in 1917, and the third in the early 1920s. The outcome was that the SMR, through its Fushun mine complex, owned the largest Mond gas facility in the world. Extensive underground coal mining was also undertaken. Electricity generated from dynamos driven by

Fig. 15.7 Postcard showing general view of a section of the Fushun coal mine, the largest open cast mine in Asia. (Author's collection)

Fig. 15.8 Postcard showing Mond producer gas plant, Fushun. (Author's collection)

Fig. 15.9 Postcard showing Oyama pit, Fushun. (Author's collection)

Mond producer gas powered an electric railway used for the haulage of sand required in a coal extraction process, known as the sand flushing system, used in subsurface mining. In November 1916, Denka of Tokyo, in collaboration with the SMR, began operating a carbide works at Fushun. There it produced calcium cyanamide, which, via ammonia, was converted into ammonium sulphate. The output of ammonium sulphate in 1917 was around 5,000 tons. This facility closed down in June 1920, and was sold to the SMR.

In addition to owning the Fushun mines the SMR controlled the new steel works at Anshan, founded in 1916 (Showa Steel Works, present-day Anshan Steel Works).

Manchuria and the SMR were described in glowing terms by the American propagandist for Japan and its Asian empire, Henry Kinney. Kinney's son, Charles Bishop Kinney, played the same role as his father.

The late 1920s saw much research focused on expanding the chemical industry at Fushun and in Manchuria. The Imperial Japanese Navy set up an experimental apparatus to investigate high-pressure coal liquefaction at Dairen in 1928. During 1928–1929, the SMR, jointly with Mitsui and Mitsubishi, investigated the Mont Cenis ammonia process for use at Fushun. Plant for the distillation of shale oil was installed at Fushun, with plans to produce 18,000 tons of ammonium sulphate annually, in addition to fuel oil and paraffin [25]. These and similar activities would soon become part of an increasingly chemical autarky programme.

On 18 September 1931, officers of the Japanese military (Kantogun) assigned to protect the SMR detonated a charge of dynamite on a short section of rail in an effort to cast the blame on Chinese soldiers at a nearby garrison for an act of sabotage. The actual damage was minimal, but this Mukden, or Manchurian Incident, provoked tensions and became the pretext for military action. Within days Japanese troops, moving along the main line of the SMR, captured almost every city on its route, as well as the connection with Korea. Though starting with an act of insubordination, these events led to Manchuria becoming a puppet state of Japan in March 1932. It was renamed Manchukuo, with its capital at Changchun, called Hsin-King by the Japanese (Fig. 15.10). The aggressive foreign policy set off a backlash. In 1933, the League of Nations censured Japan for its seizure of Manchuria. Japan withdrew from the league [26]. In 1936, following growing tensions with Japan, the Republic of China moved its research institutes in Beijing to the south.

Despite sanctions and loss of Western financial investment as a result of the intervention in Manchuria steps were taken to promote economic stability in Japan. Freed of foreign influences, Imperial Japan went its own way in following a new nationalist, and isolationist, agenda. The occupied territories of Korea and Manchuria provided vast new living spaces for Japanese citizens who were encouraged to move to both areas and serve the empire's two major industrial colonies. Increasingly, military authorities became involved in land seizures and settlement expansions, as well as providing assistance to industrial behemoths, most notably Noguchi's enterprise.

Japan expanded the SMR, coal mines, and chemical and steel industries, and built massive dams. In 1933, Manchuria Chemical Industry Co., Ltd, at Dairen was established, and two years later synthetic ammonia production began there using the Mont Cenis process. As elsewhere the ammonia was converted into ammonium sulphate. Following a failed military coup in Tokyo early in 1936, the Imperial Army gained control over national politics. Remote from its former allies it may have been, but Japan was not without new alliances, most notably with National Socialist Germany.

From the mid-1930s, Japan, increasingly under military control, engaged in tremendous expansion in production of war-related chemicals and heavy industrial products. The growing sphere of influence in East Asia encouraged and enabled autarky programmes that backed up military power. Production of ammonium sulphate and cyanamide continued to grow [27, 28]. In 1936, high-pressure coal-to-oil conversion was taken up at Fushun by the SMR in collaboration with the navy [29]. This, however, did not meet with success.

The Manchuria Heavy Industrial Development Corporation, under the control of Nissan, was created in 1937 to oversee growth in military industries. On 7 July 1937, the first shots of the Sino-Japanese War were fired as Japanese troops advanced on the Chinese capital Nanking, or Nanjing (following the Marco Polo Bridge incident). The Japanese Ministry of Commerce and Industry gave instructions that factories making nitric acid were required to operate at maximum capacity [30]. The Chinese republican government relocated to the less fertile provinces

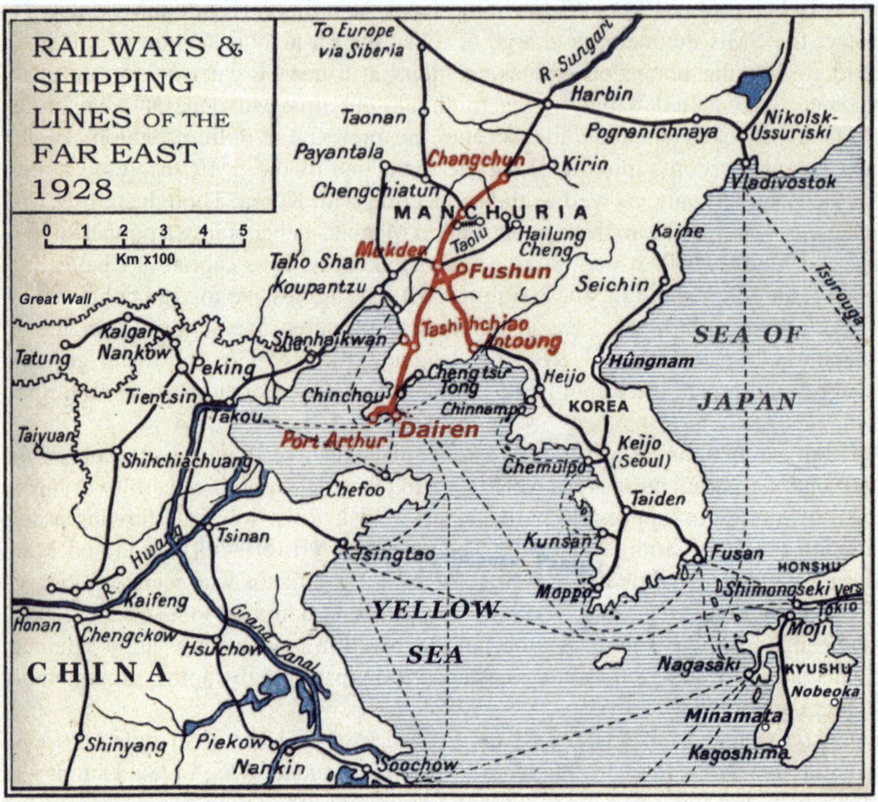

Fig. 15.10 South-east Asia, showing the northern Republic of China (including Manchuria), Korea, and Japan's main southern island, Kyūshū, with principal shipping and railway lines. Nitrogen products manufactured in this region were: cyanamide (Minamata, from 1909–1910; Kagami, northern Kyūshū, 1914; Fushun, 1916; near Hûngnam, mid-1930s); synthetic ammonia (Casale process, Nobeoka, 1923, Minamata, 1926, Hûngnam, 1930; Mont Cenis process, Dairen, 1935; Nitrogen Engineering Corporation process, Nanking, 1937; ammonium sulphate (Mond gas process, Fushun, 1916–1920; and distillation of shale oil, Fushun, 1927–1928). The main routes of the South Manchuria Railway are shown in red. The route from Dalian (Japanese, Dairen), and the nearby Port Arthur (originally Lushun; Japanese, Ryojun)—close to the southern tip of the Liaodong Peninsula—to Changchun (Japanese, Hsin-King), via Mukden (Shenyang), was 700 miles in length. In the north it linked with Harbin, through a branch of the Chinese Eastern Railway, on the Trans-Siberian line to Vladivostok. (Adapted, with additions, including Nitchitsu's Hûngnam connection, by Nick Lera, from Henry W. Kinney, *La Manchourie Moderne et la Compagnie du Chemin de Fer Sud-Mandchourien*. Paris, 1928)

in the west. Imperial Japan's invasion of China, and other nations, in pursuit of its "Greater East Asian Co-prosperity Sphere," came to an end in 1945 [31].

There was one other synthetic ammonia facility in Asia, in the Republic of China. In late 1930 the government, in part to counter fear of Japanese aggression, drew up a programme of industrialization in which manufacture of ammonium sulphate fertilizer was a priority project, administered by a nitrogen fixation committee. This body began negotiations with ICI, and soon after with ICI and IG Farben, acting together. China was an important market for both corporations; their joint interests were through the DEN group. Among visitors to China in connection with the proposed ammonia facility were Johannes Fahrenhorst of IG Farben (January 1932), and George Pollitt and Harry McGowan of ICI (in January and October 1933, respectively). Meantime, Frederick Pope of Chemico offered the Nitrogen Engineering Corporation ammonia process (which had the advantage of being outside the control of the European cartel), and the ammonium sulphate venture attracted the interest of the local entrepreneur Fan Xudong (1883–1945), who had graduated in chemistry at Kyoto Imperial University. After negotiations with the Europeans collapsed, Fan in 1934 established the Yongli Chemical Industry Company Ammonia Factory near Nanking, at first undertaking trials with the imported Pacific Nitrogen ammonia equipment. In May 1937, Yongli Chemical, with government assistance, began full-scale production using Nitrogen Engineering Corporation plant. The Yongli Chemical facility was forced to close down within a few months as a result of the Japanese invasion [32]. Prior to the founding of the Peoples Republic of China in 1949, Nanking and Dairen represented the total synthetic ammonia resource of this most populous of nations. Four decades later, China was the world's leading producer of synthetic ammonia.

References

1. Kikuchi Y (2000) Redefining academic chemistry: Joji Sakurai and the introduction of physical chemistry into Meiji Japan. Historia Scientiarum 9:215–256.
2. Tamaru Oyama H (2015) Setsuro Tamaru and Fritz Haber: links between Japan and Germany in science and technology. Chem Rec 15(2)(April):535–549.
3. Kudo A (2000) Dominance through cooperation: I.G. Farben's Japan strategy. In: Lesch JE (ed), The German chemical industry in the twentieth century. Kluwer, Dordrecht, pp 243–283, on 271–276.
4. Molony B (1990) Technology and investment: the prewar Japanese chemical industry. Council on East Asian Studies, Harvard University. Distributed by Harvard University Press, Cambridge, MA, pp 297–298.
5. Molony B (1990) Technology and investment: the prewar Japanese chemical industry. Council on East Asian Studies, Harvard University. Distributed by Harvard University Press, Cambridge, MA, pp 268–269, 297–298.
6. Nitrogen fixation in Japan (1927) J Soc Chem Ind 46(38)(23 September):860 (Chem Ind London).
7. Molony B (1990) Technology and investment: the prewar Japanese chemical industry. Council on East Asian Studies, Harvard University. Distributed by Harvard University Press, Cambridge, MA, pp 149–150.
8. Molony B (1990) Technology and investment: the prewar Japanese chemical industry. Council on East Asian Studies, Harvard University. Distributed by Harvard University Press, Cambridge, MA, pp 126–130.
9. Molony B (1990) Technology and investment: the prewar Japanese chemical industry. Council on East Asian Studies, Harvard University. Distributed by Harvard University Press, Cambridge, MA, p 136.

10. Ammonium sulphate in Japan (1927) J Soc Chem Ind 46(6)(11 February):123 (Chem Ind London).
11. Soya beans (1932) Memorandum. Institute of Pacific Relations, American Council 1(7) (28 April):1–3.
12. Molony B (1990) Technology and investment: the prewar Japanese chemical industry. Council on East Asian Studies, Harvard University. Distributed by Harvard University Press, Cambridge, MA, pp 168–169.
13. Molony B (1990) Technology and investment: the prewar Japanese chemical industry. Council on East Asian Studies, Harvard University. Distributed by Harvard University Press, Cambridge, MA, pp 288–293.
14. Suzuki T (1998) The foundation and amalgamation of Miike Nitrogen Industries Inc. and Toyo Koatsu Industries, Inc. Japanese Yearbook of Business History 4:103–125.
15. Mork RK, Nakamura M (2005) A frog in the well knows nothing of the ocean: a history of corporate ownership in Japan. In: Mork RK (ed), A history of corporate governance around the world: family business groups to professional managers. University of Chicago Press, Chicago, pp 367–466, esp. 410.
16. Molony B (1990) Technology and investment: the prewar Japanese chemical industry. Council on East Asian Studies, Harvard University. Distributed by Harvard University Press, Cambridge, MA, pp 272–275.
17. Molony B (1990) Technology and investment: the prewar Japanese chemical industry. Council on East Asian Studies, Harvard University. Distributed by Harvard University Press, Cambridge, MA, p 300.
18. Molony B (1990) Technology and investment: the prewar Japanese chemical industry. Council on East Asian Studies, Harvard University. Distributed by Harvard University Press, Cambridge, MA, pp 275–281.
19. Chemical industry in Germany (1929) J Soc Chem Ind 48(26)(28 June):650–652 (Chem Ind London).
20. Kudo A (1998) Japanese-German business relations, cooperation and rivalry in the inter-war period. Routledge, London, pp 110–129.
21. Moore AS (2013) Constructing East Asia: technology, ideology, and empire in Japan's wartime era, 1931–1945. Stanford Univerity Press, Stanford.
22. Matsusaka YT (2001) The making of Japanese Manchuria, 1904–1932. Harvard University Press, Cambridge MA.
23. Ferrin AW (1917) Development of the South Manchurian Railway. Commerce Reports, US Department of Commerce, no. 231(3 October) pp 42–45.
24. Matsusaka YT (1993) Japanese imperialism and the South Manchuria Railway Company, 1904–1914. PhD thesis, Harvard University (May), pp 161, 190.
25. Kinney AR (1962) Japanese investment in Manchurian manufacturing, mining, transportation and communications, 1931–1945. Garland, New York.
26. Schumpeter EB (ed) (1940) The industrialization of Japan and Manchukuo, 1930–1940: population, raw materials, and industry. Macmillan, New York.
27. Manufacture of ammonium sulphate in Japan (1936) J Soc Chem Ind 55(4)(24 January):83 (Chem Ind London).
28. Increased Japanese production of chemicals (1936) J Soc Chem Ind 55(13)(27 March):263 (Chem Ind London).
29. Industrial notes from abroad. Japan (1936) J Soc Chem Ind 55(7)(14 February):131 (Chem Ind London).
30. Production of nitric acid in Japan (1937) J Soc Chem Ind 56(45)(6 November):1013 (Chem Ind London).
31. Young L (1998) Japan's total empire: Manchuria and the culture of wartime imperialism. University of California Press, Berkeley.
32. Kwan MB (2017) Patriots' game: Yongli Chemical Industries, 1917–1953. Brill, Leiden, pp 77–105.

Chapter 16
High-Pressure Synthesis and Later Developments

16.1 High-Pressure Catalytic Circulatory Plants

> The commercial development of high pressure synthesis has been largely 'hydrogenation,' whether of nitrogen to produce ammonia, or of carbon monoxide to produce methanol, or of various hydrocarbons to produce more valuable products. In this latter field, by high pressure synthesis it is now possible for the first time to change or control the ratio of hydrogen atoms to carbon atoms in any hydrocarbon, and even to change aliphatic compounds to naphthenic and aromatic bodies.
>
> Charles O. Brown, "High Pressure Synthesis—Basis of New Chemical Engineering Industries," in *Twenty-five Years of Chemical Engineering Progress. Silver Anniversary Volume. American Institute of Chemical Engineers*, ed. Sidney D. Kirkpatrick. New York: Published by the institute and for sale by D. Van Nostrand Company, Inc., 1933, pp. 152–168, on p. 153.

Major developments in high-pressure "hydrogenation" chemistry after 1914 were conditional on the work achieved by Carl Bosch and his team in the ammonia synthesis. After 1918, BASF continued to investigate high-pressure technology, mainly at first related to improvement of the Haber-Bosch process. In April 1918, the BASF Ammonia Laboratory had been moved to a new building on the Oppau site; a decade later it was the largest industrial research laboratory in Germany, and probably the largest in the world [1].

Following massive international growth in the output of synthetic ammonia, propelled by the perceived need to prove the worth of the new high-pressure technology, and also by over-optimistic forecasts of demand for fertilizers, and national security concerns, there was, as we have seen, a substantial glut by the late 1920s. The situation did not improve for many years, notwithstanding formation of the international nitrogen cartel. As a result, by the early 1930s, many of the chemists at IG Farben's Oppau Ammonia Laboratory had been transferred to other departments, including those dealing with new high-pressure processes. By then, high-pressure synthesis had taken on new directions, though converters and associated equipment for the production of hydrogen remained much the same. This meant that converters could be changed from one process to another, according to demand.

A.S. Travis, *Nitrogen Capture*, https://doi.org/10.1007/978-3-319-68963-0_16

16.2 Methanol

In 1923, BASF introduced a high-pressure process for methanol. The reaction, between carbon monoxide and hydrogen, had been suggested in 1905 by organic chemist Paul Sabatier (1854–1941), though he was unable to put it into practice through lack of a means for determining the equilibrium.[1]

Research at BASF began in 1914 when Mittasch and Christian Schneider attempted to obtain a single product from the reaction, undertaken in the presence of a catalyst. As in the ammonia synthesis, thermodynamics and reaction kinetics showed that the reaction was encouraged by using a high pressure.

$$\begin{array}{ll} 2\ H_2 + CO \rightarrow CH_3OH & \Delta H = -91 \text{kJ mol}^{-1} \\ 3 \text{ volumes} \qquad 1 \text{ volume} & \end{array}$$

The two BASF chemists failed to obtain a satisfactory product, though they filed a patent for the method. The same reaction was investigated by Colonel Georges Patart, the French Minister of Explosives, who had been responsible for negotiations that led Louis Loucheur, acting for the French government, to acquire rights to Haber-Bosch technology. Patart filed a methanol patent in 1921, perhaps after learning of the studies by Mittasch and Schneider. To preempt rivals, Patart laid claim to a wide range of metal catalysts, oxides and salts included, that might conceivably favour oxidation and hydrogenation. This stimulated a response from BASF, which in 1922 assigned the problem of high-pressure methanol synthesis to recently appointed chemist Matthias Pier (1882–1965), who had studied under Wilhelm Röntgen (1845–1923) in Munich and Nernst in Berlin; he was awarded the PhD in 1908. In 1914, Pier, while working at the Centre for Technical-Scientific Research (Zentralstelle für wissenschaftlich-technische Untersuchungen GmbH) in Neubabelsberg, Brandenberg, had filed patents for an ammonia converter with an inner reaction vessel of low-carbon iron. In 1920, he joined BASF.

Pier achieved almost immediate success with the synthesis of methanol, working at 300 atmospheres and around 700 °C, using a chromium oxide-zinc oxide catalyst. A pilot plant began operating during May 1922. On 26 September 1923, the first railway tank wagon of synthetic methanol left the Merseburg works (Fig. 16.1). In 1927, the output was 18,000 tons. The success of BASF led to litigation over priority with Patart. The court ruled in favour of BASF [2–4]. ICI at Billingham began producing synthetic methanol in 1928–1929. The ICI process would remain in use for four decades, after which it was superseded by a low-temperature, low-pressure process, using a novel long-lasting catalyst. In common with the ammonia synthesis, on which they were closely based, and even made in the same converters, synthetic methanol and urea became available as the result of developments in hydrogenations carried out in high-pressure catalytic circulatory plants [5].

---

[1]From around 1900, Sabatier hydrogenated organic compounds in the presence of catalysts, a method that led to industrial processes for hydrogenating oils and fats, and in turn demand for hydrogen. However, the reaction conditions were far removed from those used in the synthesis of ammonia.

Fig. 16.1 The first tank wagon filled with synthetic methanol ready to leave Merseburg, 26 September 1923. The locomotive is of the fireless type. (BASF Unternehmensarchiv)

16.3 Hydrogenation: Coal to Oil

In terms of diversification, most relevant to the time period covered here was high-pressure research conducted in Germany during World War I aimed at the conversion of coal to oil, and hydrogenation in general [6]. This was carried out by Friedrich Bergius (1884–1949), who received his doctorate at Leipzig in 1907, and then worked with physical chemists Nernst, Haber, and Bodenstein, the latter at the Technische Hochschule Hanover, where in 1909 Bergius was appointed Privatdozent and began his research into high-pressure hydrogenation. He produced hydrogen by reacting carbon and water at a reduced temperature under pressure; this gave a supply of the pure gas, and was introduced at Chemische Fabrik AG, of Posen. Bergius speculated that coal might combine with hydrogen under high pressure to form oil, and was proved correct. He found metal oxide catalysts that improved the yield, and in 1913 filed patents for his method.

On 1 January 1914, at the instigation of chemical manufacturer Karl Goldschmidt (1857–1926), Bergius joined Th. Goldschmidt AG, of Essen, in a newly constructed laboratory at Mannheim-Rheinau, where he was appointed research leader. The hydrogenation of coal to oil received priority once the war commenced, but it encountered major technical difficulties at the pilot plant stage and consumed considerable capital. Unlike the ammonia synthesis, this process involved the reaction of hydrogen with a solid material. Bergius joined the board of Goldschmidt in 1916, and in 1918 brought in partners when he attempted to establish a consortium for coal chemistry. In 1919, he left the firm, having lost the backing of Karl Goldschmidt. In 1922, Bergius set up a coal hydrogenation plant at Rheinau, operating at 100 atmospheres and 400 °C. It produced one ton of synthetic oil, containing 25% motor fuel [7].

16.3.1 Leunabenzin

Bergius soon after began to face financial difficulties, and sold his patents to BASF/IG Farben, where Carl Bosch enthusiastically backed the conversion of coal to oil. Industrial-scale production of synthetic fuel (Leunabenzin, Leuna petrol) was first achieved at Merseburg/Leuna in 1927, mainly through the efforts of Mathias Pier, then the BASF expert on hydrogenation of coal (Fig. 16.2).

ICI had in 1926 attempted to develop its own process, for which Herbert Humphrey advised the executive committee, "speaking generally, no expense is to be spared in the prosecution of this work" [8]. However, this did not succeed. In 1927, following failure to develop a coal-to-oil process, ICI acquired British Empire rights to the Bergius process and in 1929 erected a pilot plant at Billingham for treating ten tons of coal a day. Scale-up followed in the early 1930s. The group leader was Kenneth Gordon (1897–1955).

The coal-to-oil process also interested Standard Oil of New Jersey. As a result, the Standard-IG Company was created in 1929, followed in 1931 by the International Hydrogenation Patents Company, through which patents were pooled, and which included Britain's ICI and Royal Dutch-Shell.

To encourage use of synthetic petrol, on 17 July 1933, the British Prime Minister Ramsay MacDonald announced in the House of Commons the introduction of a preference on the price of home-produced petrol through what would become the Hydrocarbon Oils Production Bill. Full-scale production at ICI began in 1934 (Figs. 16.3 and 16.4). Ramsay McDonald attended the official opening of the plant on 15 October 1935. Hydrogen was made by a coke oven process. The converters were forged from 150-ton ingots of nickel-chrome steel, as used in ammonia converters. Though the ICI product was completely uncompetitive in peacetime, it served as a valuable high-octane aviation fuel, as required by the RAF. Prior to the outbreak of World War II, ICI's fuel was in use with seven squadrons. Subsequently creosote was adopted instead of coal as raw material [9–13].

The falling price of petroleum, following the discovery of vast deposits of oil in Texas (1930) and the Near East (late 1930s) meant that production of synthetic motor fuel in Europe was hardly viable. That situation had changed in Germany when the National Socialist autarky programmes were implemented. By 1936, Leuna was producing over a quarter-of-a-million tons of motor fuel each year. In 1938, total German production exceeded 600,000 tons. During World War II, this output was tripled. Leuna produced a high-octane aviation fuel for the Luftwaffe.

At the end of World War II, high-pressure converters and other equipment at the Leunawerke were dismantled by the Red Army for transfer to the Soviet Union. The facility, rebuilt under ownership of the German Democratic Republic (GDR), was named after Walter Ulbricht (1893–1973), who had participated in formation of the Weimar-era German Communist Party and was a prominent politician in the GDR.

Fig. 16.2 Converter for the production of synthetic motor fuel by hydrogenation of coal at the IG Farben Leunawerke, almost identical with converters used in the synthesis of ammonia and methanol, 1930s. Note the two converters at left, housed in steel. (*Erzeugnisse unserer Arbeit*. Frankfurt am Main: I.G. Farbenindustrie, 1938, opposite p. 179)

Until the end of the 1980s, Merseburg incorporated 19 single train ammonia synthesis loops. Most of what remained from the pre-war and wartime period was demolished during the 1990s.

Meantime high-pressure industrial processes had became widespread, encouraging the search for novel catalysts and processes for the conversion of heavy oils into useful fuels. Reactions included amminations, ring closures and addition reactions; and in the 1940s large-scale refining in the petroleum industry, with

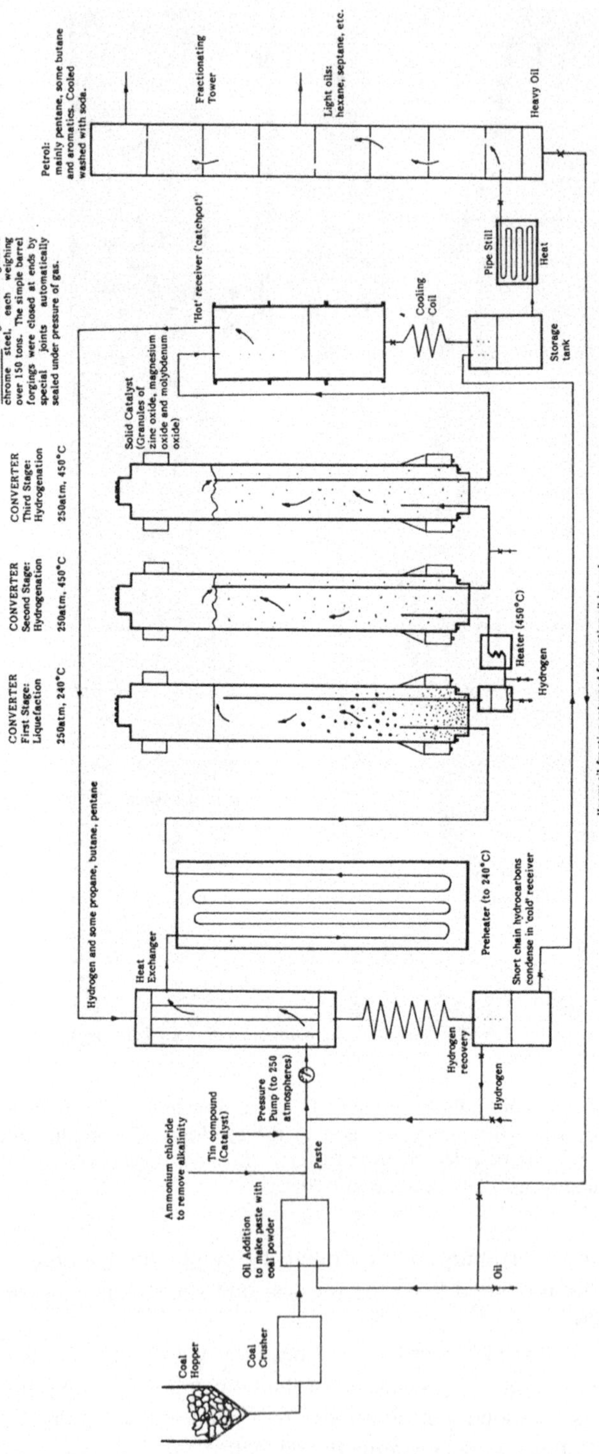

Fig. 16.3 Hydrogenation of coal, at Imperial Chemical Industries, Billingham, based on the BASF coal-to-oil (Bergius) process

Fig. 16.4 Converters for the hydrogenation of coal, mid-1930s, at Imperial Chemical Industries, Billingham. (Kenneth P. Magee/Edelstein Collection)

hydrocracking of crude oil, in which hydrogen was added to large hydrocarbon molecules to produce aviation and motor fuel. Engineering firms turned to the manufacture of high-pressure converters, compressors, boilers and autoclaves.

16.4 Polythene

The range of high-pressure investigations broadened considerably during the 1930s. The most significant outcome was the discovery of polythene, a spinoff from cutbacks in ammonia and coal-to-oil studies. In part it was a result of interest in speculative or blue sky research, as favoured by ICI research manager Francis Freeth, at the Northwich laboratories of the Alkali Group (until 1927, when he moved to London). This ICI achievement involved its dye research and high-pressure programmes, the Alkali Group, input from Du Pont, which in 1929 had begun collaboration with ICI, as a result of their exchange of know-how agreement, and an important contribution from Professor Antonius M. J. F. Michels (1889–1969) of the University of Amsterdam's high-pressure laboratory. Michels had undertaken research into the discovery of novel dye intermediates, working at over 1000 atmospheres. Following the visit of an ICI delegation to Amsterdam in 1928, Michels was provided with research funding and took on the training of two new ICI physical chemists: Reginald Oswald Gibson (with the Alkali Group from September 1926), who, through Frederick Donnan, joined the Kamerlingh Onnes Laboratory, in Leiden, in 1928 and then assisted Michels under whom he received his PhD in 1933; and Michael Willcox Perrin (1905–1988; knighted 1967), from 1929, the year in which he joined ICI [14–16].

Also joining ICI in 1929 was organic chemist Eric William Fawcett. This was at the time when coal-to-oil conversion was receiving attention at the Alkali Group's laboratories. Fawcett was sent to the United States in 1931 to investigate the constitution of petroleum at the National Bureau of Standards. While there he became interested in polymers through contact with Du Pont. By the time of his return to the Alkali Group in 1932, ammonia and even coal-to-oil research were no longer priorities. Fawcett was assigned to work with Gibson at Northwich on high-pressure studies that might be of value to the Dyestuffs Division. The apparatus was designed by research engineer W. R. Dermot Manning. With the aid of Michels, ICI at Northwich built a laboratory autoclave for working at 12,000 atmospheres.

On 27 March 1933, Fawcett and Gibson observed that the high-pressure reaction between ethylene and benzaldehyde gave a white, waxy solid that lined the inner wall of the reactor. This was achieved at just under 2000 atmospheres and 170 °C. The experiment was repeated several times. Sometimes the waxy solid was obtained; at other times a black deposit. In June they reported that the waxy solid, "appears of interest as a possible source of long chain compounds" [17]. The yield, however, was small and the decomposition reaction hardly suggested that the method could be easily controlled. From December 1935, the reaction was repeated in an improved apparatus. (A year earlier, ICI started manufacture of ethylene, by dehydration of alcohol, from fermentation of molasses, at Huddersfield.) Sufficient waxy product became available for analysis, which confirmed that it was a long chain polymer of ethylene. The material was readily moulded and cast, and showed excellent electrical insulating properties, as well as being inert. In 1936, its formation was found to have been the result of the presence of a small amount of oxygen acting as catalytic initiator.

Manufacture of the polymer, polythene (polyethylene) on a pilot plant scale began in 1938. Its properties suggested use as an insulated covering for underwater cables, as was demonstrated by a mile-long connection between the Isle of Wight and the mainland. Full-scale production began in September 1939. The novel material was soon used in high frequency radio transmitters and receivers required in radar, since it absorbed far less radio energy than existing dielectrics [18]. It also heralded a new era of diversification, as had been the case with synthetic ammonia, away from the traditional alkali product, soda ash, pioneered by Ludwig Mond.

16.5 Acetylene Under Pressure

The widespread availability of calcium carbide, as a result of the manufacture of calcium cyanamide, as well as its conversion into acetylene in welding and lighting, provides another example of an important spinoff from the development of high-pressure chemistry. After World War I, acetylene became a starting point for novel chemical reactions. As part of the interest, an alternative synthesis of acetylene was investigated in the 1920s at Oppau. This involved subjecting methane to an electric arc in a Schönherr furnace. The yield was just 14%. The unreacted methane was recycled to the arc. In the 1930s the process was used in the United States for the production of acetic acid.

The principal chemist in industrial high-pressure acetylene chemistry was Walter Reppe (1892–1969), who joined BASF in 1921. Later in the decade, under research director Curt Schumann, he was a junior member of the team that converted acetylene into the solvent butanol. He also worked on the manufacture of ethylene oxide. This started with the reaction between ethylene and chlorine. The resulting ethylene chlorohydrin was heated with lime to afford the ethylene oxide, later an important intermediate. At the same time, Reppe began work on synthetic Buna rubber. This required a viable route to the monomer butadiene. Following Hitler's rise to power, synthetic rubber (and the associated butadiene research) became a strategic priority. This was especially so after inauguration of the National Socialist Second Four-Year Plan in 1936 [19].

Reppe, from 1934 head of the new Ludwigshafen Intermediates and Plastics Laboratory, decided to develop a technical route to butadiene, starting with acetylene, through the application of high pressures, work he had started in the laboratory in 1927. Subjecting acetylene to high pressures had previously been considered far too dangerous. Acetylene, highly reactive as a result of the triple-bond between its two carbon atoms, under pressure is inherently unstable and explodes with great violence. ICI had undertaken such studies around 1930, but because of the enormous danger worked on a small scale. That did not deter Reppe. In September 1937 he applied extreme conditions, application of pressure to a mixture of acetylene and formaldehyde in the presence of the percussive copper acetylide as catalyst, to produce 2-butyne-1,4-diol (butynediol), which retained acetylene's triple bond. The butynediol was converted into 1,4-butylene glycol, which in turn gave butadiene. This three-stage process faced several difficulties, at least until the end of 1939, when Georg Niemann made substantial improvements. Then there were daunting technical hurdles to overcome. Engineers and chemists fixed the problems in much the same way that they had overcome difficulties with ammonia converters before, during and after World War I. The successful large-scale production of Buna rubber based on acetylene subsequently led to its use in the manufacture of other important organic chemicals. This in turn stimulated production of more carbide. By 1943, IG Farben's Schkopau works was the largest carbide producer in Germany. However, at that stage Reppe chemistry had not been fully developed. Some of Reppe's reactions, conducted under pressure in the presence of a nickel carbonyl catalyst in the late 1930s, were introduced commercially after 1945, including in the United States [20].

16.6 The New Ammonia Technologies

The ammonia industry during the late 1930s saw few new developments. Expansion from the end of the decade, when more than 90% of ammonia factories used non-BASF technology, was in part related to war needs. Only in the production of hydrogen were there innovations. Ludwigshafen, for example, developed partial oxidation of hydrocarbons.

It took a quarter of a century before, in the mid-1960s, the ammonia industry in western Europe, the United States and Japan underwent major technical improvements. Until then the synthesis loops, the individual trains, relied on reciprocating compressors. While reciprocating compressors are efficient, they are complicated and require frequent maintenance, and the amount of gas that they can handle is limited. ICI, for example, had a worker on hand to maintain each compressor. The improvements were based on the introduction of a single centrifugal compressor, providing all pressure requirements, such that the plant functioned on the basis of a single train. The centrifugal compressor is simple and operates at a very high speed, often driven by a steam turbine, using waste heat available from plant operations.

M. W. Kellogg and Chemico in the United States introduced single train energy-integrated plants, replacing the process technology that had changed little since around 1930. The first plant, based on Kellog innovations, was built for the American Oil Company at Texas City and opened in December 1963. Ammonia Casale (Casale SA since 2014), Uhde (ThyssenKruppUhde; from 2014 ThyssenKrupp Industrial Solutions), C. F. Braun, ICI, Toyo Engineering, Chiyoda (both of Japan), and Haldor Topsøe (of Lyngby, Denmark) were among the other firms that introduced advanced technology systems. By the late 1970s, centrifugal compressors were employed in plants with outputs of over 1200 tons of ammonia per day. (Below a capacity of 600 tons the use of a centrifugal compressor is not energy efficient, which is why some modern smaller plants continue to use reciprocating compressors.) Ammonia converters increased in size, often exceeding 20 metres in height [21, 22]. Thermal efficiencies increased threefold, and energy input dropped by almost two-thirds. Casale SA continues to innovate, and to upgrade existing ammonia plant [23].

Coke oven and water gas plants for hydrogen were on the way out in many parts of the world. In the early 1950s natural gas rapidly replaced coal as the source of hydrogen, particularly in the United States, where over 80% of ammonia was made from this source, and in Italy [24, 25]. In the 1960s, ICI developed a catalyst that enabled a move at Billingham from water gas to steam-naphtha as the source of hydrogen, and then, by 1970, to natural gas. In 1983, ICI announced the introduction of its AMV process, incorporating a more efficient catalyst, operating at a lower pressure, between 70 and 80 atmospheres, and requiring considerably less energy. In Britain at that time, ICI operated ammonia plants at Billingham, Wilton, Immingham, and Avonmouth, and Shell at Ince Marshes, on the River Mersey. Today, hydrogen is mainly prepared by partial steam reforming or partial oxidation of hydrocarbons.

In the mid-1970s, new coal-gasification processes for hydrogen were developed in response to the rising cost of petroleum. A number of countries, starting with Yugoslavia in 1953, had already introduced the direct gasification of coal.[2]

Since the 1980s, reflecting the changing geography of the nitrogen industry, national infrastructure programmes involving ammonia production have provided the most significant growth, mainly in Asia. China, the world's leading manufacturer of ammonia since the early 1990s, accounts for one-third of international production, with an annual capacity of around 67 million tons in 2015. It is the only nation that undertakes the large-scale production of ammonia based on coal (79% in 2011; with 20% from natural gas). The manufacture of synthetic ammonia continues to represent an important part of the global chemical industry, with around 80% consumed by the agricultural sector. More recent developments have focused on reducing power requirements.

In the 1930s there was a gradual changeover from ammonium sulphate to ammonium nitrate, which offered a higher nitrogen content as fertilizer. Today, three products represent the mainstay of the ammonia-based fertilizer industry: granular ammonium nitrate, liquid anhydrous ammonia, and urea [26]. Ammonium nitrate (Nitram) fertilizer is favoured in Europe. The nitrate, made from hot ammonia and nitric acid, after removal of water, is sprayed in a tower, where small lumps are formed, called prills. According to density, prills are used as fertilizer or in the manufacture of explosives. For fertilizer, particularly outside Europe, urea has come into favour, due also to its high nitrogen content. This enables more careful application, with less damage to the environment.

16.7 Successors to the Electric Arc and Cyanamide Firms

During the twentieth century many of the companies met with in this study emerged as major corporations through involvement in the nitrogen industry. Some already existed in the nineteenth century, based on other sectors, such as alkalis and synthetic dyes, and diversified or moved to nitrogen as new business strategies evolved based on prior technical know-how. After 1945, corporations in Germany, Italy and Japan were broken up, though they reappeared in various guises from the

[2]Hydrocarbon feedstocks must be carefully purified prior to reforming. In the case of methane, the catalyst poison sulphur is reacted with hydrogen over a catalyst. The resulting hydrogen sulphide is then converted, by reaction with zinc oxide, to zinc sulphide, which is readily removed. Next is the process of primary steam reforming over a nickel catalyst in tubes held in a furnace, to give hydrogen and carbon monoxide. This is followed by secondary reforming, when air is injected into the gas mixture, again over a nickel catalyst. The nitrogen-hydrogen composition is made up in the proportions required for ammonia synthesis. Steam reforms any remaining methane. The mixture then consists of hydrogen, nitrogen, carbon dioxide, and carbon monoxide. As in the earlier processes, the shift reaction converts the monoxide to dioxide. Once oxides of carbon are removed the synthesis gas is ready for conversion into ammonia.

1950s. At the end of the century, new directions, mergers, acquisitions and divestures, and political changes, in addition to movement of much of the chemical industry from the West, led to further redirection, though building on the success of nitrogen capture. The histories of better-known corporations, such as BASF, IG Farben, ICI, Solvay, Montecatini, DSM (since 2010, OCI Nitrogen), American Cyanamid, and Du Pont, have been recorded in detail. Here we briefly outline the subsequent fates of other but generally lesser known corporations involved in the nitrogen story.

Norsk Hydro, associated with the original Birkeland-Eyde electric arc enterprise, in 2004 spun off its fertilizer division, which from around 1930 was mainly associated with high-pressure ammonia synthesis, and today specializes in metals and energy. The fertilizer division was acquired by Yara International, a leading producer of synthetic ammonia [27]. Yara's Sluiskil facility (successor to the Dutch Fauser site) has Europe's largest ammonia and nitric acid capacity.

The manufacture of calcium cyanamide continues to the present-day, particularly in Germany, where the remaining facilities associated with the Frank-Caro process have undergone several changes of ownership. In 1939, Bayerische Kraftwerke AG was merged with Bayerische Stickstoff-Werke AG to create Süddeutsche Kalkstickoff-Werke AG (SKW). The state held 70% of the shares, and IG Farben 30%. From 1978 this entity was known as SKW Trostberg AG, which in February 2001 merged with Degussa-Hüls AG, and subsequently was owned by Degussa/Evonik Industries, which in turn in 2011 sold SKW Trostberg to AlzChem, today one of the firms located at the Trostberg Chemical Park. Evonik is a leading manufacturer of specialty chemicals. SKW Stickstoffwerke Piesteritz GmbH, successor to the Bayerische operated works at Piesteritz, is the main German manufacturer of ammonia and urea. The Vereinigte Industrie-Unternehmungen AG eventually reinvented itself as a major energy company, since 2000 known as E.ON.

American Cyanamid, apart from its entry into chemical engineering and synthetic ammonia, through its subsidiary Chemical Construction, diversified into organic chemistry, following acquisition of the Calco Chemical Company of Bound Brook, New Jersey in 1929, and pharmacuticals, through purchase of Lederle and Davis & Geck, both of New York, in 1930. Studies on dicyanamide had been undertaken by Caro and colleagues shortly after 1900. At American Cyanamid at the end of the 1930s it became important in the manufacture of melamine resins. Calcium cyanamide found uses as an intermediate in the synthesis of sulpha drugs, acrylonitrile, etc. American Cyanamid moved into polymers, agrochemicals and increasingly the life sciences and pharmaceutical products. In 1993–1994, what remained was split up, chemicals becoming Cytec Industries; while the rest was acquired by American Home Products (later Wyeth Corporation) [28].

Though calcium cyanamide was written off by Charles Parsons and others in the early 1920s, production increased during the following years, though its overall share of the fertilizer market declined. In the mid-1930s its agricultural applications were extended to use as an effective defoliant, based on herbicidal properties and, before planting of crops, as a seed-bed sterilizer. There was still considerable demand until relatively late in the twentieth century, partly arising from cyanamide's formulation for use as a weed killer. It remained popular in Europe until well into the 1950s, where it still represented 20% of nitrogen fertilizer (as compared with 2% in the United States). Worldwide, calcium cyanamide production peaked during 1950–1970; AlzChem at Trostberg continues its manufacture for agricultural use, for example in Turkey. Its action and advantages in application were usefully summarized by Williams Haynes in the 1950s [29]:

> Cyanamide acts in the soil very like the favored waste organic materials (tankage, blood, meal and fish scrap) the supply of which is so limited and the price accordingly so high that they become less and less prominent in the fertilizer picture. Like these natural ammoniates, cyanamide is a delayed action plant food. Its nitrogen becomes available over several months thus feeding the crop throughout its growing season. Again, like the organic materials, cyanamide is an excellent 'conditioner,' that is it improves the physical condition of mixed fertilizers. The lime in cyanamide sweetens naturally acid soils and prevents the building up of soil acidity by continued use of such high-acid fertilizers of the old standby, ammonium sulphate.
>
> With such versatile chemical talents, cyanamide has inevitably been given some special agricultural jobs. . . .But it is when used 'straight' that cyanamide shines as a specialist. This began during the 1930s when the granular form was perfected which corrected the faults of dustiness and skin toxicity that in the early days won the nickname, 'devil dust.' It now became a decent material for direct application for top-dressing orchards, pastures, and haylands or for heavy applications, broadcast and plowed in, to restore or build up soil fertility.

Around the same time that Haynes penned these words, chemists in North Korea made the synthetic fibre vinalon (vinylon), based on polyvinyl acetate. Vinalon's large-scale production made use of the extensive carbide facilities at the cyanamide works associated with Hûngnam. The manufacturing plant was the first erected in North Korea without assistance from the USSR or China [30].

References

1. Reinhardt C (1998). Basic research in industry: two case studies at I.G. Farbenindustrie AG in the 1920's and 1930's. In: Travis AS, Schröter HG, Homburg E, Morris PJT (eds), Determinants in the evolution of the European chemical industry, 1900–1939: new technologies, political frameworks, markets and companies. Kluwer, Dordrecht, pp 67–88, on 81–86.
2. Synthetic methanol controversy (1925) Ind Eng Chem 17(8)(August):859.
3. Methanol developments (1925) Ind Eng Chem 17(8)(August):772–773.
4. Lormand L (1925) Industrial production of synthetic methanol. Ind Eng Chem 17(4) (April):430–432.
5. Stranges A (2000) Germany's synthetic fuel industry, 1927–1945. In: Lesch JE (ed), The German chemical industry in the twentieth century. Kluwer, Dordrecht, pp 147–216, on 173.
6. Hughes TP (1969) Technological momentum in history: hydrogenation in Germany, 1898–1933. Past and Present 44:106–132.
7. Stranges AN (1984) Friedrich Bergius and the rise of the German synthetic fuel industry. Isis 75(4):643–667.
8. H. A. Humphrey to members of the [ICI] executive committee, 18 October 1926. Papers of Herbert Alfred Humphrey, Archives, Imperial College London, item E33.
9. Egloff G (1938) Motor fuel economy of Europe. Ind Eng Chem 30(10)(October):1091–1104.

10. Stranges AN (1985) From Birmingham to Billingham: high pressure coal hydrogenation in Britain. Technology and Culture 26(4):726–757.
11. Hayes P (2001) Industry and ideology. I.G. Farben in the Nazi era. 2nd edition. Cambridge University Press, Cambridge, pp 37–44.
12. Krammer A (1978) Fueling the Third Reich. Technology and Culture 19(3):394–422.
13. Jones WI (1934) Hydrogenation of coal. J Soc Chem Ind 53(15)(13 April):321–326 (Chem Ind London).
14. Michels MAJ (2016) Antonius Michels, his high-pressure work, and the origins of polyethylene. SHAC Spring Meeting, High pressure in the interwar period, Science Museum London, 11 February 2016.
15. Gibson RO (1933) The viscosity of gases at high pressure. HJ Paris, Amsterdam, 1933. Copy with Gerald Holten Collection, letters boxes, pressure E-GO, at Sidney M. Edelstein Library for the History and Philosophy of Science, Technology and Medicine, National Library of Israel.
16. Gibson RO (1980) The father of industrial progress. Chem Ind, London, no.16 (16 August), pp 635–641.
17. Harness A (c. 1980) The story of polythene. Plastics Processing Industry Training Board, Brentford Middlesex, p 2.
18. Travis AS (1998) Modernizing industrial organic chemistry: Great Britain between two world wars. In: Travis AS, Schröter HG, Homburg E, Morris PJT (eds), Determinants in the evolution of the European chemical industry, 1900–1939: new technologies, political frameworks, markets and companies. Kluwer, Dordrecht, pp 171–198, on 190–192.
19. Morris PJT (1983) The industrial history of acetylene: the rise and fall of a chemical feedstock. Chem Ind, London, no. 18 (19 September), pp 710–715.
20. Travis AS (2007) Unintended technology transfer: acetylene chemistry in the United States. Bull Hist Chem 32(1)(2007):27–34.
21. Smil V (2001) Enriching the earth: Fritz Haber, Carl Bosch, and the transformation of world food production. MIT Press, Cambridge, MA, pp 122–127.
22. Maxwell GR (2004) Synthetic nitrogen products: a practical guide to the products and processes. Kluwer Academic/Plenum, New York.
23. Cariati V (2009) Nata due volte. L'avventura bella della Ammonia Casale. Impiantistica Italania, no. 5 (September–October), pp 35–43.
24. Belchetz L (1955) Refining processes and plant. In: Evans RB, Samuel DL, Sell G (eds), Reviews of petroleum technology, vol. XIV (covering 1952–1954). The Institute of Petroleum, London, pp 57–87, on 75.
25. Quaderni dell'ingegnere chimico italiano (1975) 11(10)(October):155–160.
26. Pattabathula V, Robertson J (2016) Introduction to ammonia production. Chem Eng Progress (September), pp 69–75.
27. http://www.yara.com/about/history/1906-1919/index.aspxn (accessed 12 August 2014).
28. Travis AS (2004) Dyes made in America. the Calco Chemical Company, American Cyanamid and the Raritan River. Edelstein Center, Jerusalem, pp 120–124.
29. Haynes W [1957] On the chemical frontier: the Cyanamid Story. Copy held at Sidney M. Edelstein Library for the History and Philosophy of Science, Technology and Medicine, National Library of Israel, pp 80–81.
30. Kim, D-W (2004) Two chemists in two Korea's. Ambix 51(1):67–84, on 79–83.

Chapter 17
Nobel Prizes and a New Technology

The nitrogen capture story and the diversification that it spawned as told in the foregoing undoubtedly represents one of the pinnacles of human achievement. Despite the tremendous amount published on the subject it needs also to be read against a more general background. With this in mind, here and in the closing chapters we return briefly to some earlier themes, though mainly in the post-1918 period, and consider the diversity of certain social, political and economic conditions that impacted on the ways in which nitrogen technologies were taken up. We start with what many would call the ultimate accolade recognizing the success of nitrogen capture, the Nobel Prize.

17.1 Fritz Haber and the Nobel Prize

Late in 1918, Fritz Haber sent his family to Switzerland. Concerned that he would be declared a war criminal, he crossed the Swiss border, disguised with a beard, and joined his family. Fearing that his assets would be seized, and that he might also be seen as a war profiteer, Haber renegotiated his original 1908 contract with BASF, reducing his royalty income from 1.5 to 0.8 pfennig per kilogram of ammonia. The difference, it was anticipated, would be made up at a later time, by mutual agreement with the company.

In the event, no charges were made against Haber (Fig. 17.1). Both sides had used toxic gases with impunity, which would have made the case against any German irrelevant. There was, however, no case for preventing the Swedish Academy of Sciences from considering Haber as a candidate for the Nobel Prize in Chemistry. His name had been put forward in 1912 and 1913, and, jointly with Carl Bosch, in 1915 and 1916, even though few details had been published regarding the Haber-Bosch process [1, 2]. When in 1919, the academy announced recipients of the 1914–1919 Nobel Prizes, Fritz Haber was included among the five German laureates. He received the 1918 chemistry prize for the ammonia synthesis.

A.S. Travis, *Nitrogen Capture*, https://doi.org/10.1007/978-3-319-68963-0_17

Fig. 17.1 Fritz Haber. (BASF Unternehmensarchiv)

The award ceremony and the presentation speech in 1920 focused on the industrial production of nitrogen compounds as plant nutrients, even though the recent war had widely demonstrated the dual roles of synthetic nitrogen. Haber's address emphasized that the synthetic nitrogen processes, "relieve us of future worries caused by the exhaustion of the saltpeter deposits, which has threatened us these twenty years... It may be that this solution is not the final one. Nitrogen bacteria teach us that Nature, with her sophisticated forms of the chemistry of living matter, will understand and utilize methods that we do not as yet know how to imitate" [3]. Responses to Haber receiving the Nobel Prize included the usual erroneous claims that without the "Haber" process,"it is doubtful if Germany would have started the war" [4].

Similar claims persist and are cited, including in recent literature. However, estimates of the extent to which the Haber-Bosch process enabled Germany to conduct war from early 1915, when synthetic ammonia production was first expanded, until the Armistice, show that the claims are untenable. Until well into 1915 the supply of nitrogen, in the absence of Chile saltpetre, was mainly derived from calcium cyanamide and ammonia from coal gas and coke oven works, and hardly at all from the Haber-Bosch process. The sources of nitrogen necessary to provide food for the Kaiser's military and citizens and satisfy demand from manufacturers of explosives included Norsk Hydro's products. Despite claims that cyanamide-derived ammonia was inferior in quality to that obtained from the Haber-Bosch converters, it was no less suitable for munitions manufacture in time of war. France, for example, placed considerable reliance on cyanamide as a source of ammonia for explosives production. Even after Oppau was expanded considerably by late 1916, this did not make Germany dependent on the Haber-Bosch process. Nor did it help to stave off famine during the winter of 1916–1917. It was only from April 1917, with the opening of the Merseburg ammonia factory, that the Haber-Bosch process took on a leading role. That was the decisive factor in support of dependence on the Haber-Bosch process; just the last year and a half of the war. Prior to that it was one of the main contributors, but not the only one. However, as a wonder of science and technology, its importance in the war was exaggerated, even among the Allies.

17.2 Fritz Haber's Germany, 1918–1933

The end of World War I brought an end to the forty-year reign of Kaiser Wilhelm II (1888–1918). Germany from November 1918 and through 1919 was racked by chaos and a left-wing revolution. In order to restore order, the newly formed Weimar Republic, a coalition of the Social Democratic Party, the German Peoples Party and the Zentrum Party, was forced to call in the armed forces against its own citizens.

According to Article 231 of the Treaty of Versailles, Germany and her allies carried the full blame for the start of World War I. A 132 million mark indemnity, the equivalent of 50,000 tons of gold, was imposed on Germany. The conditions were ruinous. Ammonium sulphate, coal, coal tar, synthetic dyes and drugs, locomotives, rolling stock, etc., had to be handed over to the Allied nations, particularly France, Belgium, and Italy, as reparations. France alone was to receive 30,000 tons of ammonium sulphate annually for a period of three years. Its need for coal was exacerbated by the wartime destruction of some 200 mines in the north.

Haber, based in part on the calculations of Svante Arrhenius—who in 1903 estimated that some eight million tons of gold existed in sea water—decided in the hope of paying off Germany's debts to extract gold from sea water. It was not altogether a new idea. In England during 1905, the Engineering and Industrial Trust, Ltd, claimed to have similar intentions, and its proposals were endorsed by Sir William Ramsay. However its real motives were unclear, and no activity was undertaken. Haber's project began in 1920, managed by Johannes Jänicke, but was abandoned in 1927 when it was found that the work was based on erroneous calculations, as evidenced by microanalysis [5].

From 1919, German scientists returned to the research into war gases under the direction of Haber, but in order to circumvent the conditions of the Treaty of Versailles the experimentation was carried out in other countries, including Russia.

Haber promoted the restoration of scientific activities in Germany. In 1920, with leading members of the Prussian Academy of Sciences, and Prussian Secretary of State Friedrich Schmidt-Ott, he co-founded the Emergency Association for German Science (Notgemeinschaft der Deutschen Wissenschaft; later Deutsche Forschungsgemeinschaft). Haber, using his power and influence, was deeply involved in moves to bring Germany back into the global scientific community, including the removal of boycotts aimed at German scientists and their institutions. Germans had been barred from attending international scientific gatherings, notably the 1921 Solvay Conference, that Albert Einstein, in protest, had refused to attend. In the same year, Einstein gave his support to the Emergency Society for German and Austrian Science and Art, an organization founded by German-Americans in New York that during 1920–1927 promoted scientific work in Germany and Austria.

For BASF in 1921 there were problems of another kind. In September, a mixture of ammonium nitrate and ammonium sulphate stored in a warehouse at Oppau exploded, with the loss of over 600 lives, as well as 2000 injuries, wrecking much of the site. This followed the detonation of a hardened mass of the two ammonium compounds while it was being broken up with small explosive charges. Carl Krauch, the chemist who had directed the construction of Merseburg during the war, was responsible for reconstruction, which took just twelve weeks, and brought about his elevation to the managing board of BASF.[1]

The impacts of the Treaty of Versailles and wartime defeat fuelled economic and political instability in the new democratic republic. Unemployment and hyperinflation underscored the emergence of extremist political parties. One victim was foreign minister Walther Rathenau, murdered in 1922 [6]. In January 1923, the Weimar Republic announced that it was no longer in a position to fulfill the terms of the Treaty of Versailles. French and Belgian troops occupied the Ruhr region, taking control of several mining and manufacturing concerns. Strikes against the occupation created further damage to the German economy. In November, the price of a loaf of bread increased from 23 to 140 billion marks (and Adolf Hitler's putsch in Munich failed). By the end of the year the German currency was devalued to the extent that 25 billion (US trillion) marks equalled one pound sterling. For a short time, BASF introduced its own currency, the aniline dollar. Post-war Germany faced both economic and political crises.

The year 1924 marked the start of a return to relative stability and calm in Germany. A new currency, the Reichsmark, was introduced; Britain and the United States criticized the occupation of the Ruhr. However, normalization was achieved with difficulty. Germany was excluded from the 1924 Paris Olympic Games. Occupation of the Ruhr, the creation of a customs border, the crippling terms of the Treaty of Versailles (eased in September 1924) and shortages of coal created difficulties, including with power supplies, transportation and manufacture at Ludwigshafen and Oppau.

Gatherings of technical experts offered an alternative to gatherings of politicians in attempts to bring about a new world order and reconnections between former enemies. Electrical power, in particular, underscored common visions of peaceful collaboration, economic growth and stability, if not of harmony. In the summer of

[1]In September 2001, there was another major explosion, of ammonium nitrate, this time at the Toulouse ammonia factory in France; and in April 2013, a further explosion, in Texas.

1924, Haber was one of twenty official German representatives at the first World Power Conference, organized by the British Electrical and Allied Manufacturers Association. The event took place in the Palace of Engineering on the grounds of the British Empire Exhibition in Wembley, north of London, and was attended by representatives of forty nations [7]. Many saw it as a "Technical League of Nations," and perhaps a new era of normalcy in international affairs.

Haber, who represented the Kaiser Wilhelm Society (Kaiser-Wilhelm-Gesellschaft zur Förderung der Wissenschaften) and the German Chemical Society, took the opportunity to meet up with his erstwhile collaborator Robert Le Rossignol, who represented the British General Electric Company and was based at the new North Wembley laboratories of General Electric. (Le Rossignol travelled to Germany during the 1920s for family and business reasons related to his royalties from the ammonia process and met with Haber). Energy sources for synthetic nitrogen processes were discussed by various participants in the transactions (the lecture presentations were very brief, lasting just a few minutes), in particular those representing BASF, Casale, Claude, and Fauser. John Crowley, for example, promoted the Casale process [8]. Also present at the meeting was Polish electrochemist Ignacy Mościcki, who directed reconstruction of the former German cyanamide factory at Chorzów. The conference, for all its unfulfilled promise, certainly marked the beginning of a new era of scientific cooperation for Germany in its attempts to overcome the impediments of post-war antagonisms.

Later in 1924, Haber and his wife embarked on a world tour, one that would bring about a further cementing of relations with former enemy nations. Their first destination was the United States, where Haber represented the Prussian Academy of Sciences, the Kaiser Wilhelm Society, and the German Chemical Society at the Benjamin Franklin Centenary event, held in Philadelphia during September 17–19. There Haber spoke on the "Practical Results of the Theoretical Development of Chemistry." Echoing sentiments voiced at the Wembley event, he closed with a note of hope based on technoscientific universalism and internationalism: "Discontent with economic conditions arouses a strife among men, and only progress of natural science offers the certain prospect of a greater contentment than prevails in the present world" [9].

Luigi Casale was in the United States at the same time to promote his ammonia process. According to Arturo Miolati, "Professor Haber, a distinguished German chemist, said, on the occasion of a reception given in New York [1924] in his and Dr. Casale's honour, that Dr. Casale was the man who had found the best way to apply the principles which he, Professor Haber, had established" [10]. It is not possible to substantiate this statement, nor whether in fact they happened to be in New York on the same occasion. Haber did not mention it in the published version of his Philadelphia lecture. It may be somewhat of an exaggeration, but it is not without foundation.

After crossing the United States, the Habers arrived in San Francisco, from where they proceeded to Japan, to be met in Yokohama by the industrialist Hoshi Hojami (1878–1951), and the German ambassador, Wilhelm Solf. During their almost three month trip in Japan, the Habers were accompanied by Haber's former colleague Tamaru Setsuro, who acted as translator. In Kamakura, south-west of Tokyo, the Habers were guests of the Tamaru family. From Japan, the Habers, accompanied by Tamaru, crossed the Sea of Japan to visit Korea and then Manchuria, where the growing influence of Japan on industrialization was strongly evident, and then made their way back to Germany.

Hoshi during 1920–1921, had provided funds to ambassador Solf for the creation of a foundation to support science in Germany. The much delayed opening of what became known as the Japanese Institute took place in Berlin on 4 December 1926, mainly through Haber's efforts. A similar institute opened in Tokyo in April 1927.[2] In early 1926, Germany had been admitted to the League of Nations.

Throughout the 1920s, Haber's Berlin institute was associated with outstanding young chemists engaged in a wide range of mainly theoretical studies, including structural chemistry and free radicals. Meantime, Germany recovered its leading position in chemical industry. The second World Power Conference was held in Berlin during June 1930.

There, ICI's George Pollitt spoke on "The Synthetic Ammonia Industry," emphasizing the increasing role of coke oven hydrogen [11, 12]. On June 25 a party of delegates, including Pollitt and ICI engineer Humphrey—just one year before he retired from the firm—visited the Merseburg/Leuna factory that had an "annual output of 650,000 tons of air-nitrogen, 220,000 tons of which can be converted into sulphate of ammonia and 55,000 tons into nitrate of lime. The remainder of 375,000 tons of nitrogen is sent off in the form of anhydrous ammonia in special tank-cars to the plants of the I.G. trust to be converted there into various sorts of fertilizers." As for the power supply: "[E]nergy used daily at the factory is generated out of 12,000 tons of brown coal. . . .The electric current required at the factory is generated at a power plant of 85,000 KW" [13]. Forty-five gas engines, of 3000 to 4000 horsepower, drove the compressors used in the Haber-Bosch process. The gas engines were replaced by electric motors during 1936–1937.

The ammonia synthesis process was stated to convert about 20% of the gas mixture into ammonia in each cycle. The Leuna site employed around 15,000 workmen and 3000 technicians. Billingham, even allowing for rounds of layoffs, still employed around 5000 people and produced 170,000 tons of fixed nitrogen annually.

In 1931, Carl Bosch and Friedrich Bergius, for their development of high-pressure methods, became recipients of the Nobel Prize in Chemistry. It was the first Nobel Prize awarded for an industrial process [14, 15].

[2]Later in the same year, Charlotte Nathan separated from Fritz Haber.

17.3 The Death of Haber

We are not fully acquainted with the cause of [Fritz Haber's] exile but believe it to be due to the descent of his ancestors from that Semitic race from whom we Europeans learned 3000 years ago our alphabet and 2000 years ago our religion.

"Prof Fritz Haber," *Chemistry and Industry*, no. 6, 9 February 1934, p. 134.

By 1932, unemployment, poverty and the burden of reparations had taken a heavy toll in Germany. The United States had already called in foreign loans. Sectors of the chemical industry were struggling to survive—in the case of IG Farben mainly as a result of involvement in high-pressure coal-to-oil conversion. Leuna was operating at around 20% capacity in ammonia production in March, and, despite protective tariffs, both nitrogen and synthetic fuel were major loss makers. At the beginning of October it was estimated that IG Farben had shed 46% of its work-force since 1929. As with nitrogen products, IG Farben had failed to control a single market in chemicals [16]. Profits came from exports, and even there demand had fallen.

In July 1932, the National Socialists gained 230 out of 608 seats in the Reichstag. At the end of the following January, Adolf Hitler took office as Reich Chancellor. Within days Germany was in the grip of anti-Jewish demonstrations, which became violent in March and April. Now, Fritz Haber felt the increasing racialist and political tensions in Germany. His situation, and that of many of his assistants, changed almost immediately. The Law for the Restoration of the Civil Service of 7 April 1933 required that Haber dismiss workers of Jewish descent. After having done so, he handed in his own resignation on April 30. Among Haber's assistants was Ladislaus Farkas, who would within a few years found physical chemistry at The Hebrew University of Jerusalem. In May, Bosch attended a meeting with the Führer at which he pointed out the damage caused by the loss of Jewish scientists. Hitler's response, according to Bosch, was that Germany "would work one hundred years without physics and chemistry," if that became necessary [17]. Fritz Haber departed from Germany.

Support for Haber came from Harold Hartley (knighted in 1928 and director of research at the London Midland and Scottish Railway), Sir William Pope (knighted in 1919), and others in Britain, and also from Japan, including Tamaru, at the Tokyo Institute of Technology (previously the Tokyo School of Technology, founded in 1881).

In the autumn of 1933, Haber joined Pope at Cambridge. Chaim Weizmann, chemist and Zionist leader, who had visited Haber in 1932, and modelled the Daniel Sieff Research Institute at Rehovot in Mandate Palestine on Haber's institute, invited Haber to join the new desert institute, at least during the winter months. Haber intended to visit Palestine, but was suffering from heart problems, which would have made the long journey difficult. He died in Basel, Switzerland, on 29 January 1934. In April 1937, Joseph E. Coates, then Professor of Chemistry at University College Swansea, gave the "The Haber Memorial Lecture," before the Chemical Society in London. It was a comprehensive description of Haber's scientific achievements [18]. By then Haber, along with Adolph Frank, Heinrich Caro, and Nikodem Caro, had been redacted from accounts of the history of chemistry in Germany.

Nikodem Caro emigrated from Germany in 1933. He died in Rome two years later and was buried in Zurich. Albert Frank left for the United States in 1938; during 1942–1945 he worked with the Office of War Information, and after World War II with American Cyanamid.

On 9 December 1936, Herbert Humphrey gave an account of the "History and Development of Mond Gas for Industrial Purposes," before the Institute of Fuel at the Chemical Society, Burlington House, London. He drew attention to the original advantage in ammonia recovery and production of ammonium sulphate for agriculture. In 1939, he received the Melchett Award of the Institute of Fuel (the present-day Energy Institute). His award lecture was also an historical account—well suited to the time: "The story of explosives during the war, and the early history of Billingham." ICI was then producing around 200,000 tons of ammonium sulphate each year (compared with 157,000 tons in 1936), in addition to 14,700 tons of concentrated nitric acid and 64,100 tons of ammonium nitrate.

In 2011, Carl Bosch and Fritz Haber were voted the world's most influential chemical engineers of all time by members of the British Institution of Chemical Engineers.

References

1. Coffey P (2008) Cathedrals of science: the personalities and rivalries that made modern chemistry. Oxford University Press, Oxford, pp 119–120.
2. Widmalm S (1995) Science and neutrality: the Nobel Prizes of 1919 and scientific internationalism in Sweden. Minerva 33:339–360, on 343.
3. Haber F (1920). The synthesis of ammonia from its elements, Nobel lecture, June 2, 1920. Nobelstiflesen Foundation, Stockholm, 1920.
4. Alexander J (1920) Nobel award to Haber: source of resentment felt in allied countries. New York Times, 3 February.
5. Stoltzenberg D (1994) Fritz Haber: Chemiker, Nobelpreisträger, Deutscher, Jude: Eine Biographie. VCH, Weinheim, pp 487–499.
6. Volkov S (2012) Walther Rathenau: Weimar's fallen statesman. Yale University Press, Yale.
7. Meyer TH (1996) D. N. Dunlop: a man of our time. Temple Lodge Publishing, Forest Row, Sussex.
8. Crowley JF, Bergstrom EM (1924) The development of national water power resources. In: The transactions of the First World Power Conference, London, June 30th to July 12th, 1924, vol. I. Power resources of the world available and unutilised. Percy Lund Humphries & Co., Ltd, London, pp 349–368, on 359–363.
9. Haber F (1925) Practical results of the theoretical development of chemistry. Address given on 18 September 1924. Journal of the Franklin Institute 199(4)(April):437–456.
10. Quoted from Miolati A (1927) Synthetic ammonia and the Casale process. Amplified edition of a lecture delivered the 27th February 1927 at the Institute of Chemistry of the Polytechnic School of Prague. Ammonia Casale SA, Rome, p 34.
11. Tongue H (1934) The design and construction of high pressure chemical plant. Chapman & Hall, London, p 109.

12. Waeser B (1932) Die Luftstickstoff-Industrie. Mit Berücksichtigung der chilenischen Industrie und des Kokereistickstoffs. Springer, Berlin, pp 132–136.
13. Visit to the Ammoniakwerk Merseburg G.m.b.H. in connection with the World Power Conference at Berlin. Some technical data given by Direktor Dr Schneider, when addressing the visitors. Leunawerk, June 25th 1930. Papers of Herbert Alfred Humphrey, Archives, Imperial College London, item M7.
14. Steinmüller F (1993) Carl Bosch 1874–1940. In: James LK (ed), Nobel laureates in chemistry 1901–1992. History of modern chemical sciences. American Chemical Society, Washington DC, pp 198–204.
15. Barnes Z (1993) Friedrich Bergius 1884–1949. In: James LK (ed), Nobel laureates in chemistry 1901–1992. History of modern chemical sciences. American Chemical Society, Washington DC, pp 192–197.
16. Hayes P (2001) Industry and ideology. I.G. Farben in the Nazi era. 2nd edition. Cambridge University Press, Cambridge, pp 40–43.
17. Hayes P (2001) Industry and ideology. I.G. Farben in the Nazi era. 2nd edition. Cambridge University Press, Cambridge, pp 90–94.
18. Coates JE (1939) The Haber memorial lecture. J Chem Soc, pp 1642–1672.

Chapter 18
A Legacy of Synthetic Nitrogen

18.1 Homage to Inventors

The early synthetic ammonia industry became strongly associated with myths surrounding Fritz Haber's personal role in enabling Germany to wage war through his involvement with nitrogen products. Today, this is still repeated as fact. More correctly, he was associated with the use of toxic chemicals in warfare. While he was an ambitious and arguably amoral servant of his country at war, and certainly an adept administrator, he was certainly no worse than those John Galbraith described as "mostly honest men whose public and private behaviour would withstand public scrutiny as well as most" [1].

To recapitulate: Haber's main mission had been to capture nitrogen, with the ultimate aim of contributing towards resolving the world fertilizer problem. He began with applications of electrochemistry and later moved on to the kinetics and thermodynamics of gas reactions. With his collaborator Robert Le Rossignol, a capable machinist and experimentalist, this led to a benchtop method for producing synthetic ammonia. The high-pressure chemistry drew on newer directions in academic chemistry, and satisfied new and converging needs [2]. With the aid of a suitable catalyst, they showed that synthetic ammonia could be produced in amounts that some thought impossible, and others could not achieve due to the limitations and flaws of their apparatus. Then by making the reaction continuous, with unreacted gases recirculated to the steel converter, Haber and Le Rossignol developed an apparatus that in 1909 impressed the BASF visitors to Karlsruhe. The apparatus served its intended purpose. Haber's rewards were a royalty arrangement with BASF, directorship of his new Kaiser Wilhelm Institute in Berlin, and the Nobel Prize. He acknowledged in his Nobel Lecture joint ammonia studies, at pressures up to 370 atmospheres, with the late Harold Greenwood.

Synthetic ammonia arose from the combined genius of Fritz Haber and Carl Bosch. However, though BASF supported Haber's scientific work, and Haber had overcome the vexingly elusive problem of a method of synthesis that offered a route

A.S. Travis, *Nitrogen Capture*, https://doi.org/10.1007/978-3-319-68963-0_18

to commercial success, he played no part in the industrial scaling up. To go from a benchtop model to a full-size industrial plant was a big leap. While drawing on the technologies of the time, the concepts involved in design and production departed from all those methods employed previously. The enterprise took a team of BASF chemists, metallurgists, engineers, and proto-chemical engineers, led by Bosch—delving into a wide range of problems concerning materials of construction, compressors, catalysts, instruments, supply and purification of hydrogen and nitrogen, all the time drawing on prior developments in chemical and other technologies—just four-and-a-half years to realize. In the absence of powerful electric motors, the earliest compressors were driven by gas engines. As with the cyanamide process, the ammonia process was made possible by the convergence of an array of existing and new technologies.

The pursuit of a single invention dedicated to chemical manufacture had never before relied on so many skills. It created a core group of dedicated specialists for entry into a new world of chemical technology, and was driven to a satisfactory conclusion under one unyielding, single-minded and tenacious individual, Carl Bosch, whose brute force process bound nitrogen with hydrogen on a scale suited to manufacture. Bosch's role was recognized by senior management. His reward was rapid promotion, to chairman of the board of BASF, and of IG Farben in 1925 (a post he occupied until 1935). By 1931, when Bosch and Bergius received the Nobel Prize in Chemistry, high-pressure chemistry was already contributing to other developments in industry and in the laboratory.

These remarkable scientific and technological innovations should not, however, detract from other achievements in nitrogen fixation. They had started out as the activities of independent inventors, or as a desire for diversification at dye-manufacturing and other chemical firms, with the promise of profits from contributing to increased agricultural yields. Notably, Nikodem Caro was involved with Adolph Frank in the invention of a successful electrothermal process for nitrogen fixation, applied in Europe and elsewhere, particularly in Italy, Japan, Norway, and Canada. No less important was Kristian Birkeland's work. Though the original source of inspiration is uncertain, that hardly matters. His simulator of lightning was the most successful of the electric arc methods. It was, conveniently—and here was the incentive of location—invented in a country where cheap hydro-electricity was available in abundance. As for the high-pressure processes, the almost independent investigators Casale and Fauser, the gas separation expert Claude, and the consultant Uhde also deserve their rightful places among the inventors of early twentieth-century nitrogen fixation processes.

Many of these individuals accumulated great wealth or—as has been described—rose to high positions in industry or academe. Haber, for example, was particularly generous in rewarding Le Rossignol for his outstanding laboratory work, through a 40% share of royalties, but the post-1918 situation in Germany denied them of much that they might otherwise have gained. Few records exist of actual payments, but Le Rossignol considered himself highly fortunate for whatever he received [3]. A hint of Nikodem Caro's wealth became known through a 1932 court case concerning the purported amount of a dowry that he paid to Czech-born Ignaz Petschek, owner of lignite mines in Germany, when Caro's daughter, Vera, married one of Petschek's sons, Ernst, just after the war. The couple broke up several years later, leading to a dispute centred on the dowry, which was claimed to amount to 400,000 marks, one quarter in the form of shares in Bayerische Stickstoff-Werke. Kristian Birkeland, despite at times frosty relations with colleagues at Norsk Hydro—no doubt the reason for his departure in 1906—enjoyed an income from his electric arc invention which was many times that of a university professor.

Though these inventors and their enterprises are almost forgotten, efforts are being made to retain reminders of their rich heritages. At Rjukan, a Birkeland-Eyde furnace, an acid tower dating from 1911 and various buildings, in addition to the complete transport network—including railways and train ferry terminals—survive. They are included in a UNESCO World Heritage Site proposal (2014) for commemorating the impact of industries powered by hydro-electricity at Rjukan and Notodden [4]. The Norwegian Industrial Workers Museum at Rjukan includes several exhibits related to the Birkeland-Eyde process. Abandoned Frank-Caro ovens and a Linde nitrogen plant remain in place at the Odda facility. The Norwegian Museum of Hydro Power and Electricity covers historical developments at Odda and Tyssedal. The Deutsches-Chemie Museum Merseburg includes several items of equipment from the early days of synthetic ammonia manufacture at Merseburg/Leuna. The Carl Bosch Museum Heidelberg celebrates the work of the Nobel laureate on high-pressure chemistry, with appropriate exhibits. The London Science Museum has a three-crank two-stage vertical reciprocating compressor, manufactured in 1918 by Peter Brotherhood of Peterborough, originally used at Runcorn and then from 1923 in the ammonia pilot plant at Brunner, Mond's Billingham factory (and latterly for production of Drikold), and artifacts from a 1928 ICI compressor (in addition to an early Haber-Bosch converter: another one is displayed opposite the Ludwigshafen factory of BASF).

18.2 Autarky

> It is a question whether private capital would have of itself invested such large sums in the Oppau and Merseburg plants in normal times or without government aid.
>
> R. S. Tour, "The German and American Synthetic-ammonia Plants. V," *Chemical and Metallurgical Engineering*, vol. 26, no. 8 (8 March 1922):463–465, on 464.

> [The] false momentum, propelled not by consumer demand, but by the contingencies of war.
>
> Peter J. T. Morris

The electrothermal developments related to the capture of nitrogen brought about significant and far-reaching changes in the chemical industry between 1900 and 1914. At the end of that period high-pressure chemistry offered a viable, rival process to the electric arc and cyanamide processes, in the form of the Haber-Bosch process, inaugurated in September 1913. Initially, there was limited interest at BASF in the oxidation of synthetic ammonia to nitric acid. Nevertheless, elsewhere in Germany oxidation to the acid of by-product ammonia from gas and coke oven works, as well as from cyanamide at the Bayerische Trostberg facility, had been introduced on a small scale. Thus all the new nitrogen-based processes that would be required to serve the Kaiser's army and navy from August 1914 were already in place, even if the scales of operation were not great and in most cases technical difficulties still had to be overcome.

The long war of attrition, and the often desperate need for high explosives, brought about a new strategy, at first on the German side, but then adopted by the Allies. It was based, as Fritz Haber later stated, on the "internal connection" between war and the exact sciences [5]. This harbinger of change, the vast academic-industrial-military complex in which Haber was closely involved, was dependent on wartime scientific advances and technological innovations. World War I was "The Chemist's War," conducted on an enormous scale, in the scaling up and development of products and processes for capturing nitrogen, converting ammonia into nitric acid, nitrating aromatic compounds, and manufacturing ammonium nitrate and sulphate. As in Germany, the British and French merged the activities of state agencies, the military, chemical manufacturers, and universities, as did the Americans even before they entered the war in April 1917.

Research and development without economic goals was put into practice, often aimed at finding substitutes (ersatz products) for materials no longer readily available. Germany overcame the loss of nitrate from South America in order to sustain the long stalemate once it was realized that the conflict would not be a "war of movement" but of drawn out and inconclusive campaigns. Fritz Haber, by association, and conviction, remains the scientist most connected with the German war effort. Yet despite the many claims to the contrary, and even the reminiscences of Alwin Mittasch who started writing them during World War II—Mittasch was then of the opinion, correctly, that his 1914 work on ammonia oxidation contributed towards Germany's ability to continue fighting after the spring of 1915—the ammonia synthesis, it should be emphasized again, was not an overriding factor in munitions manufacture [6]. No less significant: but for the war, the enormous Haber-Bosch and other factories sponsored directly and indirectly by the Reich would not have become so readily available to provide the fertilizer required to feed humankind in peacetime [7].

BASF's appetite for risk had grown following the successes of synthetic alizarin (1870) and indigo (1897) [8]. These were peacetime triumphs, as was nitrogen capture through synthetic ammonia, though its strategic significance was recognized well before 1914. All three products demonstrated the role of science-based chemical industry in replacing natural products and replicating Nature. High-pressure chemistry encouraged even greater risk-taking in the late 1920s, as Carl Bosch envisioned new directions in product development. It drove leading autarky programmes that became common themes, political and technological, in the history of the inter-war period, as industrial nations strove for independence in strategic materials and the capture of export markets. The costs were enormous, precipitating mergers, notably the formation of IG Farben in 1925, and ICI in 1926, and acquisitions of both new technology and manufacturers by Montecatini during the early 1920s. Geopolitical and economic considerations tended not to favour foreign acquisitions in the nitrogen industry. Instead, concentration of markets through cartels—reducing competition in the industry—came about when firms had to meet the global challenge of over-production and falling prices.

For a time, Bergius high-pressure coal-to-oil conversion appeared to justify the cost. But that failed, technically and commercially, once abundant reserves of petroleum were discovered in the United States and Near East [9]. The Great Depression had not helped, nor did resistance by farmers to synthetic nitrogen, notably in the United States, and, for German, British and American exporters, the ambitious Japanese plan to build up its own high-technology chemical industry.

In Germany, it took until after the mid-1930s before the risk paid of, with the vast autarky programmes of the National Socialists. This included Walter Reppe's reactions on acetylene under conditions of high pressure, initially used in the manufacture of synthetic rubber (Buna rubber) [10]. In Britain, by the late 1930s, strategic needs supported ICI's synthetic fuel programme and the construction of new synthetic ammonia factories.

These examples demonstrate that expansion prior to (and during) World War II, was driven mainly by historical contingency—namely a "false momentum, propelled not by consumer demand, but by the contingencies of war" [11]. By the late 1920s, nitrogen was providing a low margin of profit, and soon even losses. Significantly, between 1919 and the early 1930s, in conditions of relative peace, it was the traditional products, synthetic dyes, as well as pharmaceuticals and photo products, rather than high-pressure ammonia chemistry, that generally provided the profits required to sustain the behemoth IG Farben [12]. A similar situation prevailed at ICI, as Billingham's ammonia plant increasingly drained the corporation's resources from the end of the 1920s, and were written down during the 1930s. State support by central command for a sophisticated and superior new technology during World War I at a time of strategic vulnerability did not necessarily bring the anticipated peacetime profits, at least not in the short term. Much the same argument can be applied to the German and American engagements in synthetic rubber programmes, particularly for tyres, before and during World War II [13].

In summary, before 1914, the commitment to the Haber-Bosch ammonia process was propelled by the same high-risk strategies aimed at self-sufficiency in Germany that inaugurated synthetic alizarin and indigo, and would be followed from the mid-1920s by commitment to other high-pressure industrial processes. The latter, in particular, included the emergence of Reppe chemistry, based on the availability of carbide-derived acetylene, and new polymers, notably Buna synthetic rubber [8]. Without the build-up of the German military in the 1930s there would have been no economic incentive to continue with synthetic petroleum and synthetic rubber, both based on the employment of high pressures and special catalysts (Fig. 18.1). By 1938–1939 synthetic fuel was one of the main products at IG Farben's Leunawerke. Germany's warlike ambitions caused around 60% of

Fig. 18.1 The Leunawerke (Merseburg), late 1930s. Solid coal. . .Liquid force. *The way is open to German fuel.* Adolf Hitler. (*Erzeugnisse unserer Arbeit.* Frankfurt am Main: I.G. Farbenindustrie, 1938, p. 177)

ammonia made there to be converted into ammonium nitrate, and less than 30% into ammonium sulphate. Britain and the United States responded accordingly as the threat of war increased. In the United States during the early 1940s, ten new government ammonia facilities tripled the nation's pre-war synthetic nitrogen capacity. Similar circumstances in 1915–1916 had created the drive towards the improvement and expansion of synthetic ammonia production, and the development of the modern processes for large-scale conversion of ammonia into nitric acid (Fig. 18.2).

Fig. 18.2 Artist's impression of a barge transporting tanks of nitric acid on the River Rhine with the BASF/IG Farben Oppau factory in the background. Nitrogen products were critical to Hitler's rearmament programmes. (*Erzeugnisse unserer Arbeit*. Frankfurt am Main: I.G. Farbenindustrie, 1938, p. 171)

References

1. Galbraith JK (1969) How to control the military. Doubleday & Co., Inc., Garden City, New York, p 22.
2. Lenoir T (1992) Practical reason and the construction of knowledge: the lifeworld of Haber-Bosch. In: McMullin E (ed), The social dimensions of science. Indiana University Press, Notre Dame, pp 158–197.
3. Sheppard D (2017) Robert Le Rossignol, 1884–1976: engineer of the 'Haber' process. Notes Rec R Soc London. DOI: https://doi.org/10.1098/rsnr.2016.0019 (accessed 18 March 2017).
4. Taugbøl T, Andersen EM, Grønn U, Moen BF (2014) Nomination dossier: Rjukan-Notodden industrial heritage site. Nomination to the UNESCO World Heritage List, Norway, pp 44–46.
5. Haber F (1918) Das Verhältnis zwischen Heereswesen und exakten Naturwissenschaften: Vortrag auf der 24. Hauptversammlung der Deutschen Bunsen-Gesellschaft in Berlin, 8–10 April 1918. Chem Ztg 42:197.
6. Mittasch A (1951) Chronik meines Lebens (unpublished paper, Deutsches Museum Munich), p 370.
7. Jones G (1920) Nitrogen: its fixation, its uses in peace and war. Quart J Econ 34(3):391–431.
8. Morris PJT (1982) The development of acetylene chemistry and synthetic rubber by I.G. Farbenindustrie Aktiengesellschaft: 1926–1945. DPhil thesis, University of Oxford, pp 14–15.
9. Hayes P (1987) Carl Bosch and Carl Krauch: chemistry and the political economy of Germany, 1925–1945. J Econ Hist 47(2):353–363.

10. Morris PJT (1998) Ambros, Reppe, and the emergence of heavy organic chemicals in Germany, 1925–1945. In: Travis AS, Schröter HG, Homburg E, Morris PJT (eds), Determinants in the evolution of the European chemical industry, 1900–1939: new technologies, political frameworks, markets and companies. Kluwer, Dordrecht, pp 89–122.
11. Quoting PJT Morris, personal communication, November 2014.
12. Plumpe G (1990) Die I.G. Farbenindustrie AG. Wirtschaft, Technik, Politik 1904–1945. Duncker and Humblot, Berlin, pp 550–551.
13. Morris PJT (1989) The American synthetic rubber research program. University of Pennsylvania Press, Phildelphia.

Chapter 19
Catching Up: Mainly Italy, Japan, and the Soviet Union

19.1 Nitrogen: A Strategic Asset

The synthetic nitrogen industry emerged mainly in Germany with the availability of novel furnaces, ovens, and autoclaves, improved knowledge of gas reactions, and the application of sophisticated engineering techniques to chemical processing. This and the need for fertilizers provided the stimulus for the development of heavy chemical industry at new locations, notably in Italy and Japan. In the historical literature, these nations are invariably portrayed as latecomers in industrialization. However, the fact often overlooked is that the involvement of Italy and Japan in hydro-electricity and nitrogen around 1900 established their chemical industries among the world leaders within just over two decades, and stimulated moves into high-pressure ammonia processes. As for the Soviet Union, despite the ravages of Stalinism and ideological dictates, mainly framed in starkly personal terms, tremendous advances were made in industrialization by the mid-1930s, based on the need for artificial nitrogen products required in a highly politicized system of agriculture, with its mechanized collectives. These three countries, Italy, Japan and Russia, whose representatives and entrepreneurs were unable to gain access to BASF technology, or considered the cost extortionate, demonstrate how the late growth of their modern chemical industries was based on the global circulation of rival nitrogen technologies.

The massive post-war expansion in development of synthetic ammonia processes was at first stimulated by strategic needs as much as by demand from agricultural sectors for new nitrogen fertilizers. The research and development programmes drew both on knowledge gained during 1914–1918 as well as on the new approach to goal-oriented research that was an outcome of the organization of wartime military needs. Industrial, state, private, military and academic institutions were involved, including, apart from Germany, in Japan (particularly Tokyo), the United States (Washington DC), the Soviet Union (Leningrad and Moscow), and Italy (Terni).

A.S. Travis, *Nitrogen Capture*, https://doi.org/10.1007/978-3-319-68963-0_19

The style of growth of the nitrogen industry varied from country to country, according to pressure on the land available for agriculture, energy sources, and ideological and economic needs. Adoption of nitrogen technologies mirrored political and historical events. Moreover, as a result of World War I, the type of chemical technology had changed, as had the social organization of science and technology. In the case of nitrogen, the new processes were as much based in the realms of engineering as they were in chemistry. They included new methods for making and purifying hydrogen and nitrogen, machinery for their compression, and the development of novel steels.

The scale of working and costly equipment required a massive investment of capital, which made necessary backing from large corporations and banks, as well as from the state. Monopolies on nitrogen products encouraged external investment. Fauser was backed by Montecatini which dominated a large part of synthetic nitrogen manufacture in Italy. From 1921, Montecatini, through involvement with cyanamide, and then Fauser technology, became the major player in the Italian nitrogen fertilizer industry. The same happened in Germany, as BASF increasingly controlled the Stickstoff-Syndikat.

Luigi Casale, mainly excluded from Italy by the rival Montecatini, relied on investors in Switzerland, and capital raised elsewhere. He quickly found much needed licencees in Japan, France, the United States, the Soviet Union, etc. Selling points for the Casale process were: the product was anhydrous ammonia; the relatively small size of working, with economies of scale, since it was also suited to low-volume production; and the inexpensive catalyst. There was also the ejector, replacing the costly circulation pump, which took up considerable space and had to be maintained. Ammonia Casale had a disproportionate impact until the mid-1920s, and even after, not only because of its technical improvements and certain advantages over the Haber-Bosch process, but also because BASF demanded a high price for access to its ammonia technology; the Fauser process was restricted to Italy by Montecatini until 1926; and the Claude process faced initial technical difficulties.

Enthusiasm for synthetic nitrogen in Italy arose originally from the fact that there was little in the way of coal or land suited to the large-scale production of crops. Then, between 1920 and the mid-1930s, as was the case elsewhere, the spectre, and reality, of unemployment and hunger played into the hands of a nationalistic political movement. Synthetic nitrogen fertilizer, offering the answer for overcoming the problem of food supply, was heavily backed by the fascist state. Italy remained at the cutting edge of early innovation and international expansion, which was favoured by a relatively stable political system, whatever its failings. Montecatini's near monopoly and Mussolini's autarky programme, which for the chemical industry was little more than Montecatini's autarky programme, encouraged the ongoing production of both calcium cyanamide and synthetic ammonia, the latter mainly by the Fauser process.

Just as the needs of war had encouraged the German state to sponsor the growth of synthetic nitrogen programmes during 1914–1918, the needs of agriculture favoured state sponsorship of synthetic ammonia in Italy and the Soviet Union. For Mussolini and Stalin, the new synthetic ammonia processes were essential to their economic programmes. In the Soviet Union, nitrogen fertilizers supported state and collective farms, and were integral to growth of the agricultural sector.

A similar situation existed in Japan, as the Imperial Empire introduced, through Noguchi, modern industry into northern Korea during the second decade of colonial rule. This started just as the decline in output, in particular of synthetic ammonia, began throughout much of the West, followed in 1929 by the Wall Street crash. The contemporaneous massive industrialization in the USSR led many Japanese, including among the military, to believe that the free market had come to an end and state planning and control based on the Soviet system presented the model for a modern form of economic organization. It may have been an illusion, but many Japanese had good cause to complain. They had endured considerable turmoil since 1918, including rice riots (1918), widespread hoarding of grain (1921) and the banking crises (1927), in addition to attempted coups. From the late 1920s, the growing influence of the military, which distrusted the old established business conglomerates (zaibatsu), provided opportunities for entrepreneurs such as Noguchi. While not necessarily a nationalist, he had set up in Korea in part to remove his activities far from the threats of the very same old zaibatsu that the army despised. He found favour among the military authorities there, who believed that he could achieve the sort of industrial expansion that his rivals could not. The outcome was that Noguchi through his ammonia facility and hydro-electricity was to spearhead modern industrialization in Korea: his enterprises symbolized aggressive industrial colonialization and hopes of an imperial destiny. Power was consolidated with hydro-electricity and high-pressure chemistry. Noguchi's success in Korea was what put him behind the development of hydro-electric power in Manchuria and made Nitchitsu into a leading industrial conglomerate.

Chapter 20
Conclusion

From 1918, BASF prevented access to the Haber-Bosch process and declined multiple requests for operating licenses. The clear preference for promoting its product, ammonium sulphate, rather than its technology, created a barrier that stimulated research elsewhere, based on the availability of cheap electrical power, the application of sophisticated engineering techniques equal to those of Germany, and alternative routes to pure hydrogen.

Several factors contributed towards the introduction during the 1920s of nitrogen technologies into countries outside Germany, with special reference to synthetic ammonia. Foremost, in a world still traumatized by the impact of World War I, was the need for national security, based on self-sufficiency in explosives and fertilizer. Apart from security considerations, other aspects have to be taken into account, including natural resources and location.

Then there were the contrasting approaches of Britain and the United States, in part determined by the neglect of science-based industry. Britain's chemical industry prior to 1914 was highly risk averse. A sense of fatalism had descended on most sectors, made worse by the failure to master high-pressure ammonia synthesis during the war. Not long after the Armistice, the state divested itself of any interest in ammonia, bringing in Brunner, Mond & Co., to take over the Billingham site. Eventually, with a little theft of BASF technology, Brunner, Mond created an ammonia facility that was a worthy rival to Oppau. The United States, with its great dependence on agriculture, found good reason to sponsor extensive research into nitrogen technologies at the Fixed Nitrogen Research Laboratory, a springboard for university research and source of inventions used by Allied Chemical, Du Pont, and other firms.

The rise of insurgent companies, mainly in Europe, not only threatened BASF but led to the pursuit of new challenges, none more so than technologies for the extraction and purification of hydrogen from coke ovens. This enabled the wider geographic distribution of ammonia manufacture, notably in Italy, northern France, Belgium, the Netherlands, central and eastern Europe, and the Soviet Union (and later in Japan). Security considerations, however, still played a role in plant

A.S. Travis, *Nitrogen Capture*, https://doi.org/10.1007/978-3-319-68963-0_20

location, particularly in central Europe, where potential threats posed by Germany and the Soviet Union were deciding factors, though in the long term not always the correct ones. The location of the French state-owned ammonia works at Toulouse was based on considerations of defence—that is, well away from Germany—but it was remote from coal-producing regions.

As for BASF, it is difficult to assess how it would have fared if it had not adopted a blocking strategy [1]. Part of the answer is rooted in the firm's culture, going back to the late nineteenth century. The successes in discovery and innovation in the synthetic dye industry created a sense of near invincibility. This was fully understandable, based on vast investments in research and development, and the resulting outcomes. BASF had determined both the regulatory and technical environments by its early participation in German patent law and forceful protection of its inventions outside Germany. That suited the situation prior to 1914, when BASF and other German chemical firms were still world leaders. The ability to introduce radical new technologies was a function of their size. But during the war years, other nations began to catch up in their organization of science and technology, often emulating the German model. The environment of innovation, and incentives, changed. The time gap was closing. It took half a century to catch up with German, mainly BASF, dye technology, but just over a decade to catch up with BASF high-pressure technology. Perhaps in the 1920s that came as a surprise to BASF (and IG Farben). In the 1930s, the impact brought about a complete reversal of strategy, with the sale of know-how for synthetic ammonia technology to Japan, and the pooling of coal hydrogenation patents elsewhere. Paradoxically, perhaps, in late 1933, IG Farben's lawyers emphasized the loss of substantial profits arising from the policy of withholding licenses on Haber-Bosch ammonia as rivals entered the high-pressure nitrogen industry in order to argue against introduction of a National Socialist law aimed at prohibiting the international sharing of technology [2]. Three years later, the state ordered German firms to cease sharing with or selling technology to foreign firms.

The self-imposed exclusion by BASF, through its refusal to provide licenses and know-how for Haber-Bosch technology in the belief that success could not be achieved outside Germany, had been challenged globally by around 1930. Non-BASF (IG Farben) synthetic ammonia factories were operating or under construction in the following countries: Belgium (Casale, Claude, Mont Cenis, Nitrogen Engineering Corporation [NEC]); Canada (Casale, Fauser); Czechoslovakia (Claude, NEC); France (Casale, Claude, Mont Cenis, NEC); Germany (Casale, Claude, Fauser, Mont Cenis); Great Britain (ICI process); Holland (Fauser, Mont Cenis); Hungary (NEC); Italy (Casale, Claude, Fauser); Japan (Casale, Claude, Fauser, Mont Cenis, Du Pont); Norway (NEC); Poland (Claude, Fauser, NEC); the Soviet Union (Casale, Fauser, NEC); Spain (Casale, Claude); the United States (Allied Chemical, Casale, Claude, Du Pont, FNRL, Mont Cenis, NEC), and Yugoslavia (Casale). In all cases, the basic conditions were Haber's. But each process was different, in its catalyst, pressure, design and arrangement of converters, while the methods of preparing pure gases were determined by available coal and energy sources and the scale of working. By around 1940, annual global nitrogen capacity, mainly by the high-pressure processes, was around three million tons.

When, in 1919, the Fixed Nitrogen Research Laboratory was established in Washington DC, its main purpose was to investigate the chemical methods that had been devised over the previous two decades for capture of atmospheric nitrogen. There was also a strong interest in bringing about biological nitrogen fixation, in the hope that it could be achieved within a few years. A century later, that has still not come about. In 1898, William Crookes had warned of major famines unless the nitrogen fertilizer problem could be resolved. That was achieved by chemists and engineers within 15 years. They completely overcame the technical difficulties of large-scale manufacture of stable nitrogen products, thereby offering the potential means to produce sufficient fertilizer to prevent undernourishment. However, devastating famines, the direct results of wars, drought and weather-related poor harvests, inadequate transport infrastructure, and geopolitical and regime changes brought death to over seventy million people in the twentieth century, and mostly before the mid-1960s [3]. From that time the Soviet Union, and subsequently China and India, both of which had suffered more than other countries, turned to high-pressure ammonia synthesis to support their green revolutions, and in order to remove the threat of starvation [4]. As a result, by the end of the century global capacity in synthetic nitrogen enabled 80 million tons to be applied each year, exceeding the 1940 figure thirty-fold.

Crookes's rallying call still has wide resonance, not so much about the invention of nitrogen capture technologies, but rather within a broader, more global context, concerning the policies and commitments that will enable synthetic ammonia to contribute towards food security for the ill-nourished wherever it is needed. The present challenge, as posed by the World Bank, is to increase world food production by 50% before 2050.

References

1. Travis AS (2017) Globalising synthetic nitrogen: the interwar inauguration of a new industry. Ambix 64(1):1–28, on 24–26.
2. IG Farbenindustrie AG (v.Knieriem & Brendel) to Reichsminister der Justiz Dr. Gürtner (20. Nov.1933), in Bundesarchiv Berlin, BArch, R8128/1420: Bestand IG Farbenindustrie AG, Verbindung zum Heereswaffenamt [Juli 1931–Nov. 1936], pp 93–94.
3. Food shortages (2017) Financial Times (20 February).
4. Saha M, Schmalzer S (2016) Green-revolution epistemologies in China and India: technocracy and revolution in the production of scientific knowledge and peasant identity. In: Phalkey J, Tong L (eds), Science of giants: China and India in the twentieth century. BJHS Themes 1:145–167.

Index

A.S. Travis, *Nitrogen Capture*, https://doi.org/10.1007/978-3-319-68963-0

C

E

M

Q

R

S

T

The manufacturer's authorised representative in the EU is Springer Nature Customer Service Centre GmbH, Europaplatz 3, 69115 Heidelberg, Germany. If you have any concerns regarding our products, please contact ProductSafety@springernature.com

Printed and bound by CPI Group (UK) Ltd, Croydon, CR0 4YY
05/10/2026
02228138-0017